WORLD WAR II DESTROYER SAILOR

Pelham's Revenge

Pacific War Series Volume Two

This book is a work of historical fiction. Characters and events have been developed from the author's imagination and placed within the unfolding drama of World War II. Any resemblance to living persons is purely coincidental.

Published by Pacific Series Press, Las Cruces, New Mexico

www.pacificseriespress.com

Send corrections/comments to clarelawman@gmail.com

World War II Destroyer Sailor

Pacific War Series Volume Two, First Edition

ISBN: 97817332474-5-0

Library of Congress Cataloging - in progress

Law, James W.

World War II Destroyer Sailor, a novel, James W. Law, First Edition

Editing by James Oliveri, author of "Frost Weeds, Vietnam 1964-1965," and others

Cover design by Robert Reno, WeareIT, Las Cruces, New Mexico

WWII DESTROYER SAILOR – CONTENTS

Chapters...continued

PROLOGUE

This book features the adventures of a naval officer who advances from Seaman Apprentice (E-1) to the rank of Commander (O-5) in the United States Navy during the 1930s and the outbreak of World War II. Comdr. David A. Pelham gained most of his operational experience in the destroyer navy. He teams up with Captain Harrold "Rats" Mueller, USMC, and a C-46 "Commando" transport aircraft crew to take the fight to their Japanese enemies early in the Pacific War. Mueller and his "Rat Pack" were featured in Volume One of the Pacific War Series.

It is historical fiction that takes some liberty with events, places, and people. The characters are primarily fictitious. However, the author has attempted to portray a real person truthfully. Many tactical situations, weapons, and communications are products of the author's imagination.

The story is a tribute to the "Tin Can" sailors who manned fast, lightly armored, and highly versatile destroyers during World War II.

An alphabetized glossary follows the narrative to decipher the jargon used in the narrative. Twenty-four-hour military time is repeatedly used. For example, 0630 is 6:30 a.m., 1000 is 10:00 a.m., 1200 is noon, 1500 is 3:00 p.m., 1800 is 6:00 p.m., 2100 is 9:00 p.m., and 2400 is midnight.

Hopefully, the reader will learn about the early phase of the Pacific War, life aboard a fighting "Greyhound of the Sea," and tactical at-sea real and imagined situations.

Chapter headings provide date and location information to help the reader follow the story's flow and sequence – I hope you enjoy the tale.

JWL

INTRODUCTION

Destiny commanded David Alexander Pelham to make a mark on the world. David was the second child and first son of William Pelham and Ruth-Anne "Rose" Pelham. He was born in the newly constructed Tarrytown Hospital in Westchester County, New York, in 1911. After a week in a private room, his mother took him home to their three-story house in Pocantico Hills.

His family was wealthy, comfortable, and safe. David's father and his older sister, Marie, were delighted with the new arrival except when he bawled in the night. Baby Pelham had his family, a nanny, a maid, and his father's driver, who all doted on him. He was the cutest, brightest baby they had ever seen. Grandma and Grandpa Pelham lived a half-mile away on a large estate with a swimming pool, tennis court, horses, and other farm animals. David's dad, William, was their only child, making him the heir apparent to a modest family fortune.

Davy grew up in an active, loving, joyful home, but neither parent allowed him to be spoiled or nasty. He was encouraged to explore the surrounding world and his many interests. His parents taught him proper manners, how to excel in school, and encouraged his participation in team sports. David's mom and dad often went to New York for a weekend of parties, stage shows, or to shop. He and his sister, Marie Pelham, would stay at Gramps and Grammy's house, or their grandparents would move in for a weekend.

In 1919, his father moved the family to living quarters on the lower floor of the Pelham Oil building in Manhattan. David and his mother thought it was the end of their world. They did not want to live in the city, even if Central Park was nearby. However, the family still had a nanny, a daytime maid who also cooked, and a chauffeur.

David had taken his comfortable life in Pocantico and private schooling for granted. He thought this was how most folks lived: horseback riding in the country, Easter egg hunts at the Rockefeller estate, and swimming in Gramps' pool. But, in Manhattan, he learned there was a whole different gritty world where multiple families in the Bronx, Brooklyn, and Harlem crammed themselves into small, drafty, poorly maintained apartments and barely had enough to eat.

As David prepared for high school, his father wanted to send him away to boarding school and place him in a college-prep program. However, his mother interceded for him, saying she couldn't bear to have him living elsewhere. Rose won the battle and Davy continued schooling in New York private academies. Either his father or the family driver chauffeured him to and from school.

William Pelham's business prospered, and he bought the top floor of a twelve-story building further south on West Park Avenue. During a full year of construction, an army of workers turned it into a plush penthouse that looked out on the Hudson River on one side and the Central Park Reservoir on the other.

In 1929, as a high school senior, David attempted to break free of his father's control. He declared that he was attending the University of Pennsylvania in Philadelphia or no college. There were intense father-son verbal debates as his dad argued that David could have Harvard, his alma mater Yale, Princeton, Columbia, or Brown. William contended that David should remain close to their New York home because of his mother's declining health. His mother was neutral as the two men of the family clashed over David's future.

After a particularly nasty argument when his father threatened to cut him off from family funds and stormed out to prowl the New York bars, his mother sat down with him and spoke softly, "Davy, you're a good boy, and now at eighteen, you're your own man." She asked him what in the world he wanted the most. After a long pause, he told her he had decided to be a historian and a Naval Officer. She knew much to her distress that he had always been fascinated with war and the history of warfare. From age five, he had collected Army and Navy figures from great battles and staged them in detail. He pored over books about generals and admirals and read accounts of ancient and modern conflicts. The Naval Officer career was a surprise to her.

Her son had researched each service, and the Navy seemed his best option. He could serve as a reservist while earning his undergrad degree and then apply for officer training. He had decided - the family oil business was not for him. The clincher was his visit to the University of Pennsylvania and the hours spent perusing the glorious collection of books in the main library. David saw a poster for the Naval Reserves in the Student Union Building and visited the Philadelphia Navy Base the same day. A Navy recruiter told him other U-Penn students enlisted in the Navy. They trained on a destroyer one weekend a month and cruised for two weeks during summer vacation.

David explained to his mother that he'd have to enlist as a Seaman Recruit, complete a four-week Boot Camp, and then advance on his merit. Then, during his senior year in college, he could apply for Officer Candidate School. After four months of training at Rhode Island and a commission as a Navy Ensign in the reserves, he would return to Penn for his master's degree and Ph.D. in History - specializing in the history of warfare.

His mother reached over the kitchen table, took both of his hands in hers, and said, "Listen Davy, you have to be 100% sure; there's no turning back. If we do this with your father, it will take an unwavering united front." He told her he was going ahead with the Navy enlistment papers and applying to Penn with or without his father's blessing. She asked him what he'd do for money if his dad cut him off. He told her that he had money saved from his Pelham Oil jobs, several savings bonds and that he could always get help from Uncle Tommy and Gramps.

Rose instructed him to proceed with his plans, and only she'd handle his father, saying, "He loves you very much and will come around in the end. He'll be very disappointed that you won't take over the business and that you're picking what he'd consider a lesser campus. He'll rant and rave and make all kinds of threats because this undercuts his vanity and his standing with the captains of industry. We raised you to be an independent and thoughtful young man who can stand on your own two feet. Now we, especially your father, must live with it."

NAVY HYMN

Eternal Father, strong to save,
Whose arm hath bound the restless wave,
Who bidd'st the mighty ocean deep
Its own appointed limits keep;

Oh, hear us when we cry to Thee,
For those in peril on the sea!

O Christ! Whose voice the waters heard
And hushed their raging at Thy word,
Who walked'st on the foaming deep,
And calm amidst its rage didst sleep;

Oh, hear us when we cry to Thee,
For those in peril on the sea!

TITLE PAGE

WORLD WAR II DESTROYER SAILOR

Commander David A. Pelham

United States Navy

Pelham's Revenge

James W. Law

Dedicated to my wife Clare, and my children, Ryan and Amanda

Chapter One

WAR! – Marine C-46 Flight to Hawaii

January 1942

The bright silver aircraft sliced through the cold night air at 8,500 feet, heading southwest. After leveling off and completing the cruise checklist, Co-pilot Lieutenant Bradley W. Whitlock, United States Navy, pulled off his earphones and was immediately asleep. To his left, the Plane Commander checked engine temps, oil and hydraulic pressures, and fuel quantities. The glowing red gauges and switch settings indicated all was well.

His C-46 transport aircraft was flying as designed by its Curtiss-Wright engineers. Captain Harrold C. Mueller, United States Marine Corps, put his head back on the seat headrest and let out a long, contented sigh as his body felt the rhythmic throb of the powerful Wright Cyclone engines. It was the start of a long flight - San Diego to Pearl Harbor, Oahu. He punched the autopilot on and touched the double bar railroad track insignia on his starched khaki shirt collar thinking, *Damn! In the Marine Corps for only five years of active duty and already promoted to Captain. As I expected, we've got an all-out war. It should make for great promotions – if I survive. Damn! A certified Plane Commander of a new quarter-million-dollar aircraft with a crew of five. I'm flying the most capable two-engine land-based transport in the world, and I'm the boss. Could life be any better? Crime does pay!*

Harry Mueller looked over at his sleeping Co-pilot. *That's right, kid; grab a few winks when you can. There's no telling when the boredom of transport aviation instantly changes to stark life-or-death terror. An engine quits, a prop runs away, fire breaks out, flight controls go snake shit, hydraulic pressure dumps, and it's time to earn your flight pay.*

Mueller verified that the aircraft's magnetic heading, altitude, and Indicated Air Speed or IAS matched those provided by the Navigator. The aircraft's excellent autopilot was dutifully flying the settings entered. He remembered his Pensacola flight instructor's repeated warnings, "Mister Mueller, Mister Mueller! If you want to survive Navy flying, you need to check, check, and triple-check. Take nothing for granted!"

The aircraft's Navigator, Lt. (jg) "Tex" Johnson, called Harry on the intercom and recommended he switch from the mid-wing fuel tanks to the forward ones for at least thirty minutes. Harry rogered the call and obeyed by selecting one tank, waiting for any sign of an engine problem, and then the other. He thought, *Nav, you're a nervous Nellie, always worried about fuel, weight, balance, and compliance with flight regulations. You would never have made it in the Corps.*

The outside air temperature gauge was reading 36 degrees Fahrenheit. So Mueller raised the rear cargo cabin heating switch from warm to hot. *I'll hear yowling from the boys in the cargo hold if they start freezing their nuts off.*

Harry reviewed each of the "Commando's" key flight systems as he took the pulse of the living, flying machine: engines, electrical, hydraulics, fuel flow and storage, flight controls, radios, and navigation. Minor problems could cascade into disaster in seven hours when they passed the point of no return. However, Captain Mueller's periodic instrument scan assured him that all systems, pressures, voltages, and fluids were satisfactory. The fourteen-cylinder,

1,700 horsepower engines and their four-bladed props rotated at 1,750 Revolutions Per Minute, pulling the aircraft through the air at 167 knots or 192 miles per hour.

Plane Commander Mueller glanced back at his Flight Engineer, Navy Chief Donald Watkins, sitting in the jump seat behind his right shoulder, "Chief, it looks good to me. How about you?"

"Yes sir, settings look good," replied Watkins. "All the temps and pressures were normal on take-off and climb. I'm putting on my Crew Chief hat and going aft to check the cargo and passengers."

"Go for it. We'll be fine up here."

The take-off on a crisp January evening from Naval Air Station San Diego had been uneventful. Winds were light, and the air temperature was in the low fifties. The crew had special permission to launch at night. They followed a dimly lit jeep to the apron of runway 18 and prepared for takeoff. Mueller avoided the usual 29-duty runway if the winds allowed. He worried about one engine failing on launch and plowing into the looming 500-foot-high Point Loma ridgeline.

The third time to Hawaii is the charm, he thought. *Our luck is holding. We barely got off Wake Island two weeks ago before the Japs took it. Both Whit and I have Purple Hearts from injuries caused by an anti-aircraft shell at Wake. Shit, my left cheek still is throbbing, and I bet Whit's arm hurts like hell. On the other hand, we just had repairs on the bird and pulled fourteen days of liberty in San Diego. Life is good.*

The C-46 crew had trained on the new aircraft in the Midwest and supervised its modifications for long-range flights before the Japanese attack on Pearl Harbor. With the war on, it was eerie to look out at Coronado, San Diego Bay, and downtown, seeing nothing but darkness during the second week of January.

The duty Operations Officer had told them to line up on Runway 18 and expect only dim centerline lights for take-off. The plane was near its maximum weight of 45,000 pounds, or nearly twenty-two tons. On each outbound flight, the crew agreed to add another 1,500 pounds of cargo as they learned the new transport's capabilities. Once parked at the north end of runway 18, Mueller called for the Takeoff Checklist.

Co-pilot Whitlock read from a printed list encased in a plastic bag:

"Trim tabs." Mueller responded, "All neutral."
"Controls." Mueller moved the wheel, flight control column, and pedals, "Free and clear."
"Carb heat." "Cold."
"Mixture." "Auto rich."
"Fuel pumps." "On."
"Props." "High RPM."
"Supercharger." "Low."
"Tailwheel." "Unlocked until we roll."
"Flaps." "Full up."

"Ready for takeoff Harry. She's heavy and should fly off at about ninety knots," advised Whitlock.

"Roger that."

While waiting for takeoff clearance, Mueller worried whether his estranged bum of a father would track him to San Diego a second time. Without warning, the disheveled and grubby older man had shown up at Quantico, Virginia, Dayton, Ohio, and Coronado. Howard Mueller demanded money from his son and threatened to stage an embarrassing scene at each military base.

Harry thought, *How does the bum find out where I'm stationed? How does he cross the country flat-broke and show up at the worst possible times? And this last time, he accosted Ginger while I was in Hawaii. No one else knew we were married. How did the dirty moocher find out? How did he find her in Coronado?*

Mueller's thoughts were interrupted when Lieutenant Junior Grade Ben Johnson, USN, reported the aft crew and passengers ready for takeoff. He added that they expected flames out of the engines - something that had scared the piss out of previously unwarned passengers. Mueller spoke loudly over the engine noise to his Co-pilot and Crew Chief, "OK, crew, same as before, the Chief works the throttles. If she starts to snake one way or the other, I call back on one throttle, forward on the other. Whit, you say concur if you agree or negative if I've got it backward. Chief, monitor the engine temps, RPM, and Manifold Pressure. Keep us within limits unless I scream for more."

Harry recalled his first flight in the Army's C-55, the C-46 prototype at Wright Field in Dayton, Ohio, a few months previous. He was in the co-pilot's seat, learning how the beast behaved. The Army pilot said they would use differential throttle power to make taxi turns and hold to the centerline on take-off. He said the bird flew like a dream in the air, but on the ground, it might veer in either direction or even ground-loop without warning. The C-55 was a massive transport aircraft with a 108-foot wingspan. Curtiss designed the vertical tail and rudder to provide directional control at fifty knots or more. Below that speed, they were next to useless.

At 0130 or 1:30 a.m. civilian time, the tower signaled the jeep driver, who flashed his red lens flashlight at the Marine Corps transport's cockpit. Mueller thought, *Get ready to roll; those runway center lights go on for as long as we need them, or maybe not.* It felt strange to fly at night and not use the aircraft's landing lights. The west coast was jittery over rumors of Jap attacks, and anyone with a gun was trigger-happy. Harry worried, *We'll be lucky if the Army artillery battery at the end of Point Loma doesn't open fire on us. Jesus, we make a fantastic target with the engines at full power and ten feet of blue flames shooting out of them.*

The runway center lights flashed on and went dim. Mueller called for taxi power on the throttles.

He ordered, "Port throttle up," and Whitlock said, "Concur," as they lined up on the middle of the runway. Just before the aircraft's nose centered on the long line of dim lights, Mueller ordered, "Throttles up, set takeoff power, lock tail wheel!"

Chief Watkins smoothly advanced both throttles to 2,550 RPM. The engines roared, and the plane vibrated as it picked up speed. Whitlock confirmed the tail wheel was locked and called out the increasing airspeed in ten-knot increments. At forty knots, the aircraft started to swing to the left.

Mueller held the throttle settings and put in some right aileron that brought the nose back to the centerline. With the tail up at sixty knots, he pushed gently on the rudder pedals and announced, "We have rudder."

The Co-pilot called out, "Seventy, eighty, ninety."

Harry held her down, waiting for the aircraft to fly into the air on her own accord. The runway lights shut off just before the C-46 became airborne.

The co-pilot called, "Airborne, heading 180, needle ball, climb 200 feet per minute, speed 100 knots, increasing."

His Plane Commander scanned his instruments, confirming the wet compass heading of 180 degrees, the turn and bank needle/ball centered, and the altimeter showing an increasing climb rate.

Captain Mueller warned his Co-pilot, "Standby gear, stop the wheels."

Whitlock pushed on the top of his rudder pedals, causing the main landing gear brakes to stop the spinning wheels, and responded, "Wheels stopped."

"Gear up, gear up!" Mueller ordered.

"Main gear and tail wheel coming up." A few seconds later, "All gear up and locked," Whitlock said as he scanned the instrument panel.

Harry took out some of the nose-up trim he had used just as the wheels left the ground. Now free of the landing gear drag, the C-46 surged forward and climbed faster. As they passed 500 feet, Mueller announced:

"Turning starboard to initial heading of 255. Chief, give me climb power."

Watkins pulled the throttles back and responded, "Aye, aye, sir. Setting 2,300 RPM. All temps and pressures look good."

Captain Harrold "Rats" Mueller, USMC, and Navy Lt. Bradley "Whit" Whitlock piloted the Marine transport aircraft. Lt. (jg) Benjamin "Tex" Johnson provided navigation and fuel management. Chief Donald Watkins, USN, was the senior enlisted man in the crew, and Corporal Reggie "Jinx" Jenkins, USMCR, performed the duties of Plane Captain and Radio Operator.

Jenkins was making his first combat flight after transferring from the United States Army Air Force. The Commando crew had adopted him while testing the prototype C-55 aircraft in Dayton, Ohio. His experience with the C-55 helped them modify a newer version for the Marine Corps' long-distance flight requirements. The three officers and two enlisted aviators flew together for several months. After a series of training flights, transport airlifts, and combat operations at Wake Island that grew more colorful with each retelling, they had dubbed themselves the "Rat Pack."

Mueller double-checked the flight controls with the aircraft smoothly flying at cruise altitude. He deselected the autopilot and rotated the control wheel slightly left and right - ailerons check. Rats pushed the yoke in and pulled it back just past center - up and down elevator check. He pressed lightly on the right and then left rudder pedals - rudder check. After selecting some nose-down trim, he re-engaged the autopilot. Harry eased the throttles forward slightly, increasing the Indicated Air Speed or IAS to 145 knots, ignoring the 140-knot cruise speed recommended by Nav.

Mueller smiled and thought, *Let's pick up the beat here; I've got some pussy waiting for me at the other end of this air circus.* He had a fleeting moment of guilt, having just kissed his wife of two months goodbye after several passionate days and nights in Coronado. He thought about his path to matrimony. *Last August, we flew from Quantico to Patterson Field in Dayton, Ohio, to learn more about this plane. Colonel "Dusty" White, the Deputy Program*

Manager for the C-46 program, invited me to his house for brunch. I met his daughter Marybell "Ginger" White, and my life changed in one day. I've got a cute picture of my redhead in tennis shorts painted on the nose of this plane. The "Ginger Snap" nickname has stuck for this airborne beauty. It probably is a lucky moniker. It stinks that I want to chase another chickadee. But it's just who I am. I'm like the scorpion that stings the frog midstream – I can't help myself. Life is short, and then you die.

Captain Harrold C. Mueller, USMC, stood five feet ten and had the lithe body of a long-distance runner. He kept his light brown hair closely cropped in a crew cut, and his face had a smart-aleck look with a persistent impish smile. Rats quickly accepted any challenge. His friendly demeanor, broad face, and deep blue eyes seemed to hide what he was thinking and feeling. He had a reputation in the Corps for cutting corners to get the job done.

Mueller had led the crew through several challenges to procure the C-46 from Curtiss-Wright in St. Louis, Missouri. At the factory, they met Rhonda Bessemer, a lady engineer who provided company confidential information on necessary changes to the aircraft to increase its flight range and airborne safety. After Captain Mueller briefed the bird's capabilities to General Holcomb, Commandant of the Marine Corps, the plane was purchased using Marine Corps and Navy funds. The Marines signed Rhonda as the lead engineer for the Navy/Marine C-46 procurement program.

Before Pearl Harbor, Rhonda Bessemer and the original Plane Captain for the C-46, Corporal Paul "Fitz" Fitzpatrick, fell hard for each other. It was an open secret that they were married without permission from the Marine Corps. Fitz was the PC on the first two flights to Hawaii and the dangerous hops from Midway Island to Wake Island while the latter was under siege in December 1941. On this flight, Fitz was left behind at NAS San Diego to work with Rhonda on modification plans for "Ginger Snap." In six months, new, improved C-46A aircraft were expected from the Curtiss-Wright Buffalo, New York, plant.

As Rats relaxed and monitored the autopilot, he thought, *Hate to stop flying with my friend and co-pilot Brad Whitfield, but it has to happen sometime. He's ready to be Plane Commander, and one of the VIP birds coming off the Buffalo line belongs to the Marine Corps. I need to cut Whit loose about May to get a crew together and fly the acceptance flights. I'll require a new co-pilot from the Corps or Navy. It makes no difference to me if he has a multi-engine ticket and can cut the mustard. Rhonda's moving back to New York to manage the C-46 engineering for the Navy and Marines. Guess I'll let Brad have Fitzpatrick as Plane Captain so Fitz can spend more time with Rhonda before they deploy to Hawaii. Damn! When we get two Commandos, we can deliver forty bodies or eight tons of cargo daily to Midway Island. The Japs aren't taking Midway like they did Wake, if I can help it. No siree, not over my dead body.*

C-46 "Commando" transport

Note horizontal "crease" between upper and lower cabins

Source: Google Images

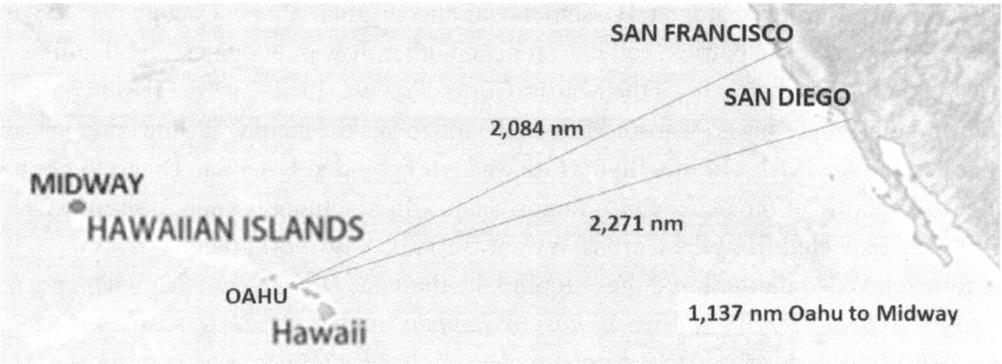

Flight distances from the West Coast to Hawaii in nautical miles

Source: Inserted on the map by the author

Chapter Two

SHUT EYE – C-46 "Ginger Snap" Passenger

January 1942

As the C-46 Commando aircraft logged its third-hour airborne from San Diego, the naval officer's awakening in the cargo hold was slow and painful. He first was aware of the stiffness of his legs and torso arranged awkwardly on the canvas web and aluminum tube seating. His back hurt from sleeping at an odd angle against the man beside him. It was dark, cold, and incessantly noisy. He could smell the strong odor of gas and oil. David felt the rhythmic vibration of the metal flying machine as his conscious mind came up for air. The repetitive drumbeat of the engines, ARRA-ARRA-ARRA-ARRA, became a slow chant in his head. *You're going to war -you're going to war - you're going to war.*

Fragments of the previous night flitted through his semi-conscious mind. He had recollections of a day out of control, memories as painful as the hammering in the frontal lobes of his brain. David Pelham moved a hand, an arm, and tilted his head to make sure they were still there and subject to his command. Slowly, he lifted one eyelid and could see that he was in an aircraft by the dim red lighting. Up slowly came the other eyelid to register a darkened cargo bay with cargo on the deck. A shadowy row of bodies was on the other side of the plane, also being tortured by the cruel web and tube seating. Next to him were a handful of row mates, all in a somnolent state of unconsciousness. David had a moment of panic and nausea - he couldn't remember boarding the plane and did not know where it was taking him.

Urgent warning messages from his 100 percent full bladder pushed these concerns aside. There was no time to lose, or the front of his white uniform would turn an ugly yellow. Pelham started thrashing his arms when he could not find the release lever for his seatbelt. A final warning came from his bursting bladder, *The moment of no return was fast approaching!* Suddenly, a human form appeared before him, unfastening his lap straps and talking, "Good morning, Commander. I bet you would like to use the head."

All Pelham could do was nod and let himself be propelled to the back of the plane while he fended off his screaming bladder.

The man spoke again. "Try and hit this piss tube, sir. You push the button on the side, or it's all gonna end up on the deck."

Thank you, Lord, David thought, letting hours of piss flow into the tube and atomize behind the aircraft. *Well done, Pelham, you just escaped one more blow to your questionable dignity as an officer and a gentleman.* Then, pieces of the missing hours surfaced in his mind and dissolved - the chance meeting with a gigantic First Class Navy Diver at the Hotel Del Coronado. He tossed back tequila shots at the Ocean View Bar and ended up in a suite with the diver and two attractive young girls.

Lieutenant Commander David A. Pelham, USNR, felt himself tremble and thought, *Good God, what is happening to me? This is not like me. I'm always in control and aware, usually two steps ahead. And now, I'm on a strange plane heading for an unknown destination after a drunken binge in a luxury hotel. Get a grip Davy, get the straight word - the gouge you need from the helpful crewman.*

In the dim aircraft lighting, he could see the man who had helped him unbuckle was checking a tie-down on some cargo. Pelham moved beside him and spoke loudly over the never-ending engine noise, "Excuse me, you are?"

"Chief Donald Watkins, Aircraft Crew Chief, at your service, sir."

"Chief, thank you for that very timely help. Could you ease my mind and confirm that we're flying to Hawaii?"

"Yes sir, the island of Oahu. Hickam Field, to be exact. It looks like you're the senior officer on this flight. Would the Commander like to get a little more shut-eye? We have a makeshift crew bunk up forward."

"Chief, that sounds like heaven. I think I had a pretty rough night."

They passed by two large metal tanks as they made their way forward. Then, when they were abreast of two more, David saw a cut-down, makeshift Army cot perched on top of them. The chief asked him, "Sir, would you like some java? It's hot and just made."

"No thanks, Chief. Just show me how to get into that bunk, and I'll be forever grateful."

Pelham stepped on a wooden stool and jackknifed onto the canvas cot.

As his head hit the pillow, he fell asleep and only partially heard the Chief say, "Enjoy your sleep, Commander, you're gonna need it."

As Lt. Comdr. David Pelham thankfully slept on the taunt Army cot in the noisy, vibrating cargo compartment of a Marine transport plane heading to Hawaii; he did not dwell on his naval reserve career or his future risks from the war. He was on active duty now for the duration and wanted only one thing – not to dream the nightmare.

Folding Army cot circa 1940s

Source: War surplus advertisement

Chapter Three

DAVID PELHAM, SCHOLAR AND SAILOR – U-Penn

1930 – 1934

David Pelham received his enlistment physical at the Navy's Great Lakes Training Center in northern Chicago. He passed the physical and breezed through the boot camp, considering it a piece of cake compared to working as a Texas oil roustabout at sixteen. He proudly returned to Pelham's spacious Central Park penthouse wearing his service dress blue bell-bottom trousers with the thirteen-button front and his dark blue tunic with a double white-striped badge. As first in his class, Seaman Recruit Pelham had already advanced to Seaman Apprentice.

Everyone was there, his mom and dad, Gramps and Grammy, Uncle Tommy, his sister Marie, and her latest boyfriend, Ephraim Goldstein, called Epi. After hugs, they all wanted to know what Boot Camp was like, who his new friends were, and whether he had met any girls in Chicago.

Uncle Tommy announced that he had made David's favorite meal, hot dogs, onions, and scalloped potatoes covered in melted sharp cheddar cheese. It was a great reunion, only more memorable when his mother handed him an official-looking envelope with the U-Penn logo embossed on the front. David's hand shook as he opened it and read the letter's first sentence to his family, "The University of Pennsylvania is pleased to welcome David A. Pelham to the class of 1930."

It dawned on him that he had just enough time remaining in the summer to serve two weeks on the Navy Reserve destroyer, spend a few days with his parents in New York, and start his studies at U-Penn.

Seaman Apprentice David A. Pelham entered the University of Pennsylvania as a Freshman in the fall of 1930. He drilled one weekend a month on a Navy Reserve destroyer based in Philadelphia or Norfolk, Virginia. During the summer months, he completed two weeks of active duty training on a Navy ship or base.

Pelham loved cruising in Navy ships; he gained at-sea experience in destroyers and cruisers. David volunteered for additional summer cruising time, especially if his assigned ship was sailing to the Caribbean, Great Britain, or the Mediterranean. He completed the required correspondence courses and advancement tests each year in less than four years and advanced to Petty Officer Second Class, Gunner's Mate. Petty Officer Pelham was fascinated with the technology and procedures for employing his ship's guns, torpedoes, and depth charges.

During his senior year at U-Penn, David applied for officer training and was selected for the Newport, Rhode Island, OCS. He breezed through the Indoc boot camp and had no trouble with the classroom training. The one thing that irritated him was being called a "Mustang" in a derogatory manner by the Navy enlisted drill instructors. He was proud of his prior Naval service. It seemed they singled him out for extra duty and held him to a higher standard. Toward the end of the course, David was surprised when the instructor staff selected him to lead one of the junior class battalions. After sixteen weeks of physical training, academics, and leadership instruction, he was commissioned in October 1934 as a Navy Reserve Ensign - Unrestricted Line Officer.

Ensign David Pelham had dated a few coeds during his four years as a U-Penn underclassman. Although they were fun and broke up the monotony of studying and classroom lectures, none of them had the right mix of character, abilities, and personality to make him consider matrimony. That changed one night as he was studying late in the basement of the main library. He was carrying an armload of history books to his carrel when he noticed a young woman slumped on the floor at the far end of a double row of bookshelves. He unloaded the volumes on his desk and reviewed one on ancient warfare.

After about an hour, he decided to check on the girl. It was past midnight; they were probably the only people left in the building. He found the row and approached the far end slowly. As David got closer, he saw she had a jacket rolled up under her head as a makeshift pillow and was probably sleeping. From twenty feet away, he said:

"Hello there. Miss, are you OK? I don't want to scare you. I need to know you're taking a break and not passed out or something."

She stirred, opened her eyes, and looked directly at him. "Who the hell are you?"

"David Pelham, first-year grad student in history. Look, I just wanted to make sure you're OK. I'm sorry to bother you, Miss. But back at you, who the hell are you?"

"Lorraine Brockman, third-year nursing school. The only one who has been crazy enough to be down here in the stacks late at night. Why are you here?"

"I need quiet, complete quiet, to do my background research. Here in the stacks, I can commune with my books without interruption."

She said, "You are communing with history books while I try to learn and memorize all the parts of the human cardiovascular system in the small hours of the night. We make a fine, quirky pair. Come closer and tell me what you're reading. I need a break from heart valves, arteries, and veins."

David walked up to her and sat down with his back against a row of bookshelves. Up close, he could see that she was about five foot eight and had brown eyes, freckles, and an adorable face. She was lying on her right side on the carpeted floor with her pleated skirt just above her knees. He thought, *Hmmm, a pleasant-looking girl, good legs, slim figure, not bad in the bust area either.*

She followed his visual assessment with an amused smile. "Before you get any bright ideas, David Pelham, know I have a black belt in karate."

"I shall always keep that in mind and make all of my manly advances deliberate and respectful. Where are you from, Lorraine?"

"Connecticut, you?"

"New York, although I plan to be here for several years getting my master's and Ph.D. Then I hope to teach at Penn."

"A master's and doctorate in history, right? She asked.

"Yes, with a specialty in the history of warfare."

"Warfare?" she questioned. "You don't look very war-like to me."

"Au contraire, Miss Brockman, sitting before you is a newly commissioned officer in our United States Navy, a real fierce sea dog - arrrrgh, ready for battle!"

She laughed. David thought, *Charming smile, good teeth, and I can make her laugh. Easy on the eyes in all regards – this one could be worthwhile.*

"I want to hear about your book; make it interesting. What do your friends call you?"

"Mostly, Dave or David. My family calls me Davy."

"I like Davy, Davy. You can call me Lorry. Now the book!"

"Yes, Miss Lorry. We shall be delving into the Peloponnesian War. A war between the Greeks of Athens and the Peloponnesian Federation led by Sparta from 431 BC to 404 BC."

"Lord, have mercy!" she wailed. "Can't we start with the Civil War or World War I?"

"No, student Brockman, each major war has its lessons. In this one, ostensibly democratic Athens and its allies were defeated by the oligarchic dictatorship of Sparta and its allies. For Navy types like myself, it's important to understand that the clash of warships from both sides was key to the fall of Athens. Once Sparta destroyed the Athenian fleet in 405 BC, Athens was destined to surrender. Athens' Navy had defeated the Spartan fleet one year earlier, but couldn't make it an utter rout because of bad weather. Athens' political leaders sowed the seeds of their defeat by blaming this on their Naval commanders. They executed six of their most able leaders and destroyed the morale of their once nearly invincible naval forces. I think…"

David looked her way; she was fast asleep. He went to his carrel several rows away and returned with his coat to cover her legs and an oversized scarf that he draped over her arms and blouse. Pelham studied for several hours and finally fell asleep, sitting in his chair with his head on the desk. When he awoke near seven o'clock, he felt a coat around his shoulders and found his scarf wrapped around his legs and tied in knots.

The next night, when he arrived at his library carrel, David found a pink rose in a small vase, a chocolate bar, and a note, "Can't wait to hear more about the Peloponnesian War. Sleeping in the dorm tonight - hope to see you tomorrow. Getting dinner at the SUB at six o'clock if you care to join. Ancient history student, Lorry."

He found her at the Student Union Building at the appointed time; they had an enjoyable dinner discussing their career plans. David learned that Lorraine first planned to become an RN and then enroll in a medical program that would lead to a Ph.D. in Psychology. She told him that her parents in Westport would prefer she marry someone respectable and start a family. They half-heartedly supported higher education for women and paid her educational expenses but not her living costs. Luckily, her grandparents had a trust set up that she could use for room, board, and incidentals.

David told her that his dad wanted him to attend a more elite Ivy League college, especially his father's alma mater, Yale. He added that his trust was his mother, who funded all his expenses, including a two-bedroom apartment and a car.

Lorraine could see a worrying change in his expression when he mentioned his mother. She asked, "Davy, what is your mother like?"

David enthused, "She's about the best person you could ever meet, charitable, caring, and a great listener. When we moved to Manhattan, men always roamed in The Park, sleeping on the benches and begging for a few coins. My dad called them bums. Much to his chagrin, my mom always gave them something. Even if they were sleeping against a building, she'd go over to them and tuck a half dollar in their pocket. She told my sister Marie and me that the Pelham family must be thankful to the Lord for what they had, and it was our duty to share with those who needed a helping hand. On Thanksgiving, Christmas, and Easter, she'd take us to Saint John's Cathedral to help serve holiday meals to the poor and less fortunate. All the while, my dad would be fuming at home, waiting for his dinner. When we got home, he'd rant and rave that the Lord helps those who help themselves. He believed there would be no bums if people worked harder and smarter."

Lorraine replied, "No doubt a Republican businessman like my dad."

"Yes Lorry, 100 percent. Anyway, all through prep school, it was clear that my dad wanted me to go into his business in the worst way. My mom and I had a heart-to-heart chat before I enlisted in the Navy. She said, 'You've done everything he's wanted your whole life. Now It's time to choose your own path. Are you sure, absolutely sure, you don't want to be in the oil business?' I told her that although my dad had strong-armed me into working for the company each summer for the last three years, it was not for me. I was an oil rig roughneck in Texas, an office boy in New York, and an explorer for new drilling opportunities in Oklahoma and New Mexico. Sometimes, my Dad had me act as a personal secretary, sitting in on high-level meetings. I was headed for upper management in Pelham Oil. Dad promised rapid promotion after I obtained a prestigious Ivy League degree. My mom held him in check and supported my service in the Navy and teaching career. She's helped so many people; it's unfair that she has to face a terrible sickness that may end her life."

"Davy, I'm so sorry. What does she have?"

"She won't tell me exactly; it's some form of stomach cancer. She has good and bad weeks where she can barely get out of bed," Pelham answered.

"The poor woman. How often do you get to see her?"

"It has been about a month now. My mom used to come down and stay with me for a day or two each month, but I don't think she's up to traveling now."

"I bet she's a beautiful lady, your mother. And I'm sure she's very proud of her son for graduating and being a handsome young naval officer."

"What about your parents?" David asked.

"What about them?"

"Where is home? What do they do? Are there any brothers and sisters?"

"My mom and dad have a nice house in Westport. Dad is always working and often goes on travel. Mom designs ladies' clothes and has a small shop in town. She's active in our church and dabbles in a few charities. I'm the oldest child. My sister is a freshman at Smith this year; my brother is a sophomore in high school."

He took a bite of his burger and said, "So you're the oldest child, and I'm the youngest. I wonder if that makes us compatible or incompatible. That would be a good study for a psychology major, don't you think?"

"While you said it, no, but thinking about it, maybe. It could be interesting if you had enough couples, say thousands, that were happily married and the same number that had divorced or separated. You need a mathematician to determine correlations based on sibling position and carefully construct the survey questions. The size of the family could also matter. Is an only child more likely to experience wedded bliss than the oldest child? Does…"

"Excuse me," David said, interrupting, "You said your mom was active in your family church. What religion is that?"

"Methodist," she replied. "How about you?"

"Presbyterian, pretty much non-practicing lately."

They finished their dinner. Lorraine returned to her dorm; David headed for the library stacks to continue his research. They had several conversations during study breaks in the library the next week. One of them would plop down at the other's carrel and start a discussion about anything. David missed seeing her and their lively discussions on the nights she didn't study in the stacks.

They met for dinner in the SUB again a week after their first meal.

David put down his tray on Lorry's table. "Hey there, you look great. How is the budding psychologist tonight?"

"You probably should say almost graduating nurse. The other is a long way off," Lorraine replied.

"Either way, I'm proud of you. You have goals. You know what you want."

She smiled and looked at him fondly. "Same to you. You've more of your life mapped out than anyone I know. Are you dating?"

"Dating?"

"Yes, do you go out with girls? Do you have a steady, or are you a monk?" she said with a wink.

"Mostly a monk. Nearly all the girls I took to movies or dances have graduated and left Philadelphia. How about you, dating or nun?"

"In three years, I've dated a few undergrads and gone steady a couple of times, but nothing special is going on right now," she responded.

He looked into her expressive blue eyes and said, "Well, at least we have our studies and conversations to rest our brains."

"Yes, agreed. How is your mom?"

"I talked to her yesterday, and she's having a pretty good week. It's such a relief!"

"You should go see her. You have a car. You could be there in no time," she admonished.

He looked at her with surprise. "Oh my God, I think I have a blinking lightbulb idea. Something we both could do that would make my Mom truly happy. It kills me to see her in pain. We could do something to lessen her worry about me."

"What's that?"

"If you could meet her, spend some time with her, and let her believe just one tiny white lie, it would lift a great burden from her shoulders."

"I'd like to meet her. Meeting her would tell me a lot about you. I want to know about you as a child and a younger man. What's the white lie?"

"We would give her the impression that you're my girlfriend, nothing serious yet, just that we like each other and we're dating. It would mean so much to her! My sister has graduated from Columbia and is engaged to her college sweetheart. He's a nice guy that we all like. So, no concerns for Mom there. She's worried sick that I'll be a hermit bachelor with my nose in a book and no family. No one else here," he waved his hand around the cafeteria, "would ever have to know, and you would never have to see me again if that's what you want."

Lorraine smiled. "Boy, you don't mess around; it must be that Navy training. Let's say I was willing to go to one of my favorite places, New York. What are you thinking? How would it work?"

"Well, we could drive up this weekend or next weekend. Leave Saturday morning, get there for lunch, have dinner on the town, and return on Sunday. What do you say?"
Lorraine said, "I'm a good sport and definitely could use a break. So tell me, what are the sleeping arrangements?"
"Oh my, we have a guest suite, and my sister's room is open. My dad and Uncle Tommy are often on the road. There are plenty of places with a private bathroom."
Lorry answered in a low voice. "And, of course, there's your room. I could always bunk in with you, right?"
David looked stunned. "I don't know that...."
She threw her head back and laughed. "Got you there, Ensign Pelham!"

Other diners saw a young woman laughing loudly and pounding her hands on the table. Her male companion had a very red face until he, too, started laughing. They laughed until they both were in tears. She went home with David for the weekend and hit it off famously with the whole family. Lorry especially bonded with his mother, Rose, and sister Marie. On Saturday night, William entertained all of them with his secret martini formula. Uncle Tommy served the Pelham family and guests, Epi and Lorry, a delicious mixture of Chinese and Japanese food. On Sunday morning, William and Rose hosted David and Lorraine for brunch at the Central Park Boat House.

It was a warm Indian Summer day. Rose was in good spirits when she demanded that David take her rowing on the pond. She reminded him that it was here on Central Park Lake that she had taught him to row a boat nearly fifteen years before. While rowing, David looked at his mom in her tailored skirt and matching sweater with a brightly colored scarf around her neck and cried. He sobbed that he didn't want to lose her. Finally, she shushed him and bravely told him that we all depart this earth sooner or later - that it will always seem too early.

She instructed him to mind his manners with Lorraine and remarked that she was pleased he had found an attractive and accomplished woman by chance. She thought a worthy companion, with the pluck and toughness to stand up to the Pelham men and their grand plans. Moreover, she saw Lorry as his intellectual equal and wise beyond her years.

As they all said their goodbyes in the Boathouse parking lot, William gave Lorry a big hug, a kiss on the cheek, and invited her to come back anytime, with or without David. Rose took her aside and discussed something with her while William slapped his son on the back and told him to come home more often and bring Lorraine. William leaned close to David and whispered that not only was she attractive, smart, and an achiever, but it didn't hurt that she was part of the Brockman empire.

In the car crossing the George Washington Bridge out of New York, David asked Lorry what his mother said in the parking lot.
"Not much," Lorry replied, "just if things work out between us, would I please take good care of her favorite son? Oh, and to watch out when you dig your heels in over a trivial matter or decision, you're just like your father sometimes with a stubborn streak."
A surprised David said, "She told you that. I'm not a stubborn person. I'm nothing like my dad!"
"Spoken like all sons who think they raised themselves," she responded.

"And what is this about the Brockman empire? Not only did you wow my dad, but he also whispered to me that you're attractive, smart, and a go-getter. Then he adds that it doesn't hurt that you're a Brockman. What's with that?"

"You've never heard of Brockman Industries?"

"Nope, what does your dad do?"

"His companies are big into the railroad business. They make the engines, the rail cars, railroad stations, the track, the switches, the coal and water stations, the trestles and bridges, just about everything needed to build a railroad. Back to you, what does your dad do?"

"He sold his company, Pelham Oil, in early 1929 and cashed out his real estate assets. He and Gramps had a terrible argument over the Pelham Oil sale. Gramps put up the initial funds to buy the Atlantic Oil stock, and my dad sold all their shares without consulting him. Dad argued if an investment group was dumb enough to pay them five times what the company was worth, they had to take their profits. He told Gramps that they still retained the East Texas oil wells and substantial adjacent land. He planned to buy up larger tracts of land around the proven wells when it became dirt cheap and repurchase Atlantic Oil stock for ten cents on the dollar. I bet that's what they're doing right now, taking back control of Atlantic, which has gas and oil distribution rights in all of Delaware and Pennsylvania. Uncle Tommy tells me that Pelham Oil has hit it big in East Texas, and is expanding to West Texas and Louisiana."

Lorry said, "So when people look at either one of us, they'll think we were born with silver spoons in our mouths."

"Yes, I suppose so."

"Do you ever think about getting your hands on your parents' money? Or find out how much they have?"

He replied, "No, I'm happy with what I'm doing. I don't need a lot to get by, and no amount of money will buy you anything on a Navy ship. You cut the mustard and advance, or you don't. If you don't, you're no longer in the race for promotion and command. I've spent most of my life unaware of our family fortune. When Dad and Gramps argued over the stock sale, I remember my dad saying that real estate and the 1929 stock market were ridiculously overvalued, so he had also sold the Pelham Oil Building."

"It sounds like your dad had perfect timing before the depression hit."

"Yes, I think he did, but Gramps yelled that East Texas oil would never pan out and wanted to know where his money was. Dad replied that one-third of the funds were in Homestake Mining stocks and bonds because he knew that they would pay a dividend no matter what. Another third was with the Kahn Brothers Investment Firm in New York because they were betting against the overpriced, insanely credit-fueled stock market. And the last third, some twenty million in cash and government bonds, was safely in the Bank of New York. He yelled back at Gramps that it had taken him seventeen years to make Pelham Oil worth ten times their initial investment, that he should be getting a medal, not a lot of grief from his father.

"That shut Gramps up. He lowered his voice and must have said something complimentary to his namesake. As they exited the office, Gramps smiled with his arm around my dad. He announced that the two of them would be celebrating at Sardi's. Since then, I've realized that I'll always be okay. The Navy has taught me that I can do almost anything and survive. It's nice to know there's family money, but it's not something I need or think much about."

Lorry replied. "It's the same for me. No amount of family fortune is going to advance me in medicine. You either know your stuff and make a positive contribution for hours and hours, maybe two days in a row, or you falter and fail, finding yourself left out of the exciting action." She looked at him intently. "Please pull over; I have something to say to you."

David pulled over on a broad gravel shoulder and stopped the car.

Lorry moved closer and took his hands in hers. "Thank you for taking me to meet your family. I really had a good time. Your mom is a peach, and your sister is too. Meeting them makes me feel like I've known you for several weeks. I want to be your girlfriend for real. What do you think?"

"If I prayed to meet someone with a personality who is fun to be with and downright beautiful, it would be you," David said.

"Then I think you better kiss me, David Pelham, buy me dinner at one of these roadhouses, and show me your digs off campus."

Lorry moved in with David a month later. They were engaged in three months and married in six from that first kiss in the car. David's older sister, Marie, forgave them for upstaging her wedding when she learned Lorraine was expecting the first Pelham grandchild's birth six months after honeymooning with her brother.

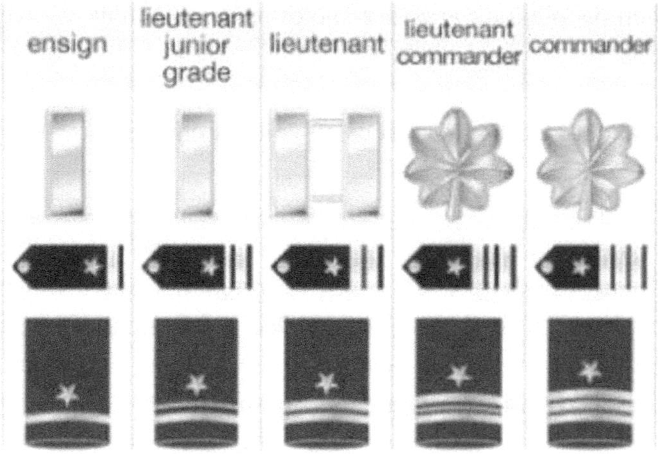

U.S. Navy O-1 through O-5 ranks.
First row, insignia worn on khaki shirt collars
Colors left to right, gold, silver, silver, gold, silver
Second row, shoulder boards worn on khaki and white uniforms and overcoats
Third row, sleeve stripes, and star for unrestricted line officers
Sleeve colors, gold braid and gold star on Navy blue, or black on aviation green
Source: Britannica.com

Chapter Four

DR. PELHAM, PH. D – Philadelphia, Pennsylvania

1935 – 1940

During his first years of marriage to Lorraine Brockman Pelham, Ensign Pelham advanced to Lieutenant Junior Grade and then to Lieutenant (O-3). He continued to drill one weekend a month and went on active duty for two weeks or more each summer. He had an extended summer cruise in 1937 on the USS Oklahoma battleship and served as Assistant Gun Director for the ship's smaller guns. David joined the University Flying Club and was ready to complete his first solo flight when another member stalled and crashed the Club's Waco F Series biplane. The accident killed both the instructor and student - and demolished the airplane. Lorraine put her foot down - no more flying for fun.

By the summer of 1939, Lieutenant Pelham had ticked off all the correspondence courses, on-the-job exams, and operational training requirements necessary to advance to the rank of O-4, Lieutenant Commander, USNR. At the University, he earned his master's degree and worked toward a doctorate in history. David specialized in military history and taught one or two courses per semester on the history of warfare. Lorry graduated number one in her nursing program and started her General Practitioner physician training. They found an excellent nanny for baby Mandy when she was one year old. When she was two, great-grandpa William bought them a four-bedroom home near the UPenn campus.

Pelham's doctoral thesis in 1939 was controversial and not well-received by his doctoral committee. Entitled "How to Win the Coming War with Germany," it traced the Nazis' rise to power and predicted that war was inevitable in Europe. Moreover, he projected a series of German victories that would eventually force the United States to declare war on Germany and Italy - predictions that did not sit well with the pacifist-minded professors on his review committee.

David used his family's experience in the oil industry to define an American air, land, and sea campaign that focused on destroying the enemy's energy resources. He argued that new technology supported unprecedented levels of aircraft, ship, submarine, and ground force mobility. He asserted that victory in the next war would go to the side that fully implemented the flexible and rapid movement of war machines, personnel, and supplies.

Pelham postulated that this increased lethal mobility depended on oil and refined oil products. Accordingly, he recommended destroying enemy oil fields, storage facilities, transportation assets, refineries, pipelines, and electrical power systems as the highest priority. It was a strategy that would reduce enemy weapons production, restrict their maneuverability, and hinder their ability to concentrate the killing power of large Army, Navy, and Army Air Force units.

The thesis also identified which oil fields in Europe, the Balkans, and the Middle East had to be controlled or destroyed. David predicted that Germany, Russia, or both would invade Poland for its oil. He enumerated the types and numbers of weapons needed to defeat Germany and Italy. Pelham analyzed the oil-producing capabilities of Europe and the Middle East. He discussed the logistics of transporting and refining it to provide fuel to the German war machine. Pelham identified vulnerabilities and bottlenecks in petroleum logistics pipelines that could be created or exploited. His skeptical review committee grudgingly accepted the 300-page guide to defeating the Rome-Berlin Axis powers. As he proudly became Dr. David A. Pelham, Ph.D. During the graduation ceremony of 1940, Pelham had no inkling of the uproar his thesis would cause in the Army and Navy's senior leadership. Nor was he aware that Japan would join the Berlin Pact with Germany and Italy late in September.

In June 1940, David was summoned to Washington, D.C. He was interrogated by Commander Roscoe Smith from the Navy Inspector General's office and an FBI agent. Pelham's predictions that Hitler would absorb Austria, the Ruhr, attack Czechoslovakia, and split Poland with Russia were too accurate and confounding for the military hierarchy. Without explanation, the Washington establishment decided to confiscate all the reproductions of his thesis. FBI agents accompanied him back to the university and collected every copy from the library, his doctoral committee, and his office, including his draft copies and working notes.

Commander Smith visited David in his campus office two days later in civilian dress.

Smith took a seat. "Sorry for this hubbub, old chap, but your book has caused quite a stir in Washington."

"I'm not sure why, Commander. Most of it was rank speculation on what Hitler and Mussolini might do and rough estimates of European oil reserves."

"Oh, Pelham, you can't be so naïve. D.C. is full of military service power structures. You opened wounds on some of the biggest ones and then poured salt in them."

"How so?"

Smith stabbed the air with his thumb and forefinger. "You claimed the battle wagons are obsolete and championed the case for aircraft carriers. You predicted unrestricted sub-warfare by both sides and undermined the Army's planned strategic bombing intentions. You pinpointed the Navy's appalling lack of landing craft. Your exposition on why the current U.S. high command structure is unworkable probably pissed off every flag officer in town."

David protested, "But all of that is factually correct. I used historical examples and logically proved my hypotheses."

"Sure you did, but nobody likes a wise-ass, Pelham. Especially if he makes everyone else look dumb and uninformed."

"So, what happens now?" David asked

"I'm here to give you a personal heads-up. First, your treatise is now classified as TOP SECRET. If you know of any other copies, you need to turn them over to our office or the FBI ASAP. You could be facing court-martial or civilian felony charges for not doing so. Second, pack your bags and be ready to be called to D.C. on a moment's notice."

"To do what with whom?"

"I'm not completely sure, but one rumor is CNO read your tome, and even though he's a battleship sailor, he directed all of the Bureau Chiefs and Fleet Commanders to read it."

David was shocked. "You're saying Chief of Naval Operations, Admiral Harold Stark, read my paper and may want a piece of my hide?"

"Yes, and on a moment's notice, but it gets worse. Vice Admiral King, Commander-In-Chief, Atlantic Fleet, may also want you flogged. Another scary rumor is that Harry Hopkins may have made a copy for FDR. The FBI plans to investigate you because of your recommendation to develop atomic weapons. Once the French embassy learns you called the Maginot Line a foolish waste of resources, I doubt you'll be welcome in Paris. Do you want more?"

"I just wrote logically and made my case," David answered weakly.

Smith lit a cigarette. "Of course, old boy, but predicting the pact between Russia and Germany to divide Poland and central Europe and then claiming that in the end, Germany must attack Russia has stupendous political implications. Your analysis that we have to rush to the aid of Great Britain or fight a much longer war with Germany, no doubt has some State Department wienies working late into the night. Your total focus on oil, petroleum refining, and energy transportation doesn't match how the Army Air Corps plans to bomb Europe."

David shook his head. "What have I done? What have I done?"

"Pelham, it's a damn good book. You will have supporters and detractors depending on their slice of the political and fiscal pie. You said many things that needed to be said, but what about the Far East? Don't you think war with Japan is just as likely?"

"Commander, it may be with the proposed oil and steel embargos, but I just couldn't handle that increase in scope. Besides, the political policies, facts, and figures are more obscure or nonexistent for the Orient."

Smith stubbed out his cigarette and stood. "Just be ready to make your case. In Washington, you're either a hero or a clown. For the Navy's sake, be prepared to stand your ground. After all the bickering dies down, your outlandish assertions and recommendations may become policy."

David was surprised. "Sir, why do you say that?"

"Because if the big brass were just pissed off that you identified many of our weaknesses, they would've classified your product as SECRET. My take on the TOP SECRET classification is that they realize that several of your arguments should change our strategy and tactics. Also, some of the weapon systems you recommend may already be highly classified. Is there such a thing as Electro-Magnetic Aircraft Ranging or Proximity Exploding Anti-aircraft shells?"

David's reply was almost a whisper. "I don't know. During my summer cruise in '38, I was in a Plymouth, Cornwall pub. A British scientist on holiday befriended me, bought me drinks, and got pretty pissed himself. He said several of these weapons were possible if the government labs would get off their bureaucratic backsides. I left several of his ideas out since they seemed too much like Buck Rogers."

"Things like what?" Smith demanded.

"Building computational machines to break enemy codes and creating long-range rocket bombs that home in on radio signals, turbine aircraft that can fly higher and faster without propellers, and amphibious tanks that can swim ashore and storm a beach. Things like that."

"Did he bring up atomic weapons?"

"No, I got that from one of our physics professors here on campus. They've been debating splitting the atom for years."

"It seems a very touchy subject in our Nation's Capital; best you stop discussing it," Smith said.

"Yes sir."

After Commander Smith left, David thought about several complications with his new notoriety. One was how the 1941 Lieutenant Commander Selection Board might react when he had a chance for early selection. Another was how he could tactfully defend his work instead of locking horns with the deep-rooted political opinions of senior Navy and Army officers. Smith's warning was accurate. Two weeks later, David was called to Washington by his detailer in BuNav and told to wear civilian clothes.

The call came on a Friday; Lieutenant Pelham was ordered to report to the CNO Duty Officer at 0900 on Monday. He had the weekend to prepare his thesis in a briefing format for presentation to several Commanders and Captains. The detailer said more orders for duty in D.C. might follow, or David could expect sea duty. After teaching his afternoon class, he went home to discuss the call with Lorraine. Neither of them could figure out the intent of the Navy orders. Would he be sent to some remote command as punishment? How long would the duty last? What could they want in Washington, D.C., from a lowly reserve Lieutenant and academic egghead?

On Monday, Pelham left Philadelphia on the first train and reported to the CNO Duty Officer in the Main Navy building on Constitution Avenue. The Lt. Comdr. Command Duty Officer welcomed him to D.C. and handed him a set of orders and a surprise. The orders placed him on six months of Temporary Additional Duty (TAD) attached to the Chief of Naval Operations Headquarters Staff. The surprise was a letter from CNO Stark congratulating him on his early selection to Lieutenant Commander. The duty officer gave him directions to the Uniform Shop and handed him a manila folder. It contained paperwork requiring his signature for the new rank and TOP SECRET security clearance. A Welcome Aboard sheet informed him that he'd be assigned to Captain Randal Banner and provided the Captain's room number. After being fingerprinted and photographed, David received a temporary Navy Department badge. He was then escorted to Banner's office by a Third Class Yeoman.

Captain Banner shook his hand and welcomed him onboard. He briefed David that he'd support the Navy Operational Planning Board, consisting of several admirals and captains from different Navy specialties. Banner said CNO Stark chaired the Board. Admiral Stark was also a member of the Joint Board that worked on the problems of the combined Navy and Army national strategy. Captain Banner told David that the Joint Board had a working staff like the Navy's planning board.

Banner gave him the rest of the day off to work on his new uniform requirements and find a place to live. Next to the Uniform Shop and the Barber Shop, he noticed a large bulletin board with several notices, sales flyers, and personal ads. One caught his eye, "Naval aviator O-4 looking for a responsible roommate to share two bedroom apartment in Georgetown."

David bought new Lt. Comdr. insignia and dropped off his service dress blue blouse at the tailor for new stripes. He called the naval aviator who worked in BuAir, and they had a congenial lunch. They agreed to share expenses; Pelham now had a place to stay on his first night in the nation's capital.

Over the next few weeks, David developed a fast friendship with Lt. Comdr. Craig "Gimpy" Adams, USN. Craig was a fighter pilot who walked with a noticeable limp. Gimpy appraised David on all the ins and outs of work in the nation's capital and warned him about the internal and inter-service rivalries for funding and operational authority. Adams cautioned David always to complement his boss's agenda and to be seen and not heard unless called upon. He recommended David work like the Dickens during the week and go home to his family on the weekend when possible. Gimpy advised David that the hours would be arduous. He said most nights, all David would want is to return to the apartment, drink a beer, and fall into bed. Months later, David learned that Gimpy's limp was from being shot down over France in early 1940. He was voluntarily flying a British Hurricane fighter against German Luftwaffe bombers.

The months flew by, and Pelham made it home to his family many weekends. He spent other weekends researching Navy/Army issues, preparing briefings on Navy deficiencies, and refining tactics for specific threats like U-boats or opposed amphibious landings. He was part of a small team working with CNO Stark and the Joint Committee on a new Rainbow War Plan scheme. They were preparing a draft memo from the CNO to the Honorable Frank Knox, Secretary of the Navy, on options for fighting the coming war. In September 1940, his CNO TAD orders were extended for three more months.

The University had grudgingly put him on a small stipend and temporarily reassigned his office to another professor. In November 1940, Lt. Comdr. Pelham, an Army Air Force Lieutenant Colonel and an Army Colonel from the Joint Board Support Staff, received TAD orders to England. They were to evaluate the contribution of new technology to winning or losing the ongoing Battle of Britain. Little did David know that being absent during Thanksgiving week would cost him everything he held most dear in the world.

Over a year later, during the afternoon of 7 December 1941, Professor Pelham was in his campus apartment preparing the final semester exam for his history class when there were loud noises in the hallway. He opened his door and heard a neighbor yelling, "The Japs have attacked us at Pearl Harbor! It's war - war in the Pacific!"

David slowly closed the door and retreated to his work desk. He sat down and put his head in his hands as he thought of the consequences for himself, his family, his Navy, and the whole country. He expected an immediate call to active duty and orders to a warship. There was a damn good chance that the war would kill him. Pelham remembered the contentious and often vicious strategy arguments of the war planning committee members, mainly focused on Europe, not Japan. The Washington planners and Admiral Stark's Rainbow Plan memo to Secretary Knox recommended a defensive posture in the Pacific and an aggressive attack against Germany in the Atlantic for war with the Axis powers.

David thought, *Now the die is cast. The Japs had taken advantage of American complacency. They combined the lessons of Taranto, Italy, with their successful surprise attack against the Russian Asiatic Fleet in 1905. Capabilities and Intent, Intent and Capabilities are the two main challenges of warfare. What does your enemy intend to do? What is he capable of doing?*

Pelham doubted that anyone in Hawaii had expected the Imperial Japanese Navy to steam 4,000 miles without detection and mount a major air attack on the Pacific Fleet. He thought *The Japs must have modified their torpedoes for shallow water attacks, similar to what the Brits did to cripple the Italian Fleet in Taranto last year. They had to maintain radio silence for weeks on end! They had to refuel at sea to pull this off, a capability we didn't think they had. And now what? No doubt, the Japs will attack Guam and Wake; maybe Midway could be next. Then a thrust south to claim the natural resources of the Dutch East Indies, Malaya, and Borneo. To do that, they will have to control the Philippines.*

The Pearl Harbor attack made it clear to Lt. Comdr. David Pelham that the U.S. was now at war with Japan, and hostilities with Germany and Italy couldn't be far behind. They must have known what their Asian Axis partner was planning and given it tacit approval. World War II would be the most demanding strategic situation of the Rainbow Plans - a two-ocean global war.

Pelham called his dad in New York and said he'd be coming home to say goodbye. Within a week, he completed his physical and received active duty orders to a destroyer in Hawaii. When he arrived at the family penthouse in Manhattan, his dad said David looked tired and depressed. William Pelham insisted that his son take the company plane to Chicago. He also booked David to Los Angeles on the Atchison, Topeka, and Santa Fe Super Chief. It was an eventful five-day cross-country railroad journey on several levels. Before boarding the train in Chicago, David learned that his dad had arranged for the Pelham Pullman car to make the long-distance run through Kansas City, Albuquerque, and Barstow. A conductor escorted Lt. Comdr. Pelham to the private railcar, where he met the butler and maid. They showed him his private bedroom and bath. They also showed him the liberal supply of liquor, wine, and beer onboard.

War is hell, he thought, *I might as well enjoy any perks that come my way. Of course, as Operations Officer on a combat destroyer, I will have trouble finding time to brush my teeth or take a decent shit.*

Santa Fe Super Chief circa 1966

Source: Wikimedia Commons

Chapter Five

HAWAII ORDERS – Santa Fe Super Chief

December 1941

Rear Admiral Chester W. Nimitz and his Flag Lieutenant moved into their sleeper berths on the Super Chief train in Chicago on their way to the West Coast. Nimitz was intrigued to see a Naval officer, a Lieutenant Commander, on the platform handing out what looked like fliers or leaflets to other men in uniform boarding the train.

The Admiral asked his companion, Lieutenant junior grade Arthur Lamar, to investigate what the officer was handing out in the adjoining berth. Then, on a hunch, he also told Lamar to ask the officer to join them for their cocktail hour at 1700. Both Lamar and Nimitz were traveling in civilian clothing. Lamar had orders from Admiral King, Commander In Chief U.S. Navy, to ensure that Nimitz got some rest on the trip to the West Coast before becoming Commander in Chief Pacific (CinCPac) in Hawaii.

Lt. (jg) Lamar returned from his scouting trip and reported that the officer invited all service personnel to a happy hour at 1700 in his family's rail car. Lamar had also observed a brightly colored car with "Pelham Oil" emblazed on the side and passed this on to his Admiral.

"So, he's traveling in style," Nimitz commented.

"He says the rail car is a surprise perk from his dad. I think I've heard of Pelham Oil up in New York. He asked if he could stop by at 1800," Lamar related.

"He sounds interesting. Let's break out the scotch at 1800 and push back our dinner until 1900. What's his name?"

"David Pelham. He's a reservist called to active duty. His dad is William Pelham, President of the oil company."

David was surprised on the railroad platform to have Lt. Lamar introduce himself and ask what he was doing with the leaflets. When David showed him the invitation, Lamar said he was traveling incognito with a senior officer and invited Pelham to join them for cocktails.

At 1800, David knocked on Lamar's compartment door. Lamar asked him to be discreet about his traveling companion. He then knocked on the adjoining compartment door. An older, tall, upright, thin gentleman with bright blue eyes and white hair came forward and firmly shook David's hand. Lt. (jg) Lamar completed the introductions: "Admiral Nimitz, Lt. Comdr. David Pelham. Mr. Pelham, Rear Admiral Chester Nimitz."

"Admiral, congratulations on your new rank. On your new job, I don't know whether to celebrate with you or commiserate with you," David said sincerely.

"Commander, we can do both at once. Please join us for a drink. Where are you heading?"

"Sir, I'm on the way to your command as Ops officer on a Porter Class destroyer, the USS Selfridge."

David saw quick, meaningful glances between Lamar and Nimitz. He thought, *What was that about?*

"A fine class of ships are the beefy Porters. Happy hunting," said Nimitz as he clinked his glass with David's.

Pelham might have been taken aback by the two-star, about to be four-star Admiral, had he not worked in D.C. for nine months with Admiral Stark, other Rear Admirals, and a host of Captains. While in Washington, David had received some choice chewing outs and debated respectfully yet forcefully on new weapons with three-star Bureau Chiefs and their staff. David now saw Admirals as mortal men who worked tirelessly to improve the Navy. Most were open to a lively yet respectful debate on strategy and tactics. He found they needed accurate, honest, and direct communications to make meaningful decisions.

"Mr. Pelham, something about your name strikes a bell," Lamar probed.

"Maybe it's my dad, William Pelham. He cuts a pretty broad swath through the New England oil industry."

"No, that's not it; it's something Navy," Lamar said, deep in thought.

"Lieutenant, maybe I ran into you last year when I worked for CNO on the Planning Board or at the Joint Planning Board," David offered.

Lamar snapped his fingers. "Now I've got it. You're Professor David Pelham, who wrote the book on defeating the Germans."

"Guilty as charged. Had I known the feathers that would get ruffled, I doubt that would've been my thesis subject."

"I spent some of my best years at Berkeley. What's your University?" asked Nimitz.

"University of Pennsylvania, sir. Professor of history specializing in the history of war."

Nimitz took a sip of his drink. "Got any tips on how to beat the Japs for this old-timer?"

"Admiral, you aren't the first who has asked. Unfortunately, Japan and the Far East were too large in scope to include in the thesis, and the factual information is insufficient."

Lamar chimed in, "Admiral, we have his book. It's part of the ten pounds of light reading Admiral King put together for you. It's now classified TOP SECRET."

"I'll have to look at it. There must be some parallels between the Pacific to the war in Europe. Lieutenant Lamar is going back to D.C. once we reach the coast. Commander Pelham, would you like to continue with me to Oahu?"

"Sir, I'm going to decline respectfully. I'm deep into a book on the history and psyche of Japan and have decided to see the author at USC. If I fight a ship against Japan or, hopefully, have command someday, I need to understand their thoughts, strong points, and weaknesses."

Nimitz held up the second drink just made by Lamar. "Gentlemen, to our brave comrades on Wake Island!"

David and Lamar, in unison, held up their drinks, "Brave comrades on Wake Island!"

"Commander, I'll be interested in what you learn at USC; plan to join me for a meal when you get to Pearl Harbor. I will read your book tomorrow; I'll have questions at 1700. Can you join us?" Nimitz requested.

"Yes sir, thank you for the drinks. Looking at the New York Times map this morning, I'm hoping the boys on Wake can hold on until we can reinforce them. If Wake falls, the importance of Midway increases by leaps and bounds."

"I couldn't agree more, Mr. Pelham. The loss of both Wake and Midway could put Hawaii in jeopardy. It also depends on the extent of damage to Pearl's ship and sub-maintenance support facilities," Nimitz replied.

"How bad is it, Sir?"

"The damage overall is shocking, the ships and men lost, and the aircraft destroyed. My beloved flagship Arizona is beyond recovery, as is Oklahoma. Nearly every other battleship and several cruisers have moderate to severe damage but are repairable. Thank God all three carriers weren't in port."

"Yes sir, if you don't mind me asking, where were they?"
"No, not at all. Lady Lex was delivering Marine dive bombers to Midway. The 'Big E,' under Halsey, was taking Marine Wildcat fighters to Wake. Enterprise would have been in Pearl the morning of 7 December except for bad weather. Saratoga was bound for San Diego to embark additional Marine aircraft and her Airwing."
David replied, "Admiral, just to forewarn you, my treatise predicts this will be a short carrier war. Older battleships that can't keep up with carriers will find other missions. The Japs have proved my point at Pearl, and the Brit's loss of Repulse and Prince of Wales off Malaya have hammered it home."
Nimitz smiled, "Oh, but us battleship admirals will argue that it proves carriers should protect the battleships. First reports are that Admiral Sir Tom Phillips had no air support when the Japs surprised him."
"Yes sir, but the current Navy doctrine to use faster carriers to cover slower battleships is a defensive strategy. It blunts the offensive application of carrier forces. As a result, I believe the era of two lines of battleships slugging it out at sea is over."
"Commander, you may be right. Certainly, Bill Halsey would agree with you. Bet I'm in for some intense discussions in the months ahead."
"No doubt about that, Admiral. Best of luck, sir. I better get back to the lads and ensure they don't damage Dad's railroad car."

The next night, David met with Lamar and Nimitz for drinks. David was ready to defend his thesis, but Nimitz nixed any talk of the war. Instead, the Admiral asked David if he was a bridge player. David admitted to being a little rusty. Nimitz had been reading all day and wanted to play some three-handed bridge. He asked Lamar to order sandwiches for dinner.

Pelham learned that Nimitz was an excellent card player as the train sped through the darkened Midwest plains. He was willing to take calculated risks in his bidding and adept at disguising his hand until playing the last card. They started each round by dealing four hands, turning over eight cards in the dummy hand, and competitively bidding against each other for the contract. The game was lively, and although Lamar and David fought fiercely, Chester Nimitz won handily. They all enjoyed the drinks, the food, and the Admiral's yarns. They were primarily humorous stories that he spun about his Texas youth. Unfortunately, there was no invitation for the next night, so David assumed the two officers had other plans.

The next day, Pelham took his meals in his dad's Pullman car and had time to complete reading the book on his new enemy. "Understanding Modern Japan," by Professor Michael Nagy, was particularly insightful from a historical perspective and presented the Japanese social structure and general beliefs. David decided he'd meet the author and discuss the book's contents. Again, he appreciated the luxury Pullman car, which had a smoother ride than most other railcars. The rhythmic clickety-clack of the wheels lulled him to sleep through the night.

Riding through the Colorado countryside at dawn, David thought about his family. His mom had passed, and he missed her dearly. His sister and husband, Epi, worked good jobs and started a family in Newark. His dad was, as always, too busy to have much time for the rest of them. David wondered how the war would affect his father and Pelham Oil. William had again bought a controlling interest in Atlantic Oil and changed the name back to Pelham Oil. Undoubtedly, there would be a high demand for oil to support the Allied war machine.

He expected his dad would be expanding the company's share of the industry and the family fortune. He could not think about his wife and daughter – it was too painful.

David remembered how his dad had lectured him on creating the family business. In 1912, William borrowed a hefty sum to accumulate Atlantic's common stock until he could control the Board of Directors. Then, he elected himself president. The Atlantic assets resulted from the government's trust-busting of Standard Oil. William Pelham took control of Standard's assets and distribution system for petroleum products in Delaware and Pennsylvania.

His dad competed against the Standard Oil companies in New Jersey and New York and was generally called a fool for bankrolling wildcatters in East Texas. He commuted from Pocantico to Manhattan daily to manage the expansion of Pelham Oil. Initially, he rented the third floor of a building overlooking the north end of Central Park near the intersection of Central Park West and Eighth Avenue. As the years passed, he leased floor after floor until he purchased the seven-story building and moved his family to the first floor. The successful CEO moved the family into a penthouse further south on Park Avenue several years later.

The Manhattan penthouse on Park Avenue was his dad's statement that he had arrived as a captain of industry. David's parents had a large master bedroom and bath combination with his and her walk-in closets. Pelham and his sister Marie had private rooms with separate bathrooms in the hallways. There was a guest bedroom, bath, and a tiny maid's overnight room. His dad's massive office looked out over the Park and had a private bathroom. The dining room, living room, and kitchen looked south on Central Park and the Hudson River. There was also a bedroom and bath suite for Uncle Tommy, who lived with them for several years. Tommy was an honored family member and something of a servant simultaneously.

As the railroad miles rolled by, David thought, *I wonder how Uncle Tommy feels now that we will fight his Japanese homeland. After meeting Professor Nagy, I should call Tommy for his advice on fighting Japan.*

Central Park, New York, circa 1940. Note enlisted sailor and girlfriend

Source: Wikimedia Commons

Chapter Six

PROFESSOR MICHAEL NAGY, USC – Los Angeles

December 1941

The Super Chief arrived in Los Angeles early on a bright sunny morning as Lt. Comdr. Pelham was packing his gear to leave the luxury of his dad's Pullman car. The butler and cook wouldn't let him depart until he had a hearty breakfast. They told him he had plenty of time because a company car was coming. David agreed to stay for another meal, realizing it was a rare gesture of affection from his distant father. While they set the table and prepared breakfast, he read through most of the morning's Los Angeles Times. It was full of war stories and speculation on the extent of the Pearl Harbor debacle.

At about nine o'clock, a tall, well-dressed man with a thin mustache entered the railcar's rear door. He introduced himself as Ronald Strong, Pelham Oil's representative in Los Angeles. David looked at the man's tan and perfect white teeth and asked if he was a movie star.

Strong replied, "I have to admit, like hundreds of other people in this crazy town, I tried it. Working for your dad is my steady gig. I still get called to double for Gable and do stunts."

"Mr. Strong, you know Clark Gable? That's amazing!" David exclaimed.

"Please call me Ron. It's not like he and I are drinking buddies. They have scenes to shoot where you don't see his face, or he might get injured. I get paid a basic union scale to fill in."

"I didn't know we had an L.A. office. What do you do? How long have you worked for the company?"

"Almost nine years now. I answered an ad your dad put in the Times: 'Wanted, intelligent, and trustworthy land broker who can work independently in the strictest confidence.'"

"To do what?"

"I wasn't sure at first. I had a real estate agent's license; your dad paid me to get my broker's license. He pays me a monthly stipend and a commission on the land I purchase for the company."

"We're buying land in California. I didn't know that." David remarked.

"Not as much lately, but quite a bit from '32 to '36. I look only at proven oil and gas acreage and buy the land and rights, or just the mineral rights. Everything has doubled or tripled in value since we started, so we've worked more in Oklahoma and New Mexico the last three or four years."

"How much land has he, uh, have you bought?"

"My territory is west of the Pecos River and north of the Canadian out in Texas. We hold about 500,000 acres. I'm not sure what your dad and Tommy have east and south of those, but it's a lot."

"Dad put lots of investment in East Texas for many years that I know about."

"Yes, we've continued that and bought acreage in West Texas, Midland, Galveston, and Louisiana. Your dad and Uncle Tommy also handle all offshore leases and anything international. They probably hold as much acreage as I have out here or more."

David was surprised. "Ron, you're saying Pelham Oil may own as much as a million acres of proven oil and gas land?"

"I'd say so, at least the mineral rights for that much."

"At how much an acre?"
"In the early 30s, we picked it up by paying back taxes, using no-bid auctions, and bankruptcy courts at $1 to $10 per acre. Last summer, most of it was worth $20 to $50 per acre. With the war on, who knows? Next year, I'd expect an average of $100 to $200 per acre."
"So Pelham Oil may be worth over one hundred million?" questioned David.
"No, no, much more than that, Commander. Some fields are pumping, then there's the retail fuel distribution system on the East Coast, and the bank is doing great. I know because I own some stock in it."
"News to me; what bank?"
"Your uncle started the Bank of the Orient in New York. I think your dad funded it and owns a majority of the stock. Tommy is the President, and your dad is the Chairman of the Board. Deposits have grown from ten or twenty million to over 200 million in five years. I rarely meet a stateside oriental that doesn't have an account in New Orient."
David sat down deep in thought. *I've learned more about what Pelham Oil has been doing in the last five minutes than in the previous five years. Dad and Gramp's sixty million seems to have quadrupled or more. Uncle Tommy has used his night school finance classes well and runs a successful national bank. Holy cow, sister Marie and I are due for quite an inheritance someday. Of course, that's if I survive this war. If it changes me, I might give the oil business another look.*
Strong said, "Not to interrupt you, but I must let you know while I remember. Your uncle will fly in tomorrow or the next day to look at some buildings and apartments."
"Why is that?"
"New Orient has been investing in income-producing buildings on the West Coast for several years. Tommy thinks like your dad. They both believe all parts of the California economy are ready to expand rapidly. So if property owners are scared by the war and ready to dump their properties at a bargain price, Bank of the Orient is a buyer."
"Ron, can you give me a lift over to USC?"
"I can do more than that. You're my number one priority. I have you booked into the company's suite at the Beverly Hills Hotel. Whenever Tommy gets here, he'll be your roommate. Where do you want to go at USC?"
"I want to meet a professor, Dr. Michael Nagy. He's an expert on Japan and Japanese history. I just read his book on understanding Japan, published two years ago, and I want to know more."
"Does he know you're coming?"
"No, I haven't had a chance to call."
"No sweat, I have some connections with the Arts Department there. I'll call and see if I can get you a meeting with Nagy. Be right back."

David gathered his bags and said goodbye to the cook and butler. He thanked them for the delicious breakfast and gave them a generous tip. Strong returned and said Nagy would be in his office until noon. They piled into a bright purple '40 Buick convertible. Ron had the top down; David noticed the seats were warm in the sun, even in December.

Ron Strong said USC was nearby, and they had time to take a little tour. He swung the car north out of Union Station and showed Pelham Chinatown. While stopped at a streetlight, David saw some vehicles racing down a side street with men firing machine guns out of the car windows at each other. Pelham was puzzled as Strong turned the Buick down the same road and poured on the gas to catch up. Above the squealing tires, he told David to look for the cameras and that it was a car chase as part of the thriving Chinese language film industry. They caught up with the vehicles and the filming crew setting up the next street shot a few blocks later. Strong pulled up to the desperados' autos, stopped, got out, and exchanged cards with the director. David could hear him say something about Clark Gable.

Strong circled the Buick back to the south and gave David a tour of downtown Los Angeles, where he pointed out several buildings and apartments owned by the Orient Bank. They finally drove down Main and took a right turn on Jefferson. Strong found a parking spot not far from USC's McCarthy Quad. They found Nagy's office listed in a directory at a nearby building.

The office door was closed, and a sign indicated the professor was out. Strong knocked loudly on the door anyway; they heard a voice saying to come in. Once inside, Professor Michael Nagy greeted them. He was clearly of Japanese descent. Nagy was about five feet ten inches tall, of medium build, light-skinned, and very friendly. He shook their hands, took both cards, and asked them to be seated as he returned to his desk chair.

David opened the conversation. "Dr. Nagy, thank you for meeting with us. I'm Navy Lieutenant Commander David Pelham, on my way to a destroyer based in Hawaii. I just read your book and hoped to discuss it with you, especially the changes in Japan during the two years after you published it."

"Mr. Pelham, I'm flattered that you enjoyed the book. How can I help you?"

"Please call me David. Believe me when I say that I mean you no harm. I'm not from the press or the government. I'm not going to quote anything you say to us. We have a lot in common. Before being called to active duty, I was a history professor at U-Penn. I've been in the Naval Reserve for ten years. I did all my training in the Atlantic, the Mediterranean, and the Caribbean. Imagine being surprised with orders to the Pacific to fight an enemy I know nothing about."

"What do you want to know that's not in the book?" asked Nagy.

"First, what is the current structure of government? Who is in charge? What are their strengths and weaknesses? Second, who's in charge of the Navy? Third, what's the quality of the Navy ships, their weapons, and their training? What kind of people will I be fighting? What is their combat spirit? Who are they fighting for, and why? Fourth, what are the military capabilities of Japan, its strengths, and its weaknesses? Fifth, what are the scandalous items you left out of the book? The things you knew to be true but couldn't publish. What are the most recent revelations in this category?"

Nagy hesitated and said, "You're asking for a lot. The hard facts for much of what you want are unavailable to me, and I don't want to mislead you. Are you sure you want to trust the opinions of a Sensei Japanese? I'm the second generation born in America. I consider myself to be American first and of Japanese descent second. My immigrant parents are Issei, born in Japan. They aren't allowed to be naturalized citizens of the USA - yet they're wholly loyal to America."

Pelham replied, "Your book tells me the whole story about trusting you. It's an unbiased history of Japan. It helps the reader to understand the culture and their way of thinking. I want to work with the man who wrote such a book to defeat Japan in this war. If that's a problem, please tell me right now. For starters, what's your real name?"

"My real name is Masahito Nagai. My father worked right here at USC for thirty years. We're proud of the Nagai family name, but my father told me to change it when I was ten. He said he had worked with white people for decades, and they still couldn't get his name right. So he believed that when I became a scholar, people would remember a Caucasian name and buy my books."

"Well, he may be right about that. It wouldn't have bothered me. Are we going to work together?"

"How much time do you have?" Nagy asked.

"This afternoon and all day tomorrow, if necessary, maybe the next day. Will I be interrupting your classes or other work?"

Nagy paused with a pained look, "No, the Dean suspended my ancient and modern Japanese history classes."

"Really? It seems we need to learn more about Japan than ever before," said Pelham.

"There's an increasing anti-Japanese hysteria here on the West Coast. Asians, even Chinese, are beaten up by bullies for no reason. Newspapers and anti-Nippon groups are calling for the incarceration of all Japanese to prevent spying and sabotage."

David nodded. "I know Secretary of the Navy, Frank Knox, has been making statements like that for years. But, surely, we aren't going to lock up American citizens, are we?"

Nagy said, "It all depends on which way FDR bends. State of California officials like Attorney General Earl Warren, the military, and white farmers want the Japanese out of all coastal areas or even the entire state. Other whites in cities and towns can't wait to confiscate Japanese-owned homes, small businesses, and land."

"Sounds like unbridled racism to me. How many people are we talking about?" David asked in an angry voice.

"It depends on the rules the military is allowed to enforce. Rumors are that they will lock up anyone with one-sixteenth Japanese ancestry. That would be anyone with a great, great Japanese grandparent. They may also include persons of Taiwanese and Korean descent because they're under Japanese control. If they lock up all the Issei, Nisei, and Sansei, it would be over 100,000 people in California, the majority of them American citizens."

"War is always an unexpected shock to a nation, its values, and social fabric. What about Hawaii? Is there a large percentage of Japanese there?" David asked.

Nagy frowned. "There's the rub. About one-third of the Hawaiians are of Japanese descent. They comprise roughly ninety percent of the island's carpenters, most transportation workers, and a large percentage of agricultural workers. General Emmons, the new military governor of Hawaii, has already stated that it will be impossible to rebuild Pearl Harbor without Japanese workers."

Pelham speculated, "So let me guess, where there is a large percentage of Japanese, it's too hard to lock them up and too expensive to lose the economic productivity. Here on the West Coast, where they are a smaller minority, we plan to invalidate their citizenship, confiscate their wealth, and exclude them from the war effort."

"Exactly, the plan is to lock up small children, even babies if they are as little as one-sixteenth Japanese, and the elderly, the mentally insane, and the disabled."

"How does that affect you and your family?"
"My wife is a doctor of Japanese descent, an M.D. We will ensure that our family goes to the same facility as my parents to care for them. Not being citizens, my parents will lose the small farm they retired to in the San Fernando Valley. It's a property that I own, and they dearly love. I'll teach school in the camp. I'm collecting books to teach high school and some college courses."
Pelham was in agony. "We're going to incarcerate a professor and a doctor, and their sons, who are all citizens? What do you mean by camp? What camp?"
"It's clear that they can't put 100,000 people into existing jails. So they, probably the Army, will have to build internment camps, no doubt with barbed wire, armed guards, and searchlights at night."
"Sweet Jesus! We'll lock our citizens in concentration camps without due process, just like the evil war machines we're fighting! I can't believe it. I can't believe it!" David exclaimed.
"War creates the ultimate in excesses, in logic, runs wild, as you said. Can I help you fight more effectively in just three days?" Nagy asked.
"I think so. I'm a little disconcerted right now. Your book was full of new information for me. I understand much about the war in Europe, its causes, the major players, and the nature of the weapons and anticipated tactics. But, on the other hand, I know little about Japan, not even its population."
"Seventy-three million, including the four major islands and Okinawa," Nagy responded.
"And we have about 132 million, not quite a two-to-one advantage."
"Dr. Pelham, the wild card is how you evaluate the Japanese-dominated populations in Manchuria, parts of China, Formosa, Korea, Inner Mongolia, eventually the Philippines, the Dutch East Indies, Malaya, and French Indochina, Hainan, Thailand, and Burma. Not to mention its influence over hundreds of populated islands in the vast region of the Marianas, Caroline, Solomon, and Marshall groups."
"Hold it, Michael, you're giving me a headache. Are you saying we are or will be on a parity regarding population?"
"Yes and no. Japan realizes her colonial ambitions and consolidates her control over most of Asia; however, this can be her undoing since she's gobbling up territory and populations like a glutton at an unlimited buffet. She hasn't conquered China after nine years of bitter fighting. She hates and fears the Soviets to the North. Once Japan invades a country or area, the Nippon administration is brutal and repressive beyond belief. Yes, they control the population and natural resources, but it's a surly, angry population ripe for revenge and resistance movements. I believe the role of Japan as an Axis power is to drive the European and USA powers out of Asia and attack the Soviet Union."
"When do you see that happening?"
"Driving out our Allies, the British, Dutch, French, and ourselves, is probably working better than they ever expected. If we don't do something drastic, they will take Wake Island soon. Guam in the Mariana Islands has fallen. Rabaul in New Britain and most of New Guinea are indefensible - they will fall quickly. Since World War I, the Japanese have controlled Tinian in the Marianas, Palau and Truk in the Carolines, and several atolls in the Marshall Islands. The Allied Mandate from World War I gave them German island possessions in the Western Pacific north of the Equator."

David's head was reeling. He had no clue where these hundreds, even thousands, of Pacific Islands were located. He asked, "Do we know where they are, and in what strength?"
"No, we don't," Michael said as he circled the map from New Guinea to Japan. "They may have many bases scattered across this vast area. It will take a massive effort to ferret out where they are and discern the maturity of their air and sea bases." He banged a pointer on the map. "They've controlled Taiwan and Okinawa since before the turn of the century; they annexed Korea in 1910. When they beat the Russians in 1905, Japan took over the southern Kamchatka Peninsula. In the early thirties, they invaded and controlled Manchuria, now called Manchukuo; that was the prelude to taking over big chunks of China. They invaded Hainan two years ago and French Indochina in 1940 as France fell to the Germans."
David held up his hands. "Stop, please stop; I think I need a break. Is there any good news on that map?"
"Maybe. I doubt Japan has the forces or the will to attack the Soviets soon. They want a big chunk of Siberia, and the Axis has probably given them a claim to all of Russia's Far East territory. China still ties up several Japanese divisions that could fight the Russians or us."
"What about MacArthur and the Philippines? We have plenty of armed forces there, right?" David said with concern.
"On paper, the Philippines looks tough with over 100,000 troops and various front-line combat aircraft. Only about 30,000 troops, 21,000 American and 9,000 Filipino Scouts, are adequately trained and reasonably equipped on closer examination. The rest are various categories of poorly armed reserves and have little training or discipline. Rumors are floating that MacArthur lost most of Asia's second-largest air force in two or three days. The Japanese can stage combat-hardened troops against northern Luzon from Formosa, only 200 miles away. With the debacle at Pearl Harbor, we'll not be able to supply significant numbers of men or equipment to MacArthur to fend them off. The Philippines will fall in six months or less."
David was shocked and subdued. "How do you know all of this?"
"My wife has relatives in Manila. We traveled there last year, and I talked to many levels of government and the defense forces. They were all bracing for war with Japan. I talked to college professors who considered war inevitable after our Export Control Act of 1940 embargoed steel and scrap iron. Japan imported ninety percent of its oil. Eighty percent of that came from the US. The handwriting was on the wall once the US and Dutch oil imports shut down last July. Japan would have to move against the oil-rich Dutch East Indies and Malaya or be slowly strangled - militarily and economically. Taking the East Indies requires an invasion of the Philippines. Otherwise, the U.S. could disrupt Japan's lines of communication to their new oil and natural resources sources."
"You aren't painting a pretty picture, Professor Nagy; do we have any positives in this fight?"
"Of course, we're far more industrialized than the Japanese. Once U.S. rationing is in place and wartime production ramps up, we'll truly be FDR's arsenal of democracy. We will pour combat personnel, weaponry, aircraft, ships, fuel, and materials into Great Britain, Australia, the Soviet Union, and China at a rate the Axis powers can't match."
"Massive aid to China and the Russians? Are you sure of that?"
"We've already sent small amounts of aid to both. It will geometrically expand until our production keeps them in the fight. If Russia can slow and stop the Nazis' armies, and China

can keep fighting and tying up multiple Japanese divisions, our war is greatly simplified. We're fortunate that we still have Hawaii as a forward base."

"You seem to be adept at making predictions. Did you expect an attack on Hawaii?" David queried.

"No, I did not. Like many, I underestimated the range of Japanese naval air power. I should've known that Admiral Yamamoto would devise a plan just this audacious and startling. Their Navy is known for surprise attacks and has a penchant for striking a knockout blow against their chosen enemy. I imagine they planned to destroy the Pacific Fleet and then negotiate a cease-fire with us, giving them free rein in the Far East."

"Why on earth would we do that?" asked Ron Strong.

"Think about it. Without our Pacific Fleet, we still have the Germans and Italians to fight in Europe. If Japan invades Hawaii, the West Coast is ripe for Japan's amphibious assaults. Given that turn of events, it might be politic for our leaders to write off Asia to maintain our continental borders."

"Do you still think that we can lose Hawaii?" David asked with an alarmed look.

"I'm sure that Yamamoto surprised even himself with the results of his daring attack. However, here's one of those weaknesses that you seek to understand. Why did Yamamoto's Admiral break off combat with Oahu in flames and defensively broken? Why did he not launch another wave of fighters and bombers, another, and another? Why was the Imperial Japanese Army not strong enough to invade Oahu and claim it for Japan?"

"Please enlighten me. I don't have answers to these rhetorical questions," said David.

"First weakness. Yamamoto had to accept an admiral in tactical command whom he didn't prefer, who lacked the aggressiveness to exploit an unexpected superior tactical advantage. Reverence for seniority in both the Navy and the Army is a weakness. To assign a major command to a more talented but junior officer is an insult to more senior officers. Their loss of face could cause them to kill themselves. Their acts of suicide would be a national disgrace for the Navy, the military in general, and even the Emperor."

Pelham was surprised. "Michael, you must be joking. So senior military brass would kill themselves because they got passed over for an assignment?"

"In Japan, a man's honor is his most prized possession. It defines him and establishes his place in a very hierarchical society. During one of the junior officer rebellions, the Army generals were sympathetic to their demands and delayed ending their occupation of several Tokyo buildings. Finally, the Emperor called the generals to his palace and played his trump card. He said if the generals didn't end the rebellion, he would lead the loyal troops to end it."

"I don't get it; how is that a threat or an inducement?" David demanded.

"If the Emperor came out of his castle to lead troops against the junior rebel officers, he'd be risking himself. It would vividly demonstrate that he did not have faith in his generals to do his bidding. Their loss of face would be complete; the warrior code of honor would require them to take their own lives.

Both Pelham and Strong were surprised.

Nagy continued, "This brings us to the Hawaii attack and another weakness. If the operational commander has struck a mighty blow against our fleet, retiring from the battle is in his self-interest. Our carriers could lurk nearby and launch aircraft, taking the bloom off his triumphant victory. His place as a national hero was secure, but not if he loses a carrier or carriers to U.S. aircraft or submarines."

David said, "So, he retires from the fray a hero. If he stays, fights, and loses ships, he loses face and commits hari-kari, right?"

"That's not the correct term. If the leader has Samurai blood, he can regain his honor by performing Seppuku. It is a ritual cutting of the stomach and partial or complete beheading performed by his second in attendance. At any level, a dishonored Japanese military leader can't help but think about the high personal cost of defeat. This brutal accountability encourages laborious groupthink. It is a time-consuming consensus approach to major decisions for the generals and admirals. For junior officers, it encourages them to commit suicide by proxy in the storied banzai charge."

"This discussion is giving me the creeps. I'm not sure I'll ever understand Japan," David said, shaking his head.

"I think you will, by listening intently and absorbing the cultural lessons. The third weakness is how the military has perverted the code of the Samurai, the Bushido. They've propagandized the rank and file to embrace death as a duty to their God, the Emperor. Surrendering is not an option. Fighting to the death is their creed. Men in the Army and Navy believe they have an extraordinary fighting spirit. This 'way of the warrior' provides a spiritual shield that will win if they freely accept death as a duty. Officers mistakenly believe this intrinsic spirit will protect them as they lead their unit against enemy automatic weapons while waving only their Samurai swords."

"I can see how this information might be helpful to an Army officer but not the Navy," said Pelham.

"I beg to differ. There you are on your destroyer fighting a submarine. You force it to the surface, and the sub displays a white flag. What do you do?"

"Well, I would send a boat to see how we could take over the submarine."

"Wrong. Wrong. Wrong!" Nagy yelled. "They will never surrender. It's all a ruse to kill more Americans. They would never suffer the loss of face inherent in turning over their ship to you, never! To them, the highest purpose in their lives would be to kill one more Yank before they die. On the battlefield, they will stab you with a concealed knife, pull the pin on a grenade, and give you a death-dealing hug. If you do capture them, they're duty-bound to kill themselves. Death is always preferred to surrender or captivity."

David shook his head, "So, is this a weakness? It scares me. The other guy has nothing to lose."

Nagy replied, "The banzai charge worked in China because the Chinese have slow-firing, bolt-action rifles. A massive attack overpowers them. The Japanese Army despises the Chinese resistance because they don't stand their ground and present a decisive battle. Instead, the Chinese operational pattern is to retreat during the day and fight at night, which frustrates the Japanese forces. If you will, the Chinese run away to fight another day or night.

"Japan's generals in China squander the lives of their men, forgetting that it takes at least two years to train a Nippon Army infantryman or a Navy seaman. More importantly, producing a fully trained aviator takes three years or more. As the highly trained and skilled soldiers, sailors, and aviators needlessly throw their lives away, wrongly following the bushido

mystique, less experienced men will replace them. As a result, training will suffer, and corners will be cut to send less capable warriors into battle.
"One example, Japanese aviators in China do not use parachutes, and their combat aircraft don't have self-sealing gas tanks. So if their time is up, they die with their aircraft and crew."
"So, they have a spendthrift, wasteful attitude about losing men in battle," David asserted.
"It starts with basic training," said Nagy. "Army and Navy recruits are beaten, starved, and marched until they drop with bloody feet. They're frozen until they get frostbite and tortured by devastating heat during war games. They're generally abused, beaten, and assaulted by noncoms and officers. This torture by fire and ice is performed in the name of the Emperor. It's justified as the means to harden the warrior into an unyielding 'sword of steel.' Estimates are two to five percent of recruits die for the Emperor in their first two years of service training."
David was agitated. "How can this happen? How can the parents and family accept such abuse? Why would they remain loyal to such a ruler, to such a harsh system?"
"Good questions here in the United States, professor. Not in Japan. Here we have checks and balances. In Japan, the Emperor is a God. It's a complete and total dictatorship. Families would rather their sons die an honorable death than dishonor the family, as defined by the Army and Navy. In the thirties, the military and their civilian government lackeys took Emperor worship to a new height. They've twisted the Bushido Code and violated the Meiji Rescript for Soldiers and Sailors to create a military dictatorship dedicated to Japanese imperialism throughout Asia."
"Excuse me, Michael, what is the Meiji Rescript?"
"It is the Meiji Emperor's instructions to all military personnel issued in 1882. It entreats each soldier and sailor to see themselves as the arms or instruments of the Emperor, who is the Commander-in-Chief or head. It encourages them to do their duty and be steadfastly loyal with words like, '...duty is weightier than a mountain, while death is lighter than a feather.' Five articles emphasize duty, loyalty, valor, respect for superiors, faithfulness, righteousness, and disregard for opinions and politics. It is read and recited continuously by the military, privately and publicly. They must memorize it and act accordingly. Taken at face value, it provides several principles of Confucius and Buddhism wrapped in a nationalistic subjugation to the Emperor and those who speak for him."
"And it's being violated by whom, how?" asked Pelham.
"Mostly by the Army, which has ignored the prohibition on politics. Generals now control the civilian government's political, judicial, and policing processes. After a decade of chipping away at the civilian side of government, you now have an Army general, Hideki Tojo, as Prime Minister, Minister of War, and Army Minister. He's running the country and authorizing military campaigns in the name of the Emperor. He controls the press, the secret police, education, or more accurately, indoctrination policies, production priorities, everything."
"I'd like to learn more about Tojo, the Emperor, and the other major government leaders," David stated.
"So you shall in the next two days. Having an Army man control the country brings us to another weakness. In Japan, Naval officers learn a broad range of academic subjects, including the strengths and weaknesses of their prospective enemies. On the other hand, Army

officers aren't educated - they're indoctrinated. They're trained to think and act as the Army defines their role as warriors, as implementers of the 'imperial way.' They've little knowledge of the outside world, making them overconfident, arrogant, and single-minded even when proven wrong. They consider themselves superior to all enemies and believe their fighting spirit will prevail in all situations. They're fearless and embrace death as a patriotic duty to the Emperor."

"They sound like mean little bastards to me. So where's the weakness?" Pelham exclaimed.

"They are mean little bastards. They kill and mutilate their prisoners and wage war on helpless civilians. They believe the Japanese are the chosen race, which justifies murdering, torturing, and raping other nationalities. We would be at a greater disadvantage if a Navy Admiral directed Japanese imperial designs. Someone like Yamamoto would be quick to finish off Wake, Midway, and Hawaii - thereby forcing the remnants of our fleet back to the West Coast; however, the Army is in charge. They tell the Navy what battle will come next and dictate how the Navy will support the Army. To some extent, they even control how the Navy uses its resources."

"Michael, let me paraphrase: you're saying that the Army is in full control of the country and the Navy. Because the Army lacks geopolitical savvy and chooses the pure fighting spirit over reason and logic, we can expect some major strategic and tactical mistakes - especially in how they employ the Navy."

"Exactly," Nagy said, "The Army hates the Navy. The Navy hates the Army – that's a strategic weakness. Neither service puts much effort into understanding the needs and combat potential of the other – weakness. The Army controls finances, combat resources, and materials to favor themselves – weakness. The Army brutalizes civilian populations in occupied territories, facilitating the creation of indigenous guerrilla fighters – weakness. The Army is insatiable for conquest and territory. They will keep expanding the empire until their supply lines snap and central coordination is impossible – weakness. The Army and the Navy expect their warrior fighting spirit to overcome severe logistical deficiencies – another weakness. They overstress offensive warfare weapons and tactics at the price of weaker defenses."

"Do you have examples of the Army overriding the Navy?" David asked.

"I'm sure it's happening right now. Yamamoto knows Midway and Hawaii are ripe for the picking, but he needs Army troops and transports to invade them. The Army is rolling south, probably based on plans put into motion for a year or more. They won't shelve their South Pacific battle plans to please the Navy, even if it's the most opportunistic thing to do. The Imperial Japanese Navy had its day of glory on December 7th. Now they must support the Army invasions. Another example, the Army sees submarines and ships as their prerogative. If they want to use the subs to insert special units ashore or resupply garrisons, they will do so. If they fear combat ships more than our merchant shipping, they will order the Navy only to attack warships."

"Michael, this is good information, but it's wearing me down. Maybe we should break for lunch," David said with a sigh.

"One more thing," said Nagy. "Like most dictatorships, there is no honesty between the government and the people. Every victory will be reported as the greatest victory ever. Every draw will be described as a victory. Every defeat will be reported as a draw at worst. This spawns another major weakness of the military. They believe their grandiose press releases and make wrong decisions because of them. The victories will seem hollow as the common people see their sons dying by the tens of thousands and hundreds of thousands. Only when food, oil, clothes, and vehicles are in short supply, and the military takes them all for themselves, will the people know the war is not going well. Only when our American bombers come and level their cities, something their leaders have told them will never happen, will they know the war is lost."

David shook his head. "What will they do? Will they rebel?"

"Never. The Japanese people are willing to die for the Emperor just as their sons are or have. They will starve. They will sicken and die. But with their last breath, they will grab a stick, a stone, a knife, or a gun to kill the despised whites who seek to defile their country. Invading Japan is our worst nightmare because the Japanese will not give up even if beaten."

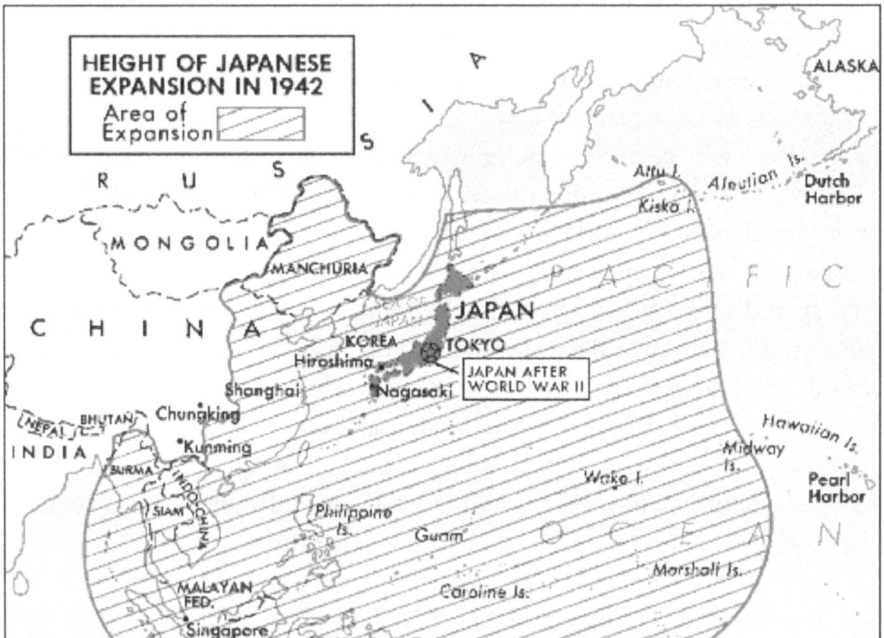

Time frame is after Japan took Attu and Kiska (Aleutians) in early June
The Allies had a toehold on Guadalcanal and New Guinea north of Australia.
Source: Britannica.com

Chapter Seven

THE NAGAI FAMILY – USC Campus

December 1941

The USC study session broke for lunch. Michael invited David Pelham and Ron Strong to his house. Surprisingly, it was on campus. Over thirty years, his father, Yoshi Nagai, built a shed for tools and garden equipment. Then he added a small shelter around the shed and, finally more rooms. His dad convinced the USC administration to allow the "Caretaker's Quarters." As the years passed, he put in electricity, running water, and a connection to the campus sewer system.

Mrs. Nagai started as a server in the faculty cafeteria and then was a cook in training. She was the head chef in the faculty dining room before the Nagais retired. Michael's parents planted an herb garden, a small vegetable garden, and a seasonal flower garden around three sides of the one-story house. The fourth side was near one of the University walkways, where his dad created a Zen garden with benches for passersby. Yoshi Nagai was the head groundsman during his last decade at USC.

Nagy's wife, Rachael Nagy, was a delightful and attractive hostess. She was nearly as tall as Michael at 5 feet 8 and had a quick, natural smile, jet-black hair, and expressive brown eyes. After serving them some iced tea, she made several sandwiches. Finally, the Nagy boys, Johnathan, age five, and Billy, nearly three, arrived home from nursery school. They all enjoyed a noisy and lively lunch at a large picnic table on the back patio. David and Ron Strong volunteered to help with the dishes while Michael took the boys back to preschool.

Rachael said it was nice to have company. Strong asked her why Michael had not hosted them in the faculty lounge. She looked indignant and said that since December 7th, no one would sit with them or near them in the faculty facility. She said faculty and students would still visit her as a doctor, but not socially. Michael returned; he and Pelham sipped some iced tea on the patio while Ron trotted off to make phone calls. Rachael went back to the university clinic.

"Tell me about your father's retirement home," David inquired.

"Not much to tell. It's a small bungalow that Dad and his friends built, surrounded by orchards, gardens, flower beds, and a couple of small forests in the making. Oh, and his latest project is planting some grapevines to produce wine in a few years."

"Sounds like a peaceful place. How many acres?"

"In the early 1930s, we bought twenty acres for a song. Mind you, there was no electricity, water, or sewer. In a couple of years, we added the adjoining twenty acres by paying the back taxes."

"And you're the owner of record? You hold the deed?"

"Yes, it's free and clear. We have all worked on it over the last seven or eight years. We took abandoned, weed-infested farmland and added a deep well, irrigation canals, a house, the gardens, and all the trees, shrubs, and plants except for an existing olive tree orchard. Mom and Dad make enough from selling their herbs, olives, vegetables, flowers, and Mom's baked bread to pay the taxes and cover most of their expenses. They have some chickens for eggs and meat and a cow for milk. Next year Dad is putting in a pond for raising catfish. He has the whole forty acres planned for development and wants to transition it to a city park when he and Mom die."

David snorted, "And now you think it will be confiscated?"

"Unless we can sell it to someone who will let us buy back in after the war."

"How much do you think it's worth? Do you own the mineral rights?"

"With the war on, I imagine it would be worth $100 per acre as farmland, but that ruins my dad's vision. They will plow everything under and rotate cash crops. Yes, we have rights to the land and all mineral rights."

"Good Michael, I might have a buyer for you. Ron Strong is my dad's land buyer in California. They're mainly interested in oil and gas land. My uncle Tommy is coming out to the coast soon. He buys land, buildings, apartments, and such if they're a good investment."

"I'm all ears, but who would be the actual owner?" asked Nagy.

"If Tommy approves, it will be me. Tell me, are your parents infirm? Do you have to go with them to a camp?"

"No, they're probably in better health than Rachael and me. I don't want them to experience that without our help and assistance. Why do you ask?"

"To tell you the truth, I'm not sure. Something is stirring in my subconscious that's not clear yet," David responded.

Ron Strong returned and told David that the office had received a message that his uncle was working on a deal in Oklahoma City and wouldn't arrive for another day.

"Michael, where's your father's farm?" David asked.

"About twenty-four miles northeast in Sylmar."

David looked at Strong. "Ron, we may be buying his father's forty acres to return it to the family after the war."

Strong responded, "Professor Nagy, do you have the mineral and water rights? What's the price per acre?"

"Yes, we have both, Mr. Strong. I'm thinking about $100 per acre."

"That sounds about right in the current market," Strong said.

"Ron, do you know Sylmar?" David inquired.

"Yes, it's mostly agricultural land and ranches right now, but it will be wall-to-wall houses, supermarkets, light industry, and services at some point. L.A. and San Fernando can only expand outward. As the population builds here, it will push through Sylmar against the San Gabriel Mountains and the Angeles National Forest. The water rights alone could be worth a fortune someday."

"My dad has hit all the water he can use with two wells at twenty to thirty feet," Nagy offered.

Strong replied, "Sounds right up Tommy's alley. But, David, it's too small for your dad and me to manage. So, if you two scholars will keep up this academic dialogue all afternoon, I can drive up there and take a look. I'll be back at about four o'clock."

Michael gave Strong directions to the farm and escorted David to his classroom. The walls were covered with maps of Japan, Asia, and the Pacific region.

Nagy picked up a pointer. "If you agree, Professor Pelham, I suggest we go hard at it for a few hours. Then we break for Cokes and a snack at three o'clock. After that, we put in another two hours before a well-deserved cold beer on our patio."

Pelham replied, "That sounds great to me, especially the cold beer. Please call me David."

Nagy pulled out one of his large map boards that displayed a topographical view of the Hokkaido, Honshu, Shikoku, and Kyushu islands. He pointed out the differences in size, mainly the much bigger Honshu and the rugged hills and mountains that limited the islands' arable land. Michael highlighted the industrial areas, ports, capital cities, and population centers. He reviewed the Meiji Restoration from 1868 to 1912, when Japan fully opened itself to trade and strove to become an industrial power while rapidly expanding its Army and Navy. The USC Professor covered the wars and treaties in the fifty years leading up to the current conflict. He identified Japan's age-old enmity with China, Korea, and Russia. Japan's so-called "Co-prosperity Sphere" in Asia, Nagy considered analogous to the Monroe Doctrine. Michael said it reflected Japan's absolute contempt for the white colonial imperialists' takeover of Asian land and resources.

Professor Nagy surprised David with his analysis of the September 1940 Tripartite Pact between Germany, Italy, and Japan. He asserted that it gave Japan a hunting license in Asia, including Siberia. He also believed that Japan and Germany had probably agreed to divide the United States, possibly using the Mississippi River as the boundary. Pelham was visibly shocked. He had not thought of his country being bombed and invaded, the men enslaved or killed, and the women raped and forced to be sex slaves for the victorious Germans and Japanese. Nagy looked at him quizzically and asked if David had forgotten the phrase "to the victor belong the spoils."

Then Professor Nagy moved on to the atrocities committed by Japan's Army in Manchuria, China, and Korea. He discussed the Army and Navy factions that pushed for a more aggressive "imperial way," assassinated government leaders, and resisted central control of their actions in other countries. Nagy used the rape of Nanking as his centerpiece of how the army had violated the true Bushido spirit and the Meiji Prescript. The details had David reeling; he felt sick to his stomach. He had wanted the facts, wanted to understand his enemy, but the depth of the Japanese Army's depravity and lack of morality was both shocking and frightening. Japan was an enemy that would routinely say one thing and do another. It was total evil and complete disregard for humanity and innocent civilians. He weakly asked for a break.

Nagy finished their afternoon session. "David, remember one more thing in your naval war with Japan. Never be taken prisoner. Another modern perversion of the Bushido Code is beating, torturing, enslaving, and starving prisoners of war. The Japanese military man believes if enemies surrender, they're cowards and deserve to be treated as sub-humans."

David excused himself and took his Coke and a box of Cracker Jacks outside. He found a bench near the building's entrance and thought, *Congratulations Davy, you've discovered more knowledge than you expected. Michael Nagy or Nagai knows more about Japan than anyone you've ever met, including Uncle Tommy. There's much more to learn, how to attack the political system, what propaganda will work, and how to demoralize the fighting forces. If they don't fear death, what do they fear? How do you shame them, insult them, and cause them to act irrationally? How do you plant false intelligence and get the desired tactical decisions from the hidebound Army? How do you wear them down until they resort to a headlong charge right into your strength?*

Pelham thought about his discussions with Nimitz and Lamar, and then the idea hit him. He walked back inside, found Michael, and they returned to the classroom.

David spoke. “Before we start, I want to know if you and your family are willing to move to Hawaii, so you can use your knowledge of the enemy to advise Admiral Nimitz and his staff.”
Nagy said, “I'm not sure I would be good at that. We're planning to be with my parents.”
“Yes, I know that. But what a waste for you, Rachael, and the boys to be shunted aside for several years. You said your parents are healthy. I think they should decide. Whenever I ask you a question, I find even more data, useful information, and wisdom you can impart. Pearl Harbor has put us in a pickle. We need to fight back hard and smart. We must completely understand our enemy's culture, vulnerabilities, and what makes him tick. I see your contribution as key to that understanding. Do you want to be incarcerated with your family, or do you want to fight for your country? We need your knowledge and expertise to win the Pacific War. What do you say?”
Nagy hesitated. “I'm having trouble seeing myself as valuable to someone high up. Who would be my boss? Do I get paid? How do we get there? Where would we live? Where would the boys go to school? What would Rachael do?”
“Those are details that'll take care of themselves. The crux of the matter is whether you want to fight for the United States and defeat Japan,” David challenged.
“Yes, I do. The Japanese military has unleashed an evil reign of destruction, terror, murder, and torture in Japan and the Asian world. We must stop them!”
“Good, so you are philosophically ready to fight back and maximize the destruction of Japan's weapons and the killing of her military forces?”
Nagy reacted with a blank stare. “It's all about killing, isn't it.”
“Yes, the nature of war is they kill you, or you kill them, take your pick,” David answered curtly.
“They will kill my family if they win.”
“Yes, all intellectuals, community and political leaders, and potential leaders will be executed or starved as slaves. Plus, on the German side of the Mississippi, there will be no mercy for any Jews,” David replied.
“If you take care of the details, we'll go. We will move to Hawaii,” Nagy said.
“Excellent. Let's get started on our next session so we can earn that cold beer.”

Professor Nagy spent the next two hours briefing Pelham on all he knew about the Japanese Army and Navy, their recruiting, training, operating procedures, officer ranks, and his best guess on the current order of battle.

Ron Strong tracked them down at about 4:30 pm, and they went back to the Nagy house. The three men grabbed a beer from the fridge and relaxed on the patio. For once, they didn't talk about the war. Instead, David and Michael compared their two universities, one public and one private. Strong gave them the current Hollywood gossip and talked about his favorite pro-football teams. Rachael came home from the clinic and asked Michael which one of them would pick up the boys. He got up, and they went together across the campus.
“So, Professor, what do you think of this USC professor?” Strong asked.
“He's the real deal, Ron. He's cramming my head with so much information, I'm not sure how much I'll retain. Michael is truly an American scholar of ancient and contemporary Japan. He knows far more than he realizes. I plan to take his family to Hawaii to augment the Pacific war plans staff.”
“Just how are you gonna do that?”

"Honestly, I don't know. It will come to me. Any news about Tommy's flight?"
"Yes, they're supposed to land at the Los Angeles airport at about 12:30 pm."
"Please call them up and ask them to land at Burbank, the Lockheed Air Terminal; I think it is. Send him a message that it's a personal favor for me. We'll pick him up and look at some land in Sylmar."
Strong got up to leave. "Consider it done."

David had some quiet time sitting on the patio. He could hear a bell tolling in the distance, birds chirping, and the buzzing of insects collecting pollen from the trees, flowers, and shrubs planted by Mr. Nagai. *I wonder what it was like for them to live here for thirty years. Did they earn the respect of the whites, or were they treated like immigrant servants? Did Michael and Rachael go to school here? I forgot to ask.*

His thoughts were suddenly interrupted by shrieking and laughing in the house. The Nagy boys and their parents were back. He took a long swig of his beer and realized Rachael was about ten feet away, staring at him with tears rolling down her cheeks. *Oh boy, he told her. She's either mad as hell, scared to death, or crying out of happiness. It could be any of them.*

She walked over to David, took his free hand, and kissed him on the cheek. He could feel some of her tears on his face.
"He told you about Hawaii."
"Yes, do you think you can pull it off?"
"I think we have a chance. But, we have to move quickly."
"I was dreading going to a camp and having the boys shut-in for years. If we can go to Hawaii and help the war effort, that is what we want."
"I've been thinking about it. You can ship your household goods, and we'll fly you to Oahu on a Pan Am Clipper flight out of San Francisco," said David.
"But not before Christmas, I hope. I want the boys to see my parents before we leave."
"Where are they located, your parents?"
"Las Vegas. They operate a small chain of dry-cleaning stores."
"We will have to talk it over with Michael. It will probably be after the New Year, but if we wait fifteen minutes too long, it won't work."
"How so?" Rachael inquired.
"I'm assuming the Army and the FBI are making up lists of Japanese for relocation inland. Next, they will probably freeze your assets, bank accounts, and such. Then, they will ban any non-approved travel and finally give you orders to a staging area. My uncle is a banker who can help with financial accounts. He'll probably purchase the Nagai farm so we can sell it back to your family after the war. He arrives midday tomorrow in Burbank, and we'll see the farm. You mustn't mention this scheme to anyone, not a soul! Please go tell Michael that right now."
Strong returned to the house, liberated his beer from the fridge, and came out on the patio. "Your uncle says he can't wait to see you in Burbank tomorrow."
"Thanks, Ron. That thing I told you about the Nagy family and Hawaii. Forget it; you never heard it. You won't share it with anyone, my dad and Tommy included, got it?"
Strong put his fingers to his mouth. "Yes, my lips are sealed."

Michael invited them to dinner. David accepted, but Strong begged off, saying that someone should use the company suite and the hotel pool. When David asked how he'd get to the hotel, Michael invited him to stay overnight in their guest room. Dinner featured what the boys called Japanese meatloaf. It was delicious, made primarily from beef and pork with baby shrimp, rice, and spices. Rachael put the boys to bed at 7:30 p.m. The two professors retired to the patio with a scotch on the rocks. Their talk turned again to the campus life they both loved. It struck David that they were sitting outside in mid-December - temperate weather made USC look attractive.

Rachael joined them and sipped a red wine when Michael announced he'd show David the guest room. He took David around the west side of the house to a sliding door, where they took off their shoes and entered a medium-sized room. There were low tables and chairs. Nature paintings, ceramic pots, and some live plants gave the walls a feminine touch. Rectangular bamboo mats covered the floor. David did not see a bed. In the far corner was a good-sized soaking tub. Michael removed the top cover and found some swim trunks in a cabinet. He told David to change and dive in while he went for fresh drinks. David put on the swimsuit and tested the water, which was nice and warm but not too hot. The tub was big enough for five or six adults. He felt his whole body resting and relaxing from the soak, the tasty dinner, and the after-dinner drink.

Michael returned, wearing a swimsuit and carrying drinks. He handed David a whiskey soda and slid slowly into the warm water.

"How is this tub heated? I don't see any heating equipment," David asked.

"My dad built the whole shebang. This room was originally the tool shed. He put a long black water hose on the roof, and an electric pump pushes the sun-heated water into the tub. So, is it all right if Rachael joins us in a few minutes?"

David grinned. "Sure, will she be wearing a swimsuit?"

"Hard to predict. We usually go in as a family in our birthday suits. She's hoping we can talk about Hawaii if we keep our voices down."

"I don't see why not, although it's just in the half-baked stage. Do you have passports?"

"Yes, we used them to go down to Mexico City. Ironically, we had planned a trip to Japan someday. Can you tell me what I'll be doing in Hawaii?"

At that point, Rachael entered through the sliding doors and said, "Wait, I want to hear this also." She was wearing a knee-length sundress. She sat on the tub's edge and held her dress down as she slid into the warm water.

David answered, "I'll give you my best guess. You could make an important contribution to the intelligence and mission planning people. You've got a feel for the culture and morality, or lack of the latter, that most Caucasian staff officers don't have. In just a few hours, you convinced me that your knowledge of Japan's history and politics could predict their current decisions and likely reactions to our strategy and tactics."

"You used those last terms a couple of times today. Are they the same?" Michael queried.

"Not at all. A strategy is a high-level set of goals and actions to achieve a specific objective in military terms. For example, I bet that our Joint Staff has decreed that we must support fortress Great Britain and our friends in Australia at all costs. Let's say that's true. Tactics would be what we're doing with our combat forces daily, weekly, and monthly to achieve the strategic objective."

Rachael pouted. "This conversation is not holding my interest. Will Michael's job in Hawaii be full-time?"

"I think so. Do you both read, write, and speak Japanese?"

"Yes, we both do. So, how does that help?"

"For starters, you could be interpreters. You could listen to Japanese radio stations to transcribe what they're saying and explain the contents of newspapers, captured documents, and maps."

Michael said, "You won't get much from radio stations; they only broadcast propaganda."

"Yes, that's true, but the intelligence value is in what they don't say or how they lie about a situation where we know the truth. So we must listen to what their leaders announce, decide what the Japanese want us to believe, and modify our tactics for the best outcomes."

"I have relatives on Oahu who will welcome us and help us settle," Rachael interjected. "I'm pretty sure I can work as a doctor for the military, in a civilian hospital, or even start my own small clinic."

David swirled his arms in the warm water. "Which do you favor?"

"Frankly, the clinic intrigues me. There are plenty of non-English speaking Japanese in Honolulu who need modern health care. Of course, it wouldn't make a lot of money, but providing the whole spectrum of family health care is an exciting challenge. There could be a book in it."

They talked more about Hawaii and escaping the years in the internment camp, the war, and how the Nagy boys would adapt to life in the islands. Michael described the main features of his parents' forty-acre farm. David yawned and remarked that he had not seen a bed in his guest room. Michael and Rachael laughed and said that it would be no problem. Rachael returned in a few minutes, rolled out what she called a futon duvet, and provided him with a bean pillow.

Pelham peacefully slept until he could see sunlight through the sliding doors and hear the boys in the house running and laughing. David exited the guest room, walked into the house, and found Michael trying to persuade the boys to sit down and eat their breakfast. Nagy said Rachael had left for the clinic, and Ron Strong had called to say he'd pick up David at noon in front of Michael's building.

"What is our plan this morning?" David asked.

"Have some of this coffee and pick out some cereal. We saved you some cream off one of the milk bottles delivered this morning. We drop the boys at pre-school at about 8:30. Then we can go to the classroom for a three-hour session before Ron picks you up."

"Picks us up," David replied. "I hope you'll come with us. We would buy the land from you, and we don't know who will take care of it until the war ends."

"That makes sense. My mom is expecting us for lunch at one o'clock," remarked Nagy.

"She doesn't have to do that. So I don't want to put her out."

"Nonsense, she loves company, and lunch will continue your education in the Japanese culture. So go light on your breakfast. You can expect a multi-course lunch that will knock your socks off. I'm hoping for some sushi, tempura shrimp, grilled steak, noodles, and fish."

"That does sound like quite a spread; my uncle will be especially pleased. Did I tell you he's Japanese?"

Nagy was surprised. "No, where is he from?"

"New York now, but originally from outside of Kyoto. It's a long story. He's from a poor Christian family. My dad brought him to the States and adopted him in his late teens. Later, he became a citizen and helped my mother manage our flat and my father's oil business. Ron Strong has just informed me that he's managing a bank in New York, the New Orient."

Nagy said, "It's a small world. Our family banks with New Orient of New York."

"Could you tell me why?" David questioned.

"New Orient started in 1933 after the banking panic and passage of the Emergency Banking Act. It promised responsible management, prudent investment policies to avoid the stock market, and larger cash reserves than other commercial banks. They opened branch offices in every US city with a significant Asian population. Asian community word-of-mouth has fueled eight years of rapid expansion. It is one of the few American institutions we can trust."

"I bet my dad is proud as punch with his other son. I wanted the academic life and adventure with the Navy. Business and money were never my primary drivers. If it's OK, I'd like to shower and shave before we go. What's the dress at your father's farm?"

"Anything goes. Most often, Dad's working outside shirtless with shorts and sandals on. Lunch company is special; Mom is going to make him dress up. Plan on slacks and a sports shirt. I better get word to them about your uncle. Mom will tan my hide if we just show up with a bank president."

"Do they have a phone?"

"No, I have to call a nearby rancher, and he'll walk or ride over to tell my parents. They don't have electricity, running water, or a city sewer line. To some extent, you could see them as living in rural Japan, enjoying the simplicity of pure well water, kerosene lanterns, and a septic tank."

"What do they do for refrigeration?"

"They have a root cellar, and the iceman comes and stocks their large icebox. Excuse me while I make that call."

David took his shower, shaved, and dressed for the farm visit. He had a tinge of jealousy as he thought, *I wonder what part of the family fortune goes to Uncle Tommy. His bank stock and personal property must be worth a lot, but what about Dad's bank stock and Pelham Oil stock? Who inherits that someday?*

David and Nagy dropped the boys at preschool and returned to the USC classroom. Michael said they would cover the trade history of Japan and their deep resentment of the Caucasian race. David reminded Michael that this might be their last session and that he wanted an update on the current government leaders, their backgrounds, and history.

"I've thought about that. You should get that information from my dad. He reads the Japanese newspapers, keeps track of who is in charge, and probably knows their family histories. My dad may surprise you. He's a well-read scholar. In his last twenty years at USC, he attended one course each semester on an audit basis."

Michael gave David a brief synopsis of Western European influences on Japan, starting with the early Portuguese arrivals in the mid-1500s. The Spanish ships came not long after that, then the Dutch, British, and Americans. He highlighted the critical voyage of Englishman William Adams as a navigator on behalf of the Dutch to Japan in 1600, resulting in 200 years of beneficial trade relations. However, during that period, the constant bickering and conflicts between Western merchants and Jesuit priests from different countries were observed by Japanese leaders with contempt.

Nagy covered the American attempts to open trade with Japan. In 1846, Commodore James Biddle and two warships arrived to negotiate a treaty. Biddle was treated with disrespect by a samurai guard. When he turned the other cheek and didn't severely punish or kill the man who knocked him down, the Japanese decided the Americans were weak and refused to deal with them. It was a different story seven years later when Commodore Perry sailed his four heavily armed warships up the forbidden Edo Bay toward the capital. Perry's gunboat diplomacy forced an agreement that Japan would respond to his demand for a trade treaty within a year. He returned in six months with ten warships. Japan signed the Convention of Kanagawa because she lacked the ships, trained sailors, and long-range cannons to defend herself. Great Britain, Russia, and other Western powers followed suit and demanded what the Japanese considered unfair rights and privileges. Unrest over the Shogun's unwillingness to expel the Western barbarians led to the Boshin Civil War in 1868.

In the late 1800s, Japanese oligarchs controlled the government in the name of the fifteen-year-old Meiji Emperor. They hired hundreds of Western advisors to accomplish industrialization, expansion of railways, installation of telegraph lines, and adoption of Western science. The realization of a national education system demanded a new reverence for the Emperor as a living god. The 250-year Shogun or warlord domination of Japan ended, and the rigid class hierarchy made it more flexible. The oligarchs modernized and expanded the Japanese military during the forty-four-year Meiji period. They created a seemingly democratic government, waged war against China, and later annexed Korea. They fought and won a war against Russia, establishing Japan as a significant power in Asia. In World War I, they sided with the Allies and received Germany's possessions in the Marshall, Caroline, and Mariana island chains.

During their last hour of discussion, Nagy turned to religions in Japan. He attempted to explain the widespread Shinto spiritual practices of honoring multiple essences, human spirits, or gods. Michael emphasized how the Meiji period government tried to mold Shinto into a more standardized nationalistic system of Emperor worship. He covered the influence of Buddhism in Japan and discussed the decentralized nature of thousands of Shinto and Buddhist temples and priests. Christianity was banned under the Shoguns but allowed by the Meiji Constitution of 1889. Current Christians, he said, were a tiny minority and suspected by the government of disloyalty. He explained that the military, civil police and secret police controlled every aspect of life in Japan down to the slightest village.

At noon, they walked out to meet Ron Strong.

Before Strong arrived in the convertible, Pelham asked his mentor, "Michael, how do we beat Japan? How do we win this war?"

"Food."

"Food?"

"Nippon can't feed herself and support large armies and ships at sea if you cut off her food imports. A starving population is not as productive or supportive of the military government. So stop her trade with neutral countries and sink the ships bringing food to the islands."

During the drive to Burbank, Ron Strong asked David how Tommy became his uncle.

"Uncle Tommy joined the family in 1921 when I was ten," answered David. "My dad and I had sailed from San Francisco to Hawaii on a business trip. I liked Hawaii; I'm looking forward to returning there. I remember it was warm, both the air and the water. We enjoyed a luau and watching attractive dark-haired girls dancing in grass skirts with coconut shells covering their breasts. They served pineapple with everything, and I was in seventh heaven when I got guided rides through the Waikiki surf on longboards. After that, we sailed on to the Philippines, Hong Kong, Korea, and Japan. My dad said you never know where to find oil or a new customer to cultivate. In Japan, there was a lot of bowing, tea, and rice. The meals often included unusual vegetables and various parts of fish or animals. The Japanese stared openly at my sandy hair and exclaimed over my bright blue eyes everywhere we went.

"My dad met with several men high up in the government who wore coats with tails, stiff white shirts, and stovepipe hats. One of them insisted that we journey from Tokyo to Kyoto to see the beautiful gardens and shrines. We boarded the train with an interpreter in tow. The further we traveled from Tokyo, the less our English or even signed language was understood. The Kyoto hotel room was luxurious, and the staff was very attentive, but their Japanese language was nearly impossible to understand. On our third and last day in Kyoto, we entered the hotel when a young man approached us on the entrance stairs. In broken English, he tried to convey that his parents would like to speak to Pelham-san, the Prince of America. Our interpreter broke in and attempted to drive the young man away in a nasty tone. The young man held his ground. My father dismissed the interpreter, saying he'd like to find out what the youngster wanted. You could see that the boy's father and mother were both Japanese. The father wore a worn suit and tie, the mother a modest dress with a sweater - both were shorter than their son."

"The young man introduced his parents as Christian missionaries and stated that they were all Christians in their family, mother, father, three sisters, and two brothers. He asked if Pelham-san and his son were Christians. My dad replied that we were Presbyterian. This response elicited smiles from the family, and the father stepped forward with his hands in the air exclaiming, 'blue sky, blue sky.' The minister said that God had provided them with five children but not the means to shelter, clothe, or feed them. Dad asked if they were requesting money, and the father protested that he did not want charity. Instead, he wanted the Pelham Prince to take his oldest son, Takenaka Yoshido, to America.

"The father explained rapidly that the boy was nineteen years old, the smartest in their family, spoke excellent English, had several years of schooling, and was a hard worker. Takenaka would belong to the Pelham dynasty and make himself invaluable such that he could send back modest funds to sustain his family and keep their ministry afloat. Our interpreter glared at the Yoshido family from inside the lobby. Finally, my dad looked at the boy's mother and asked if she was ready to give up her son. She said it was best for him, necessary for their family, and they would meet again in heaven with tears in her eyes. Dad shook his head and looked at each family member before telling them to return at nine o'clock in the morning for his decision.

"That night in the Kyoto hotel was the first time my father talked to me at length about a serious subject. He paced the floor, weighing the pros and cons of getting involved with the Japanese family. He stated that in the 20th century, it was flat wrong to take on an indentured servant. He doubted the young man could legally leave Japan or even enter the United States without a visa and a sponsor. Periodically Dad would stop and request my opinion, or what would my mother and sister think? I wondered how long the boy would live with us. My dad started pacing again, stating that this was a life-changing move for the boy and the Pelham family. He said the lad would probably be part of our family until he left for some reason or we kicked him out. He mused that he'd like to help the family, help the boy find a better life. He said maybe this was fate or God asking the Pelham family to step up and be bold, to do the Christian thing.

"Dad worried that living in America would not be easy for the boy, saying there was a lot of prejudice against Orientals and a new immigration ban that was considered outrageous and insulting in Japan. Finally, the pacing stopped. My dad snapped his fingers and said it probably couldn't be done anyway. He'd call some of those arrogant stovepipe ministers to confirm that fact. That would be the end of it. Dad disappeared downstairs to use the telephone and get assistance from our interpreter. He returned in about an hour and told me there was a way to take the boy to America. It involved doctored travel papers, and there was a price to pay for the authorities to look the other way.

"Dad also had the interpreter call the local Police Chief, saying the Pelhams were considering donating to the Yoshido ministry if it was on the up and up. The report said that although the family was poor, they were honest, had never been in trouble, and helped their neighbors. Dad told me all this and said, 'Davy, we're going to do this if we can. We will make a better life for this boy and his family. They're struggling to live amongst pagan idols and animal worshipers. I think it's why we've come to Kyoto to see the wonderful gardens, the elaborate Godless shrines, and then find out that good Christian people are starving. So I shall sleep well tonight with confidence in our family's future. Thank you, David. It has been so helpful to talk to you. Goodnight!'

"The next morning, at nine o'clock, the entire Yoshido family was waiting outside the hotel's main entrance. The young man approached my dad, bowed, and introduced himself as Takenaka Yoshido, or "Taakee" for short. Dad returned the bow and firmly shook his hand. Taakee was holding a small cheap suitcase. My father bowed to his parents and made it clear that Taakee was to come with him and the parents should say goodbye. They hugged their oldest son and spoke to him in Japanese, probably telling him never to forget them, their love for him, and to make them proud. His mother took a small packet from her jacket pocket and placed it over her heart. While looking up into his eyes, she pressed it to his heart and put it into his hands. Taakee's father came forward to William, made a short bow, and said he'd be forever grateful for the Prince to look after his son with Jesus's blessing. The minister gestured for everyone to hold hands and bow their heads. In perfect English, he led us in the Lord's Prayer. It was on a busy street with the hotel staff and many surprised glances from passersby. With tears in their eyes, each parent kissed their son, bowed to him, to my father, and then they were gone.

"The four of us rode in silence to the train station. Taakee sadly realized he might never see his family again. At the train station, the interpreter bought him a third-class ticket. When Dad caught on to that, he told the sulking interpreter that Takenaka would travel with us from now on. He's part of our family for better or for worse. On the ten-hour train ride to Tokyo, Taakee and I tried our English on each other and became fast friends. I drew him a map to explain where the West Coast and New York were located, but the place names, oceans, and distances only confused Taakee. I showed him how to play Tic Tac Toe, and he pulled out a small chess set.

“In Tokyo, it took four days for my father to work out the final details of Takenaka’s travel papers. Dad groused that the so-called fees and taxes were ever-increasing, and he hoped he’d never have to do business with these Japanese crooks again. We both showed Taakee how to eat with a full array of Western silverware, first practicing in our room using room service and then in the dining room. We bought him a new wardrobe like mine. Leather bags replaced the shoddy suitcase with decorative straps and gold-toned spring snaps. I noticed the bags were embossed with the initials TYP and asked my father what they meant. He explained that after we cleared US customs, we would call Takenaka, Thomas, Tom, or Tommy. His middle name would be Yoshido, and he’d take Pelham as his last name. We boarded a passenger freighter that sailed nonstop to Seattle with the phony paperwork in hand. Takenaka’s papers portrayed him as a son of Japanese diplomats in New York who returned after completing the Japanese equivalent of High School.

“The long train trip to New York added another page to our Pelham family history. Tommy couldn’t believe that one country could be so big. I had him try hot dogs, burgers, and Coca-Cola for the first time. Dad decided that Thomas was too old to be my half-brother, so he became Uncle Tommy to my sister Marie and me. When we took him home to my mom, she took one look at Tommy and asked him if he believed in God. When he said yes, and Jesus was his savior, she hugged him and welcomed him to the family. That was that. Mom knew Tommy and his family needed help and could not turn them down.

“Uncle Tommy quickly became essential to the efficient operation of our home. He helped in a hundred ways, cooking, doing dishes, cleaning, and doing the laundry. Tommy traveled with Dad on the Pelham oil business, handling the bags, ordering food, keeping track of appointments, and taking notes. He attended night school, earned his high school equivalency degree, and then completed college extension courses in accounting and finance. And here he comes, twenty years later, a New York banker no less,” David said as he pointed toward the end of the Burbank runway where a shiny, silver DC-3 with Pelham Oil markings was touching down on the blacktop.

Douglas DC-3

Source: Orange County Register

Chapter Eight

UNCLE TOMMY – Sylmar, California

December 1941

After the Pelham Oil plane taxied to a stop and the props stopped turning, Strong drove the convertible over next to the passenger ramp. The first passenger deplaning was David's uncle, who wore a double-breasted dark blue suit, a white shirt, and a striped tie. He looked the part of a successful businessman and banker. At forty-one years old, he was trim and had a natural tan that Ron Strong must have envied. Smiling and waving, Tommy approached David and gave him a long, affectionate bear hug.

"Oh, Davy, it's so good to see you! I'm so sorry I was away when you came to New York. Was your trip out here OK? Did you like the Pullman?"

"It was a great trip, Uncle Tommy. Thanks for arranging that. Let me introduce Professor Michael Nagy from the University of Southern California. He's teaching me about your native country, so I can kick their ass in the Pacific."

Tommy ignored David's cut. "Professor, so nice to meet you. I hope my honorary nephew has not been too much of a burden to you."

"No sir, he's kept me on my toes and ingratiated himself with my whole family."

Tommy shook Strong's hand, "Good, good. Ron, what is it we're looking at today?"

"You better ask your nephew while we drive to Sylmar," Strong said.

David got in the back seat with Tommy and started explaining. "Tommy, the short story is that the Army may put Michael and his family in a camp for the duration of the war."

Tommy sighed. "It's a sad state of affairs. I'm afraid Roosevelt has so many fires to put out that this crime against our democracy will come to pass."

David continued, "We're going to see the retirement farm of Michael's parents. They're Issei, so Michael owns their forty acres and all the mineral and water rights on their behalf. I want to buy their land at a fair price and sell it back after the war."

"Ron, what do you think?" Tommy asked.

"I think it's a sure thing. Sylmar land will greatly increase in value in the next twenty years. I went out and looked at the property yesterday and met Michael's father, Yoshi Nagai. He's improved the neglected farmland and has a plan for every acre - even hopes to make it a city park someday."

"David, if you want to buy it, just do it. Why do you want the Orient Bank involved?" Tommy questioned.

"Uncle Tommy, I know I can count on you. Who knows what will happen to me in this war? I'd have to leave it to Marie, and then she'd have to change the deed and sell it back years from now."

Tommy patted David's shoulder. "For you, anything. If it's a good investment, the bank will buy it and hold it in trust for you. Mr. Nagai or his son can recover the same property at the same price after the war."

"Thank you. How's my dad?"

"He's great, never slows down. He's organizing the oil pumpers to deliver whatever volumes the government wants at fair prices."

"So, how did you become a banker?"

"Your father and I got tired of going to the East Coast bankers and selling them on financing new projects. So we decided to start our own bank. Your dad asked me to manage it. I put Orient in the name because I wanted to help Asian people who come to this country. They're subjected to all forms of prejudice, including economic discrimination and denial of banking services."

"It seems to have worked out. You're looking very prosperous."

"Thanks. We started with a big chunk of your dad's and granddad's money. We now handle all of Pelham Oil's financial transactions and hold many of their triple-A bonds. The deposits keep rolling in because we're not going to freeze any assets of American citizens without a valid court order."

"Do you have a branch in Hawaii? Maybe I should open an account," David said.

"We do. You and your sister Marie already have letters of credit for a large sum. Just go to any branch and use your Social Security number and identification."

When they arrived in Sylmar, Ron Strong turned off Polk Street onto a dirt road, and the ride got bumpy. After about half a mile, he turned left into a fenced area. The open gate had a sign saying, "Welcome to Nagai Farms." They drove past a large flower garden and vegetable patch. Strong stopped the car on the circular drive of crushed white rock in front of the small bungalow.

Michael Nagy pointed to a couple coming out of the front door. "Here they are, my mom and pop. My kids' grandma and grandpa."

Michael jumped out of the car, approached his father, and bowed. His father returned the bow and then shook his son's hand. He then hugged his son and kissed him on the cheek. Michael repeated the same sequence with his mother, except he kissed her cheek.

Michael gestured at Tommy. "Mom and Dad, Yoshi Nagai and Youko Nagai, may I present Mr. Thomas Pelham, President of the Bank of the Orient."

Tommy stepped forward, and Yoshi bowed to him, which Tommy returned. Then, after a few seconds, Yoshi repeated his bow and said, "Mr. Pelham, you do us great honor to visit our home."

Tommy returned a less formal bow and said, "Mr. Nagai, the honor is mine. Thank you for your invitation."

The two men shook hands. This sequence was repeated with Mrs. Nagai, who wore a white linen kimono with a bright red obi sash tied in the back. She had white split-toe socks on and was wearing wooden sandals. Michael then introduced David and Strong to his parents. After all the bows and handshakes were complete, Yoshi invited them to enter the house for a modest meal. Inside the door was a step where they placed their shoes and stepped into slippers. Youko led them through the entry and to the right. They then removed their sandals and entered a room with a tatami mat floor and decorations.

There were six cushions on the floor with back supports arranged in a rectangle. Youko led Tommy to the far side of the mats and asked him to sit at the head of the table. She had David sit to his uncle's right, Michael next to David, and Strong to her husband's right side as Yoshi sat opposite Tommy. Each had a sturdy, low wooden table with chopsticks, small sauce bowls, drinking glasses, and a fresh flower arrangement.

Youko brought out a serving tray with hot rolled hand towels. Michael told David that the cloth was an oshibori and only cleaned the hands. Once done, fold it up and place it on the table. Youko stopped at each man and let them return their oshibori to the tray. David noticed that she always started with Tommy.

She next brought in a large pitcher and announced, "Gentlemen, ice-cold Japanese beer."

Michael whispered to David to wait for the toast. After their six glasses were filled, Youko knelt at her table. Tommy raised his glass to Yoshi and Youko and said, "Itadakimasu!" Michael translated that it was like saying grace, Bon appetite, or thanks for preparing the meal. Yoshi held his glass up to Tommy and said it was an honor for his humble home to have such notable guests; a bank president, a naval officer, a stand-in for Clark Gable, and his no-account son. They all laughed and raised their glasses before taking another drink of excellent beer.

Youko got up, left the room, and returned with another large pitcher. She poured water into their other glass and asked them to try the ice-cold water from their first well. Then, she moved her table back and started arranging her cooking stoves and utensils inside the rectangle formed by the low tables. Uncle Tommy complimented her on the exquisite flower arrangements placed at each inside corner of the dining area.

She bowed to Tommy and said, "Pelham-san, you are our most honored guest. With your permission, I want to defer to my youngest son on the food sequence. I don't get to spoil him as often as I would like."

"Of course, he needs to choose for all of us," Tommy responded.

"Mom, what are we having?" Michael asked.

"Tempura prawns, chicken Yakatori, sirloin steak, and sweet and sour catfish."

"Please serve it in that order. I can't wait for the yummy prawns!" Michael exclaimed.

Youko turned up the gas on her three portable cookers and served them all rice. Then, she dipped six large prawns in batter and started cooking them. David followed Michael's lead on eating the food, where to place his chopsticks, and when to sip beer or water. Everything was delicious, including the side dishes of rice, boiled vegetables, and noodles. Youko efficiently prepared the food, served it, ate her portion, and cooked the next serving. She had all the main courses and ingredients organized in stacked lacquer bowls. It took only a minute or two to cook the food in oil, boiling water, or the portable hibachi grill. David thought he'd burst as he finished his last bite of the tender and spicy fish and his beer.

Tommy thanked the Nagais, "Such a wonderful lunch! The best food I've had in a long time. Gochisosama-deshita."

Youko bowed to Tommy. "Thank you so much. It was all my pleasure, osomatsusama-deshita."

Michael whispered to David, "He said thank you, that meal was a feast. She said with false modesty that the meal was not a feast."

David thought, *When will I understand these people, and this culture? When do we get to talk business?* He had tried to discuss buying the farm during lunch, but Michael had cut him off, saying the family did not discuss business at meals. The satiated men went outside in the sunshine. Youko was picking up dishes and stacking them in the kitchen. Yoshi took Tommy by the arm and set out to tour the property. Strong and Nagy discussed Hollywood celebrities as they strolled toward the large flower garden. David went around the north side of the house and discovered a Zen garden similar to the one at the USC Caretaker's Quarters. He sat in a shaded lawn chair and immediately fell asleep.

When David awoke, he found Youko sitting quietly beside him, shelling some pea pods. She had changed into khaki shorts, a light blue blouse, an apron, and leather sandals. Simple wooden sticks held her hair in place instead of the elaborate ivory ones she wore at lunch.

"Did you have a nice nap?"

"Yes, thank you. Your son has worn me out, trying to teach me to understand the Japanese culture."

"He can be pretty intense sometimes. I understand he may have a family move coming up?"

"That must stay a family secret unless you and your husband disapprove."

"I approve, and I'm sure Yoshi will also. We don't need them to go to a camp on our account. We would feel guilty the whole time, several years, right?"

"Yes, four to six years, at least. Do you mind going?" David inquired.

"We'll miss the farm, but if this is what our government wants us to do, so be it."

"You seem pretty passive about this illegal, rotten policy. Why is that?"

"You have probably learned that the Japanese Empire is 2,000 years old. Our culture is a mixture of influences from Confucius, Buddhism, Taoism, and even Christianity. We roll small parts of those beliefs into our traditional Shinto practices. In this case, going into the camps must be fate. It does one no good to battle with fate. We must accept our lot, make the best of it, and move on."

"So you aren't afraid? You're not angry?"

"No, I'm concerned. It's a bad policy, but it would be a waste of my energy to nurture anger. Yoshi is working on ideas for self-governing of the camp, and I'm sewing warm padded clothing. It will be cold in the winter and hot in the summer."

Pelham replied, "I'm surprised you're not scared of such a large disruption. Your son has taught me that the Army teaches everyone in Japan to lay down their lives for the emperor. So, it makes me wonder what makes the Japanese afraid. What do they fear?"

Youko laughed. "Like children everywhere, the bogeyman."

David smiled. "You mean the scary man in the closet, up in the attic, or under the bed?"

"Yes, every mother threatens her kids with the bogeyman and bad Kami."

"Bad Kami, as in evil spirits and shady influences?" he questioned.

"That's right, and the prospect of being trapped forever in a dark, gloomy underworld somewhat like your Christian hell."

David paused for a moment. "So even though a Japanese soldier or sailor is ready and willing to die for the Emperor, he may be scared of spirits, evil ghosts, and supernatural events. He is also frightened that after dying, he might get stuck in this hellish underworld."

"Yes, Pelham-san, the land of Yomi is a sad, cold, dirty place beneath the earth where dead people might wander forever in the shadows," she replied.

"Youko, you've given me much to think about and a great meal. Thank you so much!"

Uncle Tommy and Yoshi returned from their tour of the farm. Tommy said, "Davy, you should see what these two have done with this land. It's amazing! And Yoshi's plans for each of the forty acres are ambitious and well thought out. I see it as an excellent investment."

"So, can we now discuss the purchase to keep it in the family?" David responded.

"Yoshi and I have shaken hands. The Orient Bank will buy the farm and place it in a trust for your family. Your family, meaning your dad, you, and Marie, will sign a directive that the property will be sold back to the Nagai family after the conflict. The price will be the same as the sales price plus expenses for taxes and insurance."

David waved his arm in a circle. "Who will look after all of this?"

Yoshi replied, "We have friends, a Mexican family of six on some adjacent land. They are hardworking Christians that we trust. They will care for the farm and can sell many of the same products."

He bowed to Tommy. "Pelham-san, I'm sure Youko would like to pour you another glass of beer or some of our wonderful water while I speak to your nephew."

David sat back down and looked at Yoshi expectantly.

"David-san, thank you for helping our son's family. If you can keep them out of the camps, I will be forever grateful."

"Your wife says that you two will make the best of it, that it is just another phase of your life. What did you want to talk about?"

"David-san, what are you willing to risk to win the war?"

"Well, I guess everything. My life will be on the line once I get to my ship. I'm sure we will see a lot of combat."

"I understand. Based on my son's description of your character, I would expect that from you. But are you willing to risk your reputation? Are you willing to work with the devil himself to win the war?"

"Yoshi, I'm not sure what you're asking. Are you speaking of spirits or real people?" Pelham questioned.

"A little of both. Every country has secret groups. Japan is no different. Here, you have the Mafia gangs, the KKK, religious sects, and the gypsies. In Japan, we have the Yakuza. They keep to themselves and often commit crimes to fund their organization."

David shook his head. "For the life of me, I don't see how this relates to myself or the Navy."

"My son told me the kind of questions you are asking and how you want to know the nature of the Japanese soldier and sailor to confuse them and defeat them. It struck me that you would also like to know about their ships, aircraft, and plans. Things like will they come back and attack Hawaii, and when?"

"Yes, those answers would interest me greatly. How would I get them?" David asked with a puzzled look.

"You have your Uncle Tommy; I have a Yakuza uncle in San Francisco. We have never met, and I do not wish to meet a criminal relative who brings great shame to our family. I think he is very high up in this secret group, maybe even the family head, or Kimicho, here on the West Coast."

"Yoshi, do you think they know what is happening in Japan?"

"In a way, they know everything that happens in Japan. They control prostitution, gambling, illegal liquor, and many small businesses like food stands and boarding houses. I don't personally know what they know. The Yakuza gang here in America extends from their family of criminals in Japan. Just trying to contact them could be dangerous. They could kill you; you would disappear forever. If they've information that you want, you would be dealing with the devil. They will make you pay. They're criminals. Nothing would be for free."

"Whose side are they on? Do they want Japan to win or the United States?"

"That is hard to say," Yoshi answered. "Being crooks, they will lie, cheat, and steal with both sides if it benefits them. In Japan, they hate the Army and the police who have total power now that we are at war."

"Why would they deal with me? This seems to me to be more of a spy operation," said David

"I can't say they will deal with you, but they might because it's clear who and what you are. You are a naval officer who wants to win the war. They have a strange sense of chivalry and honor. They probably would respect you. Anyone else, especially any civilian, can never be trusted to be who they say they are."

Pelham questioned, "If I wanted to make this connection, how would I do it?"

"David-san, we call a number in San Francisco and give them the number of a payphone here and a time to call back. They call if they're interested. You make your pitch. If they like it, they give you further instructions."

"My pitch. What is my pitch?"

"It's up to you. I make the first call, and the rest is your ballgame."

David got up. "Let me see if Ron knows a payphone number at the hotel. Please wait for me here and put your thinking cap on. Michael said you could brief me on the current players in the Japanese government. I'll be right back."

Pelham found Strong and Michael sitting on a bench near the flower garden, still discussing Hollywood starlets.

"Ron, would you, by chance, know a payphone number over at our hotel?"

"Yes sir, I do some of my best work out of the lobby phone."

"Good. Second question: do you still have the Chinese movie director's card? Perhaps you could write the number on it?"

Strong wrote on the card. "Will do; here ya go."

David returned to the house, used the bathroom, and asked his uncle if he wanted to listen to a Japanese government discussion. Then, they went to the Zen garden, where Mr. Nagai drew a diagram in the sand.

Yoshi pointed at the drawing. "We start at the top with Emperor Hirohito, who began his reign on Christmas of 1926. His name translated means 'abundant benevolence,' and the period of his reign is the Showa Period. Showa translates as 'enlightened peace.' His father had mental and emotional problems, so Hirohito was the Regent of Japan for about four years before his reign. On paper, the power of the Emperor is complete. The Meiji Constitution states that he's the supreme commander of the Army and Navy. He can cancel, alter, or issue all laws and orders. Therefore, Hirohito is the head of the Empire and the supreme authority in all government matters."

"Does the military report to the civilian government?" asked David.

"No. They report to the Emperor via the Imperial General Headquarters and the Supreme War Council. In many cases, the Chief of the Army General Staff, General Hajime Sugiyama has conferred directly with the Emperor. The Army has taken over the government. They're controlling the people through propagandizing Kodoha, or the 'Imperial Way' that fuses the population, the land, and higher morality to the Godliness of the Emperor."

"Who is on the Supreme War Council?" David queried.

"General Sugiyama, for one; the Chief of the Navy Staff, Admiral Osami Nagano; the Minister of the Navy, Admiral Shigetaro Shimada; Emperor Hirohito; General Hideki Tojo, the Prime Minister; the Minister of War, also General Tojo; and the Minister of Foreign Affairs, General Shigenori Togo."

"There's also a cabinet and a bicameral legislature, right?" David queried.

"There is the Diet, composed of an elected lower house called the House of Representatives and an upper house called the House of Peers, made up of mostly nobles. They were established during the Meiji period."

"So, there is some form of democracy in Japan, correct?"

"Not really. The Meiji Emperor and oligarchs only extended a small role to the government, but the Emperor could overrule their actions and laws at any time. For example, the Diet can veto the yearly budget, but the previous year's budget is adopted if no budget is approved. The Diet serves as the legislative branch of government with limited powers."

"What are the cabinet positions, and who are the current people?" requested David.

"You have General Tojo as Prime Minister, Minister of War, and Army Minister. General Togo is the Minister of Foreign Affairs, Admiral Shimada is the Navy Minister, and Okinobu Kaya is the Finance Minister. Nobusuke Kishi is the Minister of Commerce and Industry. Shimajiro Matsuura is the last Minister of Education that I know of. It could be an Army officer now - consider it the Ministry of Propaganda. Also, Takeshi Terashima serves as Minister of Communications, and Ino Hiroya as Minister of Colonial Affairs. The Prime Minister and his Cabinet are the Executive branch of government."

"I see what you mean by military domination; we have the Military reporting directly to the Emperor, and the Prime Minister and his ministers reporting to the Emperor. So, who else is important?" asked David.

"Yes. The Emperor and Empress have a large staff to run the imperial castle complex and their public and private lives. They have their imperial guards and access to nobles and the oligarchs for advice. There is also the Privy Council that has lost influence and political power with the rise of the military."

"What is the Privy Council?"

"It is twenty-four counselors appointed for life by the Emperor to advise on important matters of state. The current Chairman is Hara Yoshimichi. The Lord Keeper of the Privy Seal also advises the Emperor, controls his schedule, and administers the Royal Prescripts. Kido Koichi currently holds this highly influential position. Finally, there is the Grand Chamberlain, Hyakutake Saburo, who has the Emperor's trust and handles many of his relations with the nobles and public officials."

"All state power comes together with the Emperor in charge, right?"

"In theory, yes. In practice, many military officers or government executives exercise political and economic power in the name of the Emperor, sometimes with his blessing and sometimes not. A strong, forceful, and vindictive Emperor can rule the country with an iron will."

"Is Emperor Hirohito strong and forceful?"

"No, he's not. He is well-educated, has traveled Europe, and has symbolic Army and Naval officer commissions. Upon assuming the throne in 1926, he became the Grand Marshall and Supreme Commander in Chief of the military. His style is nonconfrontational, mostly passive, and seemingly pacifist. Yet he has let the military commit all sorts of aggression against other nations. He has permitted them to lie to his face and refused to punish them for their gross atrocities against innocent civilians and prisoners."

"So, he's not averse to seeing his empire expand? Yoshi, do you think he ordered the Pearl Harbor attack?"
"No doubt he knew of all the plans and approved them. Declaring war on the British, Dutch, and Americans is not a minor decision. There would have been many meetings and planning sessions. No one, not even the Army generals or Admiral Yamamoto, would dare make such a move without the Emperor's consent."
"Yoshi-san, thank you for all of this information. How do we beat Japan in this war? How do we bring the country of your birth to its knees, begging for surrender?"
"Iron ore."
"Iron ore?"
"Yes, that and other natural resources that Japan must import. We must cut them off from trade with other countries wherever we can and destroy their shipping so they can't make steel."
"Thank you, Yoshi. I think I'm ready to make that phone call to San Francisco when you are."

They drove the Nagai truck to a small Sylmar grocery store. Yoshi dropped some change in the payphone and asked the operator to place the call.

Yoshi talked to the party in San Francisco, "Yes, hello, this is Yoshi, Y-o-s-h-i Nagai, N-a-g-a-i, in Los Angeles. I have a friend that wants to talk to you."

David took the phone receiver and leaned forward to talk into the mouthpiece, "This is David, D-a-v-i-d, Pelham, P-e-l-h-a-m; I'm a Lieutenant Commander in the Navy. I want to come by and see you so I can pay my respects." A man told him to wait. There was a pause, and another voice asked when he'd like to come by.

"I can be there tomorrow afternoon. I have a number for you to call at eight or nine o'clock tomorrow morning."

David provided the phone number, and they returned to the farm. After several goodbyes and much bowing and handshaking, David, Uncle Tommy, and Strong drove to the hotel. David asked Tommy if the company plane could take him to San Francisco and back the next day, starting with a pickup at Burbank at 10:30 a.m. Tommy asked no questions and replied in the affirmative. They moved into the company suite, had a few drinks, and ate dinner on their patio, which looked down on the pool. David was hovering near the lobby payphone when it rang the following day.

"Good morning, Lieutenant Commander David Pelham here."
There was a young female voice on the phone. "Good morning, Mr. Pelham. Could you please tell me who is Secretary of the Navy?"
"That would be the Honorable Frank Knox."
A voice on the phone, "And Chief of Naval Operations?"
"Admiral Harold Stark"
Voice on the phone, "Thank you. What does it mean to swing on a hook?"
"That's when a ship rotates around a single anchor."
"What is an anchor ball?"
"It's a round flag that's raised to show that a ship is anchored."
"What is the binnacle list?"
"It's a list of the men on a ship who aren't fit for duty."
The voice on the phone with a giggle, "What is brass monkey weather?"

"A day cold enough to freeze the balls of a brass monkey," he answered.
"Who uses the Acey Deucy club?"
"Enlisted sailors, usually second and first class."
"What is a bluenose?"
David hesitated. "Hmmm, bluenose, bluenose, someone who has sailed over the Arctic Circle."
"What are knee-knockers?"
"A passageway door opening where the bottom bangs on your shins."
"Who is Oscar?"
"He's the dummy thrown over the side for a man overboard drill."
"Good, be standing on the southwest corner of Geary Boulevard and Parker Avenue at two o'clock today, wearing a blue ballcap."

U.S. Office of War Information Poster

Source: Wikimedia commons

Chapter Nine

BOSS KIMICHO – San Francisco

December 1941

David went to the hotel gift shop and bought a blue Los Angeles Angels ballcap. Ron Strong drove him to the Burbank airport. He met the pilots and slept on the three-hour flight to San Francisco International in the Pelham Oil aircraft. David told the pilots to plan on a five o'clock departure and took a cab to the Geary and Parker intersection.

He had an hour to kill, so he asked the cabby where to get lunch. The driver dropped him off at the University of San Francisco entrance, and Pelham ate in the student cafeteria. He read the San Francisco Chronicle and the latest Time Magazine with Admiral Husband Kimmel on the cover. He walked west on Golden Gate and north on Parker at a quarter to two. At precisely two o'clock, a dark blue Chevy sedan stopped at the corner, and the driver waved him into the front seat. David kept his hands in the open and pulled down his hat. The driver made a series of turns, most to the right. When the car stopped about ten minutes later, Pelham had no idea where they were.

The driver instructed him, "I'm only gonna ask you this once. Remember, the penalty for lying is very severe. Do you have any weapons on you?"

"No."

"Are you in any way connected with law enforcement, local, state, or federal?"

"No, I'm a history professor and a naval officer."

"Raise your hat, go through the door in front of the car. You'll be checked for weapons. If you put up any resistance, the response will be severe."

Pelham entered a stairwell of the multi-storied building and met a burly man covered with tattoos on his neck and arms. After thoroughly patting him down, the tattooed man directed him to go to the top of the stairs and knock three times. David climbed the long flight of stairs, thinking they must go to the third or even the fourth floor. At the top, he paused, caught his breath, and knocked on a metal door. The door opened, and Pelham walked into a dimly lit room. The door shut, and he felt the presence of men on either side of him. Another man was seated at a table at the far side of the room with the only lighting source, a naked light bulb hanging behind him. The man gestured for him to come forward and spoke:

"Welcome, Mr. Pelham. I trust you had a safe trip from L.A."

"Yes sir. What should I call you?"

"Mr. Kimicho will do. Boys, check him over again."

David was patted down again, bowed to the seated man, and said, "Mr. Kimicho, it's an honor to meet you."

His bow was returned by a forward movement of Kimicho's head. David bowed again and received a slight nod. The man was wearing a hat of some sort that had a hanging veil hiding his face.

"Why are you here, Mr. Pelham?"

"A mutual friend told me that you and your organization might have information of value to the U.S. Navy. I'm on my way to a fighting ship in Hawaii. The more I learn about Japan and her military, the better."

"And what do you have to trade for this information?"

"Sir, I don't have anything to trade now. If your information is timely and accurate, I'm sure someone in naval intelligence could pay something."
Kimicho spoke to the two bodyguards. "Leave us. Mr. Pelham, please sit down."
David took off his hat. "Thank you, sir."
"Ask me something."
"What? I don't understand?" David responded.
"You say you want information. What information is that? Go ahead and ask."
David thought for a moment. "OK, why did they attack Pearl Harbor? What was the purpose?"
"Mr. Pelham, Nippon's leadership believed war was inevitable with the U.S., so they might as well make an unannounced attack to cripple the Pacific Fleet, break the will of the American people, and get a hands-off treaty in the Far East. It is all the result of stupid Army leaders drinking their own bathwater. They think that iron will, faith in the Emperor, and death with honor trump Uncle Sam's warfare capabilities. They thought if they sank three or four battleships, it would demoralize and unnerve the American people. Then you Yanks would sue for peace and accept Japan's demands."
"Thank you, that's a lot to take in. Did Hirohito order the attacks?"
"Yes, he must have been briefed over months or years about the Army and Navy war plans and approved the final decisions for war. He released an Imperial Rescript, a proclamation declaring war on the U.S. and the British Empire. He's in this war up to his neck."
"Will the Japanese attack again and even invade Pearl Harbor, and if so, when?"
"Admiral Yamamoto was beside himself when the Commander of the First Naval Air Fleet didn't launch more waves of aircraft and destroy every military target in Oahu. He wants to return, destroy the military in Oahu, and invade in strength. He must take Midway Island first, and then Oahu and the less populated Hawaiian Islands to do that. Wake Island will fall soon. Unless the Army stops Yamamoto, the Combined Fleet will return in May or June to finish the Oahu job. He has great influence in Japan, but the Army calls the shots."

David was stunned as he thought, *He could be making this all up. There is no way I can verify what he's saying. Ask him more questions.* "Who was the leader of the attack? The Commander of the First Naval Air Fleet?"
"Vice Admiral Chuichi Nagumo, who is now a national hero in the Nippon newspapers."
"What did he miss? It sounds like his planes did a pretty good job."
"My spies tell me that he missed all the aircraft carriers, failed to destroy the Fleet's fuel supplies, over a million gallons in plain sight, and left the ship and submarine repair facilities nearly intact. The planes sank two operational battleships, and none of the cruisers were destroyed. They could have done much more damage."
"Mr. Kimicho, the carriers were spared not by design, just by their schedule. Enterprise was supposed to be there on the 7th, but heavy weather delayed her."
"What about the other two?" Kimicho demanded.

David hesitated. *Was this a test? I'll tell you what you want to know if you share your knowledge with me.*
"Saratoga was in San Diego, and Yorktown delivered planes to Midway."
"Ah so, we were not sure about Yorktown. Do you know that your Army and Navy shot down several Enterprise planes when they tried to fly into Oahu on the 7th?"
"No, I've not heard that. How many carriers did Nagumo have?"

Kimicho answered, "Six."
"Really? Six carriers?"
"Yes, six first-class carriers and over 300 aircraft."
David whistled. "Holy cow, maybe it's better that we didn't find them!"
"If your battleships and carriers had engaged Nagumo, they probably would've been sunk. He could have then returned to destroy a million gallons of fuel. Your fleet then has to retreat to the West Coast, leaving Hawaii defenseless."
"Not a pretty picture. Could I please have some water or a Coke?"
"Of course, please excuse my manners."

Kimicho rang a bell, and a man appeared instantly from a nearly invisible door. The man returned with ice water, a cold Coke for David, and hot tea for Kimicho.
"Lieutenant Commander Pelham, please forgive me. I thought you might be in a hurry. I should've started with serving you tea and cookies."
"I plan to fly back to L.A. at 4:30 or five o'clock. That should be enough time for both of us. Who will lead the air and amphibious attack on Midway?"
"It has to be Admiral Nagumo, of course. Who else? He's the hero of Pearl Harbor. To pick anyone else would dishonor him and the Navy."
"How would it dishonor the Navy?"
"In Japan, much is learned by what is not said. Putting Nagumo aside would be the Navy admitting that he did not get the job done on Oahu. They would lose face with the public and especially the Army. Nagumo, in that situation, would probably protest by publicly committing suicide."
"So Nagumo, is he a good choice for the job?"
"No, not at all. He's a terrible choice. He is old and somewhat infirm with arthritis. Most important, he's not an aviator. His experience is with surface ships. He is an expert in torpedoes and gunnery, not in the aggressive application of naval aviation. He was initially against Yamamoto's plan to raid Oahu. Several Army and Navy flag officers told him to launch his one-two punch and immediately retire."
"So, if we had a choice, if we could select the commander, it would be Nagumo, good. How many carriers will he have?"
"Four," the gangster replied.
David challenged, "Four, how do you know?"
"It is simple. If the U.S. has three carriers, he won't engage without superior numbers, so it must be four carriers or more. But the Army will not let him take all the carriers again. They need them to attack the Philippines, New Britain, Java, Burma, and Malaya."
"Will Yamamoto direct the battle for Midway and Hawaii?"
"You can count on him planning the whole shooting match. Destroying the Pacific Fleet is his strategy and the standard strategy of the Imperial Japanese Navy, the IJN. He wants to strike one huge blow against the opposing U.S. Navy. He destroys the Pacific fleet's fighting ability and then sues for peace on favorable terms. He'll control the operation from his super battleship Yamato and be close to the action."
"Please tell me a weakness of Japanese naval aviation."

"None of the planes have self-sealing tanks. One bullet through the wings, and they're flying grenades. They require tremendous maintenance because the materials and components used aren't of the highest quality."

"How about the Army? What is their weakness?"

"The major weakness of the Army is its arrogance. They said they would defeat China in a few months. It is now many years later, and no sign of victory there. The Manchukuo Army took on the Soviets along the Manchurian and Mongolian border in the summer of 1939 and got their asses kicked. They will take over much of Asia in this current offensive and then have trouble administering it."

"Whose side are you on? Are you Nisei or Issei? Do you expect to be locked up?"

"Mr. Pelham, now you have come to the most difficult of questions. I shall try to answer them frankly. My crew and I are swinging over to the side of America. The Army and the secret police now completely control Japan. That is very bad for our business. Young men who would have worked for us are drafted into the military or forced to work in industry and agriculture. The secret police seize our property and finances whenever possible. We need peace to be successful in what we do.

"There will be no peace for us with the Army in charge. Hence, our support for the U.S. Most of my men are Issei, but we won't be locked up because we aren't officially here. We identify ourselves as Chinese, Indonesian, Malaysian, or Burmese. We keep a low profile and run several prosperous lines of business here on the West Coast."

"Mr. Kimicho, can you tell me about those businesses?"

"Let's just say that local police and the FBI would prefer us gone. We make short-term loans, run black market lotteries, make a book on sports events, and control non-approved gambling rings. We supply ladies of the night to the finest hotels and the grittiest flophouses. If someone needs muscle, we provide it for a price, as long as it is not against our fellow Japanese. We consider ourselves to be good Yakuza. We are Bakuto, or gamblers. We do not sell drugs, murder people, steal, or sell damaged or shoddy goods. We live off of sin and gin."

"How can I get back in touch with you, and what is the quid pro quo of getting more information?"

"You'll be given a phone number to call, spell your name, and provide a payphone number for a return call."

"I'll be busy on my ship; can I send another Navy person?"

"No, this meeting today is very unusual. It happened because my nephew has never asked for anything, not a favor or one red cent. How is he, by the way?"

"He's good. His farm is prospering. He and Youko hosted us for a grand feast of a lunch. They're both preparing for life in the camps and consider it a new life phase. I'm buying their farm and will sell it back to them after the war. Neighbors will look after the property."

Kimicho showed irritation. "Did you also meet my great-nephew, the professor with the white name?"

"I more than met Michael Nagy and his family. He tutored me in Japan's history for three days, and I stayed at the campus house one night. He can be very useful to Admiral Nimitz for strategy and tactics. I'm taking him to Hawaii. Could he act as a go-between for us?"

Kimicho paused and considered the request. "Possibly, yes. I will take that under advisement. Tell Nimitz and his crew that you need a short-wave radio and a one-time code pad. My deepest sympathies on the impending loss of Wake Island."

"Thank you, sir. Please tell me, how do we win this war against your country?"

"Oil."

"Oil?"

"Yes, as you probably know, Japan must import ninety percent of her oil. Destroy her tankers, storage facilities, and the Dutch oil fields – you will have her on her knees."

David got up and bowed to Kimicho. "Thank you so much for your time and hospitality. I believe this alliance can be beneficial to both of us. Please have someone contact me in Hawaii. Aloha."

The next day in Los Angeles, David took Ron Strong into his confidence about his ideas to fight the Japs at sea. They met with the Chinese movie director, his half-Chinese brother, who owned a fireworks company, and a sound engineer. Deals were made, and payment was promised. Pelham asked Strong to return to New York with Tommy and visit the Pitcairn Aviation facility in Willow Grove, Pennsylvania. As they drove to San Diego, David wrote down the specifications for Ron to deliver to Harold Pitcairn. He also wrote a note to Tommy authorizing Strong's use of his line of credit. Finally, he asked Strong where his dad would stay when in San Diego.

"Either the US Grant Hotel downtown or the Hotel Del Coronado on the ocean," Ron replied.

"Make it the Hotel Del. I've heard a lot about it but have never stayed there. My orders say I'll fly out of NAS San Diego, but I still have a couple of weeks of freedom left."

"Sure, no problem. We can take the ferry over to Coronado or drive around the Bay."

"I choose the ferry, being the nautical person that I am," David said.

When they pulled up in front of the hotel, Strong asked, "Is payment needed for what you're ordering from the aviation company?"

"Yes, please have Uncle Tommy pay for it out of my letter of credit and for any charges from the people we talked to in Chinatown. This note approves your charges up to $20,000."

Strong saluted. "Yes sir, your wish is my command."

They said their goodbyes and shook hands as a bellhop took David's bags from the trunk. David followed him to the front desk, where a clerk told him that no rooms were available. He put his name on a waitlist and tipped the bellhop, who agreed to lock up his bags.

It was a brilliantly sunny day, so Pelham strolled along the Orange Avenue shops and then had a leisurely lunch in the hotel's main dining room. The "Del" was sporting gorgeous Christmas decorations at the outside entrance, foyer, and every public space. After lunch, he walked outside on the ocean boardwalk. It was great to smell the sea air again and hear the seagulls cawing. David returned to the front desk at three o'clock. The clerk said he didn't have a private room but could double him up with another Naval officer. David paid half the rate for four days and got a key.

During the next two days, Pelham never saw his roommate. Each morning after breakfast found him walking north on Orange Avenue, stopping in several small shops, buying ice cream, a nonfiction book, and some magazines. David had no plans, requirements, or near-term deadlines on Christmas Eve. He relaxed on a bench overlooking Glorietta Bay and read his book, "Escape From Freedom" by Erich Fromm. Next, he had lunch at the Coronado

Yacht Club and noticed a few sore spots in his mouth. Later, there were small red spots on his feet during a shower. After an afternoon nap, David felt hot and sweaty; red marks on his chest started to itch. He called Rachael Nagy to discuss his symptoms which now included fever and a pounding headache. She advised him that it could be measles, chickenpox, or shingles and directed him to see a doctor immediately or go to a hospital emergency room.

David informed the front desk of his condition and asked the clerk to send a doctor to his room. Instead, the desk clerk told him to report to the bellhop station. They were sending him to the Balboa Naval Hospital. He spent Christmas and the following week in isolation while they tried to bring his fever down from 103 degrees, put calamine lotion all over his body, and asked him not to scratch his chickenpox blisters. He had a duty Yeoman at the hospital inform Operations at NAS San Diego that he'd not be ready for transport to Hawaii for a couple of weeks.

When David Pelham recovered, he checked out of the hospital and took a cab back to the Hotel Del.. The desk clerk said he'd find him a room later in the day. David wrote a letter to Admiral Nimitz, urging him to move the Nagy family to Hawaii as soon as possible, and provided several general reasons for how Michael and Rachael could be helpful. The note also stated that there were other, more sensitive and compelling reasons why Nimitz should trust his judgment. He sealed the envelope and wrote on the outside, "To Admiral Chester Nimitz – eyes only. From: Lt. Comdr. David Pelham, USNR."

On his way to the Seaside Bar, he bumped into a large Navy First Class in uniform carrying a seabag over his shoulder. After introductions, they decided to head to the bar for a few drinks. David bought the first round. He asked the bartender if anyone at the hotel had flown from San Diego to Oahu. The bartender said there was a regular rotation of PB2Y "Coronado" pilots in and out of the hotel and that he'd point them out if they came by the bar.

Three officers at a nearby table waved Pelham over and introduced themselves. "Commander, I'm Captain Harrold Mueller, pilot of a C-46 Marine transport. This is Lt. Brad Whitfield, my trusty co-pilot, and Lt. (jg) Ben Johnson, our esteemed navigator. I heard you say Oahu, are you trying to get there?"

"Nice to meet you, gents. No, not today. I wanted to send this letter to Admiral Nimitz and sought someone to take it."

"We can take it for you; rumor has it we're launching for Oahu tonight. I'd love to meet the Admiral," Harry replied.

David noticed that the three officers were drinking Cokes, not alcohol. "That would be great. You may only get to drop it off. I can't say that you'll meet him. He's a great guy, really down to earth. Many thanks."

Pelham returned to his table, where Petty Officer Ricky "Tiny" Wright was an entertaining companion with an unending collection of Navy diver stories. They had another round of drinks, and David ordered a couple of burgers. After more drinks, Tiny challenged him to a game of darts, with the loser buying the beers.

Later in the afternoon, two attractive young women walked into the bar and took a seat. Like a flash, Ricky Wright was up and waiting on their table, getting their order and telling them they were about to meet a real Navy hero. David wondered who the hero was until Wright returned with the drinks and waved him over to meet the women. *Oh, oh,* thought David, *This could be dangerous. Wright is going to introduce me as the hero!*

Hotel Del Coronado, Coronado, California

Source: Wikimedia Commons

PB2Y "Coronado"

Source: Google Images

Chapter Ten

SENIOR OFFICER PRESENT – Flying To Oahu

January 1942

In the droning Marine C-46 Commando aircraft, Lieutenant Commander David A. Pelham slept soundly on an uncomfortable Army cot lashed on top of two sizable 100-gallon gas tanks. Little did he know he was flying in the same plane as his letter to Admiral Nimitz.

His deep sleep was only disturbed by a troubling dream about daughter Mandy's first day of school and Lorraine's inability to get her ready to go out the door. He awoke to someone shaking his shoulder and mentioning coffee. David's mouth tasted like brittle cardboard. Coffee sounded like heaven. Crew Chief Watkins again said good morning, offered a mug of steaming coffee, and reminded him where the John was. He took the coffee, walked aft through the cargo bay, and used the head. Sunlight was streaming through the aircraft windows, and several men were awake, reading books or playing cards. The constant noise of the engines and the air passing over the plane's surfaces were nearly deafening.

Pelham ducked under the top of the small rear door and re-entered the cargo bay. His brain was operating at half-speed as he remembered why he was in San Diego in the first place. The shock of the devastating December 7th surprise attacks on Pearl Harbor led to his active duty orders a week later. He had thirty days to take a physical exam, purchase extra uniforms, extract himself from civilian life, and report to his new duty station. As surprising as the Jap sneak attack, BuNav had assigned him, an east coast sailor, to a destroyer in Pearl Harbor. He sent several less than respectful complaints up the chain of command, stating that he had spent nearly a decade training for sea duty in the Atlantic and expected an East Coast ship billet. The answer from his Navy Detailer was one that now absolved all seniors from revisiting dumb decisions, "Quit bitching Pelham. Don't you know there's a war on?"

Chief Watkins joined him as he was making his way forward. Watkins pointed down at a huge sailor sleeping on top of several seabags. "Sir, this First Class bullied the Air Station Ops Officer into putting you onto the manifest and more or less carried you onboard."

"I'm afraid to ask, how did he get me on the flight?"

"He had numero uno priority. I'm guessing he's a top-notch Navy diver, and they want him in the worst way in Pearl. He told the Ops Officer that you were gonna coordinate all of the ship repair diving efforts in Hawaii, and he was not going without you."

David grimaced. "Umm, the way I feel now, I'm not sure he did me any favors."

The First Class Diver was still sleeping atop his mattress of other sailors' seabags, no doubt crushing their uniforms and possessions. David thought, *God, he's enormous, must be 300 pounds or more. I guess no one is going to argue over his makeshift bed. What's his nickname? Something to do with his size. Tiny, that's it. The big paw reached out to me in the Hotel Del Coronado lobby, and the smiling face said, "Just call me Tiny; everybody in the Navy does."*

David eased past the large tanks and sat on the wooden stool, sipping on the hot, strong coffee. He thought, *Thank God for coffee - the Navy runs on coffee. The Krauts and Japs may destroy all our beans, oil, and bullets, but look out if we ever run out of coffee. The Navy would come to a grinding halt.*

His thoughts turned to his Uncle Tommy. It had been over twenty years since David and his dad had met nineteen-year-old Tommy in Japan. He became a permanent Pelham, in the New York clan; and an indispensable advisor and confidante to William Pelham, the President of Pelham Oil. Tommy always had time for David to play with him and listen to his thoughts and dreams. He also bought his honorary nephew the best presents for his birthday and Christmas. David's father had pulled some strings during the nineteen-twenties, and Thomas Yoshida Pelham became a bona fide American citizen.

Seeing him in L.A. as a successful banker and businessman was good. Too bad he never had time for a family of his own. So here I'm flying to Hawaii to wage war on his native country, to kill Japs while I love my uncle like a second father. Life is complicated.

Pelham was roused out of his reminiscing by the Crew Chief joining him. The Chief looked at him. "Sir, do you have a few minutes to chat?"

David thought, *No Chief, I'm busy here drinking my coffee for several more hours.*

"Sure Chief, what can I do for you?"

"I hope you're feeling better, Commander because I've got some news for you and a bit of a difficult situation."

"Shoot Chief, with that nap, this coffee, and some chow in my future, I'm feeling in the pink."

"Commander, you should know you're S-O-P, Senior Officer Present," stated Watkins.

"Chief, are you sure? You keep calling me Commander, but you can plainly see by these shoulder boards that I'm just a lowly Lieutenant Commander and a reservist at that."

"Sir, I'm sure that both of those designations will change quickly as the Navy expands by leaps and bounds. The heavy brass don't like to fly at O dark 30, so you're our SOP. You are flying in a new aircraft with a fresh-caught Marine Corps Captain at the wheel, assisted by a junior Navy Lieutenant, a Lt. (jg) Navigator that bilged out of flight training, me as your able Crew Chief, and a Marine Corporal plane captain."

"What's the tricky situation, Chief?"

"Here's the deal, Commander. Lieutenant Johnson, our navigator, is chewing his fingernails off because his dead reckoning plot shows us arriving at Pearl with no fuel reserve, or worse, going into the drink short of Pearl."

David's face showed alarm. "Chief, are you kidding me?"

"No sir, Johnson may have had some problems with carrier quals, but he's a smart lad, double checks everything, and works the navigation desk nonstop from takeoff to landing. He's good!"

"What's the basic problem here?" Pelham demanded.

"Sir, Nav thinks we have more headwind than forecast and that the pilots are burning more fuel than necessary. You should know they're no longer speaking to each other."

"And the plot thickens. Why is that?"

"Lt. (jg) Johnson tried to get Captain Mueller to make some adjustments, starting back at NAS San Diego. Unfortunately, the Plane Commander Marine doesn't believe the Navy Lieutenant's calculations. He's finally told him to shut up and get off the intercom."

Pelham spoke sternly, "So you want me to fix this, how?"

"Well sir, you're the Senior Officer Present. I'm trying to give you the straight skinny. The Navy may blame you if we have to ditch and lose the plane and several lives. I don't want you to feel like you got stabbed in the back."

Pelham was agitated. "Christ almighty, Chief. I think you're informing me the knife is already half in. What aren't you telling me? Spit it out, now!"
"Sir, I don't like to talk out of school, so this is just between you and me, right?"
"Right, get to it," David snapped.
"Captain Mueller, or "Rat Face" as he's affectionately known to his fellow Marines, has a new squeeze on Ford Island, and he wants to get there when she gets off work. Since last night, he's been pushing Nav to find a reason why we should land at Luke Field; read that Ford Island instead of Hickam across the bay. He moved our scheduled launch time from NAS San Diego up by an hour last night. We spent thirty minutes with both fans turning, waiting for the Duty Section to man the tower. He got very terse with Johnson over the intercom. Nav pleaded with him to at least shut down one engine while we were waiting and strongly recommended returning to the flight line to top off the tanks."
"Good grief, Chief, I met your officers earlier today at the Hotel Del. Why do you think I should get in the middle of this cluster fuck?"
"Well sir, you're no dummy. I saw your orders, Ops Officer on a heavy destroyer, Porter Class. You don't get that job of being next in line to the CO/XO by sitting on your hands. Besides, if Nav is right, we have twelve souls onboard that are your responsibility as SOP. Flying a plane is not much different than conning a ship; it's just that everything happens ten times faster. We're coming up to the halfway point of a fifteen-hour flight. If we need to make some changes and corrections, we must do them now."
"Chief, I'm a ship driver. I've done some flying in private aircraft and a larger one owned by my dad's company, but I was more of a passenger than a crew member. Where do I start?"
"Commander, I would start with a Reader's Digest update from Nav. Then introduce yourself to the pilots and make them aware that you're the SOP. After that, I'd tell them you have to check Nav's figures, but in the meantime, they need to throttle back to normal cruise speed."
"What kind of plane are we in?"
"A Curtiss-Wright C-46 Commando, twin-engine transport, just came off the assembly line."
"What's the normal cruise speed, and how fast are we going?"
"Sir, normal cruise is 157 knots True Air Speed; right now, our TAS is closer to 165 knots."
"So, we can get to Ford Island before quitting time?"
"Yes, sir."
"Chief, what is the range of this aircraft at normal cruise?"
"Normal cruise is about 2,500 to 2,600 nautical miles, assuming no headwinds."
"And how far is it from San Diego to Pearl?"
"A little less than 2,300 nautical miles as the crow flies, sir."
David calculated, "So, with no wind on the way, you could fly for another hour or two once we reach Hawaii?"
"Yes sir, probably about an hour to an hour and a half."
"And you say winds are the problem. What were the forecast winds?"
"From the southwest right on the nose at twenty knots," replied the crew chief.
David scratched his head. "So if a twenty-knot wind is on the nose, as you say, for fifteen hours, it puts us 300 miles behind our position without any headwind. Then our 200 to 300-mile reserve shrinks to, let me see, um, to next to nothing and less than a few minutes of flight time!"

“Yes sir, you’re getting the picture,” Watkins replied.
“And Chief, you’re telling me that the pilots burned fuel with both engines while sitting on the deck in San Diego, and now we’re reducing more of that 2,500 to 2,600-mile range by flying faster than necessary?”
“Yes sir.”
“What does Nav, er Lieutenant Johnson think the real winds are?”
Chief Watkins paused. “Sir, he thinks the headwinds are about thirty to forty knots.”
David replied in a pained voice, “Why does he think that?”
“There’s a beacon ship stationed about halfway to Hawaii. We should’ve picked up its signal about an hour ago.”
David was even more alarmed, “Jesus fucking Christ, Chief. If he’s right, we will not make it to Hawaii! If the wind is forty knots and we used twenty knots for our calculations for the last seven hours, we may be 140 miles behind our projected course.”
“Yes sir, I’d say that sums it up perfectly.”
David said harshly, " Why did you let me sleep for several hours?”
“Sir, I hoped they would work it out, and frankly, you were in pretty bad shape, not very likely to provide outstanding headwork.”
“OK, OK, Chief, water under the bridge. I’m going to get educated by the Navigator and go forward to see if the pilots are truly trying to kill us. Here is what I want you to do. Roust out the passengers and see if you have any Yeoman or Admin types. Have them list each man’s last name, first name, rate or rank, branch of service, and special skills. And Chief…”
“Sir?”
“If we get out of this mess in one piece, you’ll buy me mucho drinks at the Chiefs’ Club.”
“That would be my great pleasure, Commander,” replied Watkins.
Lt. Comdr. Pelham stood just forward of the large tanks in the cargo bay to collect his thoughts. *OK, Davy, think this through, get the facts, make the right decisions, and be calm and collected. They may be youngsters, but they’ve had the best training. You just need to get them back on track.*

He went forward in the Cargo Hold and saw the young Lt. (jg) bent over his map table on the plane's starboard side. David took a couple of deep breaths and walked toward him, mouthing the words, “Once more into the breach.”

When David stepped up to the navigation station, Lieutenant Junior Grade Johnson was calculating numbers and jotting down answers in a spiral notebook.
David interrupted him, “Excuse me, Lieutenant, the Crew Chief told me you know your stuff and have some concerns about our flight.”
Johnson turned away from the charts. “Yes, er, sir,” with an inquiring look.
“Sorry, let me introduce myself. I’m David Pelham, Lieutenant Commander, USNR. According to the Chief, I’m the Senior Officer Present. He wants me to understand our operational situation.”
“Yes sir, we may not make land. We should be declaring...”
Pelham held up his hand. “Stop right there, Johnson. Let me tell you what I think I know, and you can provide more details and any recommendations. I know we burned up some precious fuel back at San Diego, and that our airspeed may be greater than an efficient cruise speed. I know that you believe the headwinds may be as much as thirty or forty knots, and if so, we

may have no reserve fuel at Pearl Harbor. The Chief said you believe this because we have not yet received the radio beacon of the navigation ship. Is that all correct?"
"Yes sir. This is our third flight to Hawaii, and the only time we have not received the ship's signal when expected."
"Good, Mr. Johnson, I have some questions for you. One, are we flying a Great Circle route? And if not, why not?"
"No sir, we aren't. The Plane Commander does not believe a Great Circle route is any shorter."
David looked surprised. "The Chief told me that we're about halfway. How many miles do you think we would save if you were to fly a Great Circle from here to Pearl?"
"Well sir, we're flying between thirty-two degrees of north latitude at San Diego, southwest down to nineteen degrees at Honolulu, and halfway there would put us in the mid-twenties, so with about 1,200 miles to go, I think we would pick up about ten to twenty miles."
"Good man. Can you estimate our course if we want to fly the Great Circle route?"
"Sir, off the top of my head, I'd say about twenty degrees starboard and take off a degree every sixty miles."
David thought, *I like this kid. He took 1,200 miles, divided it into sixty-mile segments for each degree of longitude, and developed a rough estimate off the top of his head.* "Lieutenant Johnson, what are your recommendations?"
"Sir, we need to slow down and quit wasting fuel. We need to fly this bird on its max-range profile. If my numbers are even half right, we need to divert to the Hilo airport on the Big Island of Hawaii now!"
"I don't understand; going to Hilo gives us what?" Demanded David.
"Sir, Hilo is a hundred miles closer to San Diego than Pearl Harbor, and it's to the left of Pearl about 200 miles."
"Very clever, Mr. Johnson, so although we have not been flying the shortest Great Circle route to Pearl, we may be close to one that ends in Hilo?"
"Yes sir, exactly sir!"
"And what speed change are you recommending?"
Johnson pointed to a small instrument cluster. "Sir, you can see on my airspeed gauge we're going 150 indicated airspeed. We should be going 135 to 145 indicated. That's the longest-range speed for this aircraft, depending on the weight."
"So we're burning more gas than we should, and it may cause us to run out of gas before we can land. Is that the tactical situation?"
"Yes sir, we need to slow down as long as we keep her on the step."
David was curious. "The step, what is the step?"
"Commander, this plane is sort of like a boat that you push through the water faster and faster to get the bow down. When you get to a speed where it's planing on the surface, you can ease back on the throttle. For us, pushing through the air is like a boat pushing through the water. Once we get her on the step and planing, we get a boost of about five to ten knots and some fuel saving. You can almost feel it."
"And right now, our speed over the ground is 150 knots. Is that correct?"

"Oh no sir, that's our Indicated Air Speed or IAS. Since we're at 8,500 feet in altitude, the air sensed by the IAS pitot tube is less dense than at sea level. We must add roughly sixteen percent to get our True Air Speed."

David used a pencil and pad. "So our ground speed is 150 plus 24 or 174 knots, correct?

"Yes sir, if there was no wind, but with a twenty-knot headwind, our travel over the ground, or the sea, in this case, is 174 minus twenty equals 154."

"And if it's thirty knots of wind, we have a 144-knot ground speed, and if it is forty knots, we have a 134-knot ground speed?"

"Yes sir."

"Lieutenant, you're scaring the bejesus out of me. What's your first name?"

"Benjamin, sir."

"Can I call you Ben?"

"Certainly sir, or Tex is my nickname."

"OK Ben, I'm going to go forward and reduce the airspeed, get us on a max range profile, and start heading us to Hilo without saying much about it. You've some work to do. First, figure out a course for a Great Circle route to Hilo. For starters, I plan to have the pilots turn five degrees to starboard. Second, determine if the beacon ship should pass down our port or starboard side if we're on that route. We need a Nav fix from the ship to get a handle on the wind. Is that correct?"

"Yes sir, it's critical to whether we make land or not. Good luck with the pilots. I could use one more piece of information from them."

"What is that?"

"I need the fuel flow rate at the current speed before they slow down and then the new flow rate."

"OK Ben, you got it, but educate me; what type of fuel flow indications do the pilots have?"

"They have a readout for each engine that shows the current burn rate in gallons per hour."

"I think I understand. The higher that rate, the faster we run dry, right? I expect you've been keeping a time log and can back into fuel used from engine startup?"

"Yes sir, fuel flow is the most accurate measurement of fuel used. We know the tanks were full when we started the engines, and I've kept time logs for fuel usage since then."

"Keep doing what you're doing, Ben. You may be saving everyone on this airplane, and I will make sure you get the recognition you deserve."

"Thank you sir!"

Right Davy, he thought, *Good old Navy leadership. Pat them on the back and kick them in the ass, whichever is necessary.* David took a few steps forward and stood behind the pilots. All the levers, knobs, switches, and gauges amazed him. The sun streamed in through the side, front, and top windscreens. Although not as loud as in the cargo compartment, the engine noise continuously drummed on the fuselage. Both pilots wore khaki uniforms and crushed campaign hats with large earphones pushed back on their heads. The Marine Captain in the left seat noticed him first.

"Good morning sir; what can we do for you?" said Harry Mueller.

"Good morning, Captain. I just came up to have a look-see. Wow, this really is a new bird, bright and shiny."

"Yes sir," Rats enthused. "This is the largest twin-engine transport in the world. It has twice the cargo space, carries three times the cargo weight of a C-47, and flies double the flight range."
"Well, it looks like we have good weather so far. Your Crew Chief asked me to come forward and meet you. He's informed me that I have the dubious honor of being the Senior Officer Present."
The Co-pilot said, "It's good to see another Navy type, sir. Are you on your way to Pearl Harbor?"
"That's what my orders say. I will report aboard the USS Selfridge as Operations Officer by next week."
"We should have you in Pearl by 1700 local time if all goes well," Lt. Whitlock responded.
"That's why I'm up here, getting in your hair. I understand there is some debate between you and the Navigator on the adequacy of our fuel reserve at Oahu. Is that the case, Captain?"
"Sir, Lieutenant Johnson means well, but he tends to be overly conservative and gets his panties in a knot over the small stuff."
David spoke in a loud voice, "Captain, are you telling me that running out of fuel, having to ditch, and the inevitable loss of some or all of your crew and passengers' lives, let alone this aircraft, is a trivial matter?"
"No sir, not at all. It's just that Johnson is always looking for the worst case. I doubt we would get anywhere on time if we listened to his objections."
"Co-pilot, what is the most efficient cruise airspeed for this C-46 aircraft?" David asked.
"Sir, do you want it in miles per hour or knots?"
"I'll take knots since I've been briefed on distances in nautical miles," David answered.
"135 to 145 Indicated Airspeed sir, depending on the bird's weight."
"Co-pilot, are we five to ten knots higher than our maximum range speed? Is that correct?"
"Sir, I think you better ask the Plane Commander. He sets the speed."
"How about it, Captain? Are we running a little fast and burning more gas than we have to?"
"Yes and no, Commander. We have a briefed twenty-knot headwind, so we fly a little faster than if we have no headwind or a tailwind."
"I don't understand," David said.
"Think of it this way, the longer we're in the air, the further back the twenty-knot wind pushes us. So we speed up to reduce the total time it's slowing us down. In this case, the rule of thumb is to speed up half the speed of the headwind or ten knots."
"And if the headwind were thirty knots, you would speed up fifteen knots?"
"Yes sir."
"This is all new to this sailor, so bear with me. Would that be ten knots faster using indicated airspeed or true airspeed?"
Mueller hesitated and pointed at the instrument panel, "Well, I usually just apply it to IAS 'cause it's right there. Which is it, Whit?"
"I think he's got you, Rats. It seems to me it's TAS, not IAS," said Whitlock.
"I'm not trying to trap you. I'm trying to understand the safest cruise speed for all of us. Tex told me this aircraft is up on the step, much like a speedboat planing. Is that correct?"
Both pilots replied, "Yes sir."

"So, with my new infantile grasp of aviation, for twenty knots of headwind, you would add ten knots of true airspeed greater than the speed that keeps us on the step?"
Mueller replied, "In theory, that's not too bad, but the weight, drag, and thrust components vary during a long flight. We're about halfway through our fifteen hours and have burned off tons of fuel. We could pull a few revs off, stay on the step, and go a bit slower, but the wind affects us longer in the wrong direction."
"What would be a good indicated speed with no wind at this point of the flight?" David asked.
"135 to 140," said the Co-pilot.
David figured on a pad, "OK, let's see 140 IAS is about 162 TAS, right?
"Yes sir, that's what I've got on my A-N-C," answered Whit.
"Excuse me, A-N-C?"
"Air Navigation Computer," replied the Co-pilot, holding up a metal device. "I work one up here, and Tex has another one."
David worked quickly with a pad and pencil. "OK, if we want to compensate for twenty knots of headwind, we want a TAS of 162 plus ten for 172. If we take off sixteen percent to solve for IAS, we get about 148.2, not far off our current 150 indicated speed."
"Sounds about right," Rats said smugly.
"Captain, it looks like through your infinite wisdom, or for all the wrong reasons, we may be flying at an optimum speed, do you agree?"
"Yes sir, when I'm good, I'm really good!"
"So it appears that your speed is OK. We may need to increase it if we get a navigation fix and find higher winds. In the meantime, as SOP, I'm directing you to turn starboard five degrees while Nav and I review the fuel figures."
Mueller protested. "Sir, what's with the course change?"
"It's just a hunch. Nav believes it will give us a few extra miles in range if we use a Great Circle route - end of discussion. Co-pilot, put your ears on and monitor for the beacon ship signal. If Nav is right about higher winds, we may have a real fight to make land. If he's wrong, we can increase our speed later. Maintain the settings I've given you until I say otherwise, and let me know if you get a bearing on the ship."
Captain Mueller shook his head and replied, "Yes sir."
David leaned forward and spoke quietly into Mueller's right ear. "And Captain, just so you remember, if I come back up here and find even five revs too many on the engines or a degree off course, I will provide some negative input to your next FITREP. Do you understand?"
"Yes sir! Thank you sir!"
David turned away. "Carry on."
He stopped at the radio station aft of the pilots and spoke to the Marine. "I take it you're the radio operator?"
Jenkins replied, "Yes sir, radio, Plane Captain, cook, and bottle washer. Corporal Jenkins, or "Jinx" if you prefer."
"OK, Jinx, tell me what you're doing right now."
"Sir, I'm trying to raise the Nav ship by voice or pick up its radio beacon."
"Very good, keep at it until you get them. Are you able to talk to Nav?"
Jenkins pointed to a radio panel. "Yes sir, got him right here on the intercom."
"Good, Corporal, carry on."

Pelham went to Johnson's navigator position and stood behind him. "Ben, the Co-pilot should have the fuel flow data you need. It turns out that the speed may be OK in response to the headwinds. I'll be back in a few minutes, looking for an update using Hilo as the new destination and winds of twenty, thirty, or forty knots. Just in case, I need an estimate of the increase in flight range for every thousand pounds of gear we dump. One more thing, I could eat a horse. Have someone fix me a sandwich or something edible. Got all that?"

"Yes sir, I made a list, aye, aye, sir!"

Smart kid, thought Pelham, *Maybe I can keep us out of the drink with his help. Christ, the Pilot and Co-pilot look like they're barely old enough to have a beer. What is it going to take for us to make land? Facts, guts, and a little luck. Get on the stick, Pelham. It won't do your career any good if you never get to your destroyer.*

C-46A drawing from Pilot's Manual

Note the double rear doors that opened upward

C-46A cockpit circa 1944

Source: National Museum of the United States Air Force

Chapter Eleven

NAVIGATION FIX – C-46 Point Of No Return

January 1941

Lt. Comdr. Pelham moved aft of the cargo hold fuel tanks to locate Chief Watkins. The Chief was looking over some material on a clipboard with three or four other men and motioned for him to join the discussion.

"Commander, we have your list of personnel. What do we do now?"

"What about other officers or noncoms onboard?"

"We have an Army Captain ground pounder, a Chief Bosun's Mate, an Aviation Electrical Chief, four more passengers, and our C-46 crew of five."

"Chief, how much cargo do we have by weight?" David asked.

"Sir, about two and a half tons, 5,000 pounds, give or take a few pounds, distributed in three storage bays."

David was confused as he looked around the bay. "Three storage areas?"

"Yes sir, two below the floorboards can hold up to 3,000 pounds total."

"OK, how is the cargo distributed now?"

"About one-third forward and two-thirds in the aft bay to keep the C-G within limits."

"Excuse me, what's the C-G, Chief?"

"Center of Gravity sir, the balance point, or a range where the bird is balanced forward and aft. The plane doesn't fly if the CG is wrong, and we fall from the sky."

"Who in the crew is responsible for the correct CG?"

"The Plane Commander, sir. I'm the loadmaster, calculating all the weights and CG force moments of everything added to the aircraft depending on its location. I pass my results to the co-pilot for review, and then the Pilot reads the weight and balance sheet and signs it."

"OK Chief, you have my head spinning. Now that I think about it, we have similar ship calculations. New subject, do we have any radio types onboard?"

"Let me see," he looked over three lists, "we have a Radioman First Class."

"Have him meet me over by the forward cabin door," David said as he pointed at the low door leading to the forward cabin.

Pelham took a paper from the clipboard and wrote:

1. Your job is to help Corporal Jenkins, the radio operator, up forward in locating the beacon ADF ship.
2. At the same time use all possible frequencies to call the Nav ship and determine if it's on station.
3. Also you are to use any and all frequencies to obtain this afternoon's weather forecast for the Hilo airport on the island of Hawaii. If possible, talk directly to the Hilo Tower.

After shaking hands with the Petty Officer, Pelham gave him the instructions and directed him to report to the radio position on the starboard bulkhead behind the pilots. He turned back to the Crew Chief and motioned for him to come over.

"Yes sir?"

"Chief, I got a short briefing from Johnson. I think he'll agree that until we get a Nav update from the beacon ship, we'll be unable to estimate the winds and be 100 percent sure what our situation is."

"Yes sir, I felt a course correction, maybe?"
"Here's the latest, Chief. We are flying a maximum range configuration at the same speed, about 150 knots indicated. Nav and I agree on using a Great Circle route to our new destination, the Hilo Airport on the Big Island. Johnson is running the numbers now for twenty, thirty, or forty knots of headwind. Hilo is 100 nautical miles closer to San Diego than Pearl Harbor."
"How did this go over with the pilots, sir?"
"Chief, they're unhappy and aren't clued in on the divert field. It turns out that the Plane Commander was flying an OK speed profile. With headwinds, you fly a bit faster to fight them."
Watkins gestured at the passengers in the cargo bay. "What can we do back here?"
"I think we need to tell the men what is happening. I don't want to scare them, but they need to know that this could be a dire situation for what I may have to ask you and them to do. So if you agree, I'd like you to form them up on each side of the cargo in the aft bay, nothing military, just get them in position so I can talk above all the noise."

Chief Watkins boomed out to the passengers to form two lines on each side of the cargo on the deck and to stand by for an announcement from the Senior Officer Present.

David thought, *Nice try, Davy boy. You don't want to scare them, but you know down deep you're shocked. It may be too late to turn this mess around and keep from ditching. We destroy a brand-new aircraft and don't even think about the men that will be injured or killed on impact, drowned, or eaten alive by sharks. Steady breathing now, calm and collected; give them just enough information to be part of the solution.*

He stood just aft of the line of gas tanks, faced the two rows of men, and said loudly, "Everyone at ease, indicate if you can hear me!" The number of nods and thumbs-up gave him the go-ahead to continue. He noticed that the heavy Navy Diver, Tiny Wright, was back among the land of the living and smiling at him.
"Men, we have what may become a dangerous operational situation. We're about halfway to Hawaii on a fifteen-hour flight. When we took off from San Diego, the winds were forecast to be twenty knots on the nose. If that turns out to be correct, we barely had a safe fuel reserve when we reached Pearl Harbor."

He thought, *No use bringing up ditching at this point; keep it positive.*
David continued, "Here is what your talented flight crew is doing. One, they're now flying an airspeed that gives us the best flight range for the fuel burned. Two, they've made some navigation changes that should shorten the miles we must fly. Three, we've started down a path that has us diverting to Hilo. It's on the east coast of the Big Island of Hawaii and 100 miles closer than Pearl. Four, we're approaching a navigation aid ship. We can figure out our ground speed and the headwinds when we determine our exact position. Then we can calculate our fuel reserve at Hilo. If it's not a safe amount, we may have to take more severe measures during our final 1,000 miles.
"Crew Chief Watkins has asked me what all of you can do. I want him to form working groups to evaluate what cargo we can throw over the side. Make a list of the type of cargo, the estimated weight, and its location. Use your best judgment to prioritize it as first to go, second to go, third to go, and fourth to go. Plan for about 1,000 pounds on each dump. My use of the word cargo is broad; everything in the aircraft is fair game. Hopefully, our navigation

calculations will show we don't have to take these extreme measures. Let's stay ahead of the problem, work as a team, and know our best options. Who is the Aviation Electrical Chief?"
A Chief Petty Officer raised his hand.
"Chief, if Chief Watkins agrees, I want you to form a working party and comb this aircraft from stem to stern for loose aircraft items and removable nonessential structures or components that we can dump over the side. Prioritize them from first to go, second to go, third to go. Note the item's type, current location, and estimated weight. One more thing, if we must take it apart, estimate the time to do that. Carry on!"

Pelham moved ahead of the auxiliary gas tanks, drew his usual cup of black coffee, and sat down on the wooden stool, sipping the hot java. He mused, *What's next? We have people working, and things are getting done. What am I missing? Think, think, it's just the same as taking a ship to Hawaii, only much faster and with no chance to refuel. We must have the Nav fix one way or another; that's key. It will give us the average wind for the last seven or eight hours. We can expect the wind to be about the same for the rest of the flight. We must know how much fuel we've burned and how much remains. Fuel! We must squeeze extra miles out of every last gallon of fuel. If we can't make Hilo, we must declare an emergency early in the game, and ask for help from shore units and possibly ships at sea.*

There was a whooping noise up forward; men were yelling excitedly and celebrating. The Radioman Petty Officer started to pass David, heading aft a few seconds later.
"Hey there, Sparks, what's going on? Looking for me?" Pelham asked.
"Yes sir," the radioman said, beaming, "we have a solid ADF radio signal on the Nav ship!"
"Where does she bear, ahead or behind?" David asked.
"Sir, she's about fifteen degrees port of the nose."
"Good. Why do you think she's ahead of us?"
"Sir, we had a weak in and out signal but no bearing. Now we have a firm bearing, and the volume is increasing."
David motioned the radioman to follow him to Johnson's station and shouted, "Outstanding Petty Officer, ADF off the nose about fifteen degrees to port. Let's ask Lieutenant Johnson, our trusty navigator, which side of the bow we want the ship to pass down on our new course?"
Johnson grinned from ear to ear. "The port side would be the best, sir!"
"Thank you, Sparks. Have you talked to the ship and confirmed she's on station?"
"Yes sir, I got them on the key, and they reported on station."
"Very well, please go aft and tell the Crew Chief the good news."
"Yes sir, I will sir. Thank you sir!"

David went back to the front of the cargo bay and sat on his stool, *One piece of the puzzle was falling into place, except that ADF gear only provides a bearing to the transmitter, a magnetic direction at that, and no range. We know the ship's position in north latitude and west longitude. If we stay on our Great Circle route and pass north of the beacon ship, we can get our west longitude position when we cross their true north radial, a line drawn from the vessel to the north pole. Then we need our latitude, our distance north of the Equator, to get our fix.*

As Pelham finished the coffee and put the mug into the tray by the big, stainless steel pot, he felt the aircraft tilt to the left. *We're turning, that dirty son of a bitch. He's willing to trade miles at the end of this flight for a warm, cozy Nav fix now.*

Pelham strode forward to the Nav desk, put his hand on Johnson's shoulder, and asked if he had given the pilot a heading change.

"No sir, I just asked the Plane Commander why we were turning, and he said he wanted to fly over the navigation ship. I recommended that he stay on the previous heading until we cleared it with you. He told me to fuck off. He's in charge of the aircraft, and you…"

David cut him off. "Thank you, Ben. What course will put us on the route to Hilo?"

"Sir, now that we've found the ship, I think we need to pass about forty to fifty miles north of her."

"Give that to me in heading degrees."

"Commander, about fifteen to twenty degrees further starboard from our current heading."

"Thanks again, Ben. One more thing, use the local magnetic variation to determine the ship's ADF magnetic bearing when we fly through her true north radial. Our west longitude at that point will be the same as the reference ship. We may have to estimate the latitude, but it should be close enough for government work."

"Aye, aye, sir!"

David moved forward behind the pilots and spoke loudly so both could hear him over the engines' din. "Captain Mueller, why have we turned to port?"

"Commander, I figure if we pass directly over the Nav ship, it will give us the best possible fix. Sir, with all due respect, I'm the Plane Commander; the safety of this aircraft is my direct responsibility."

David responded in a firm low voice, "Captain, you've had your head up your ass most of this flight. If you wanted to change course, you needed only to run it by me first. I made that clear enough, even for a jarhead to comprehend. First, you burn up fuel in San Diego for no good reason. Then you fly a speed above max range cruise speed into a mostly unknown headwind to meet some mythical schedule known only to yourself. You lucked out on that one.

"Now you're turning to burn precious fuel and miles that could mean the difference between life and death at the end of this flight. You, Captain, have jeopardized this aircraft, crew, and passengers with poor decisions and shoddy headwork. Unless we get this disaster waiting to happen squared away, I see a court-martial in your future and excellent potential for you to join the infantry. Take your hands off the controls, now! That's an order! Co-pilot, you have the aircraft."

Both rage and fear showed on Mueller's face. "Sir, you've no right, you can't…"

"Nonsense, I have every right. I'm senior here, and you're hell-bent on killing us all. Co-pilot, take us to starboard twenty degrees; Captain Mueller, get out of that seat and follow me. Lieutenant Whitlock, I'll have Johnson come up here and cover the controls while you move to the left seat. Have him brief you on diverting to Hilo. I'll be right back, and we all will be working on developing a Nav fix to assess the headwinds."

Pelham took Mueller in tow and went aft to the Nav station. He instructed Johnson to go forward and assist the co-pilot. Johnson rolled his eyes at David meaningfully as he passed by going forward. David looked longingly at his coffee cup and thought, *No more, Davy boy. You're pretty caffeine-wired as it is. I better eat soon if I want to survive this marathon.*

He took the abashed, sullen Mueller to the front of the gas tanks, leaned in, and spoke to him eye-to-eye. "Mueller, you get some sleep. We'll discuss whether you can return to crew status when you get up. Stop thinking with your dick, and start being a team member, now hit the rack."

David returned to the flight deck behind the two junior officers and spoke, "Lieutenant Whitlock, are you going to have any trouble flying this aircraft?"

"No sir, I will say that Rats, er, the Captain has far more landing experience and that I may need his help if we have an emergency like an engine fire or such."

"OK, I'll see if I can find you some help. Be aware that Tex has us heading for Hilo on the island of Hawaii because it's 100 miles closer to San Diego.

Whitlock questioned, "Being halfway through our flight, does fifty miles make that much difference?"

"That plus flying a Great Circle route may give us sixty to seventy more miles and keep the sharks from feasting on your lily-white balls, Mr. Whitlock," David replied sarcastically.

Whitlock looked scared and distressed. "I'm all for avoiding sharks, believe me."

"Johnson, can you shoot the sun for latitude as we cross the ship's 360 radial?"

"Great idea Commander; I just happen to have the sextant out and ready. I may be able to do it from this cockpit, but it might take some turning to line it up."

"Good. Plane Commander give him whatever he needs. Another question Ben, isn't there a way to estimate your distance from an ADF radio station based on the rate of bearing drift as you pass by?"

"Yes sir. Whit, I need you to start your clock, note the current bearing to the ship, and then write it down at three-minute intervals. I have the formula in my navigation school notes."

"Here's something for both of you. After talking to the Crew Chief, I want a full briefing on the aircraft's fuel system. Include managing the fuel, gas consumed, and what remains in the tanks."

"Sir, is that before or after I take the fix?" asked Johnson.

"You might as well time your sun shot for the ship's 360 true radial crossing, develop our fix, and then start the fuel calculations."

"Mr. Whitlock, how is your autopilot?" asked David.

"Like a rock, sir, no problems."

"Good. See if you can spot the ship with the glasses. It may help us estimate our latitude."

"Will do, sir."

"One more thing, when I return, I'll want to know if there is fuel-saving from flying at half of this altitude, what accessories on the engines we can safely shut down, and if this aircraft burns less total fuel per mile on one engine than on two."

After Pelham left, Whitlock muttered, "Just one more thing, my ass, he's gonna hound us until he ends up flying the Goddamn plane!"

Johnson nodded. "No doubt he's playing catch-up, and we're just navigating and flying the route. He's working with the entire crew and passengers to prevent us from ditching."

Whit looked alarmed. "You think there is a possibility of that? Rats thinks you're nuts."

"I'll know more when we get the Nav fix and figure the winds, but I'm guessing we would've never made it to Oahu."

"No shit?"

"No shit, Dick Tracy!"

As David moved aft, he noted that Mueller was snoring loudly. *Poor kid thinks he has luck on his side, so what others think are risks won't happen to him or the people with him. His ass is probably whipped to boot since he has ten-plus hours into this flight, including briefing and preparation. For now, let him sleep and see if I can put the fear of Jesus in him later.*
Moving aft in the cargo bay, he nearly ran into the three chiefs going over handwritten lists.
"How's it going, Crew Chief?"
"Sir, we have three crews checking the cargo and aircraft components; we're making up rules while we go."
"Rules, rules like what?" David asked.
"We have two large bags of mail for Pearl. We hope they're the last thing we toss out for morale reasons. We've found some ship parts marked with the highest possible priority and figured they're critical to putting a ship back in service, so we gave them a later dump priority. We found lots of things for the first dump, like stacks of teletype paper, mess kits, and carbon paper that should never have made it onboard."
"Outstanding, chief. Do you think you can have 500 to 1,000 pounds ready to dump in less than an hour?"
"Well sir, it depends. Chief Norris, the aviation type, has some ideas to run by you."
Pelham shook the aviation chief's hand, "Chief Norris, nice to meet you; what have you got?"
"Commander, I formed up a team of three, and we first swept the aircraft looking for heavy stuff that we can do without."
"Sounds good, Chief. Is there a question in there somewhere?"
"Yes sir, we aren't sure how much damage you're authorizing to the aircraft. If we go full tilt, it may not fly long-range again for months."
"Give me some examples."
Norris walked over to the tanks in the forward cargo bay. "Sir, each of these is a 100-gallon auxiliary fuel tank. There are four; I estimate their empty weight, including cradles, is about 1,000 pounds."
"Chief, I see them, but I'm missing the point."
"Commander, some of the tanks are part full, but if the pilots empty them, we could cut them up, ditch them, and lose about 1,000 pounds. New tanks might be hard to get, so this plane could not fly to the mainland."
"I see," David said.
The Aviation Electrician Chief continued, "And Commander, forward from the port side tanks is an auxiliary oil reservoir. We can pump forty additional gallons of oil to the primary oil tanks in either engine - especially for long flights like this one. At some point, if the engines aren't burning too much oil, we could top off the main oil tanks, plug the lines, pull the whole assembly off the bulkhead, and lose ourselves 300 to 400 pounds. Tossing the freshwater tanks, the honeypots, the coffee urn, and redundant electronics like backup radios sheds more weight. Deciding to toss life vests and rafts overboard is above my pay grade. We probably should run these choices by the pilot-in-command."
"Chief, chiefs, I agree, but Captain Mueller has been relieved of his Plane Commander duties; he's bunked out sleeping like a log. After he has his beauty rest, we'll revisit his motivation to be a functioning crew member."

This news caused raised eyebrows and more than one questioning look amongst the chiefs that communicated, "What the hell is going on?"

Pelham continued, "Look, men, here is our status. We're about to get a workable Nav fix to reveal the headwinds. Lt. (jg) Johnson can then figure out the distance to Hilo. Next, he updates the fuel used to reach the nav fix, which gives us the best guess on the remaining fuel. Then he computes the fuel we'll need to reach Hilo. I'll double-check his figures. Then we'll know if we're in deep shit and must make these difficult choices. You're off to a good start, and I commend you. Chief Watkins, was there anyone on the manifest with flying skills? Johnson is now in the right seat and has aviation experience. My preference is to keep his nose down in the navigation. If we have that luxury, Lieutenant Whitlock could probably use some moral support."

"Yes sir, the Army Captain. Captain Jordon put down that he's a pilot. Go figure, the Army calling him up as a rifleman."

"Outstanding, connect me with him. For the rest of you, keep up the good work. I'll see if we can safely drain these gas tanks, and we should have the answers we need very soon. Carry on."

He met the Army Captain and immediately liked him. Jordon explained that he had flown several aircraft types, including multi-engine mail planes. When the Army called him up, they gave him a more stringent eye test and flushed him out of aviation into the infantry.

David took him aside and spoke in a low voice. "Listen Captain; we're just about to determine how much trouble we're in, if any. Our navigator, Lieutenant Johnson, is bright and thorough. He thinks we might not have enough fuel to make Pearl Harbor or even Hilo, which you heard me say is 100 miles closer. This mess is due to higher-than-forecast headwinds and poor fuel management by the pilots during the first half of the flight. As SOP, I've relieved the Marine Captain Plane Commander from his duties, and he's getting some sorely needed sleep. You'll be playing alter ego with a wet-behind-the-ears Navy Lieutenant with only a handful of flights like this one. Do what you can to make him feel comfortable and in command. Between you and me, I will make up with the Marine Captain before we enter the final phase of this flight. He may have to lead us through a ditching at sea or a dicey landing at Hilo."

Now, Pelham had the man's full attention. The Army Captain looked at him with frightened eyes and an open mouth. "Sir, did you say ditching like crashing into the sea?"

"I did, but we don't know what will happen. So let's go forward and meet the lads."

After introductions on the flight deck, Johnson, having shot the sun for latitude, returned to his charts, and Captain Jordon settled into the right seat.

Standing behind the two pilots, Pelham commented to Lieutenant Whitlock, "Whit, it looks like you have everything under control. Is there any reason we can't finish draining all of the auxiliary tanks in the cargo bay?"

"No sir, we just pull a couple of valve rods down here. Can I ask why?"

David replied in a loud voice. "Good. I want you to ask why, and all of us to check each other and check again. We have an aviation Chief who has pointed out that if we're really in extremis, we should empty the tanks, dismantle them, and throw them out the rear cargo door."

A startled Whitlock said, "Good God, sir, are things that bad?"

"Johnson will tell us soon. After I double-check his figures, we'll go to general quarters if necessary. In the meantime, can you switch over to the rear tanks?"
"Yes sir, makes sense to me, no problem," Whitlock replied.
"Were you able to see the ship?"
"No sir, the visibility was good, but I never picked her out."
"Good, that means we're further north than ten or fifteen miles, which is where we want to be. Question: If we slowly descend to 5,000 feet, will it add miles to our range during the next 1,000 miles or not? I'll come back for your answer in a few minutes."
Turning aft, he spoke to Jinx and Sparks, who were working the radios. "Any luck with getting the Big Island weather or raising Hilo?"
Both answered together, "No sir."
"Well, keep trying. Nav is determining the severity of our situation. You two think about how we'll declare an emergency on the radios if need be."
Both together, "Yes sir."

As he walked back to the cargo bay, Pelham thought, *The war is going terribly for us so far. Guam surrendered in two days, Wake fell on 23 December after a sixteen-day fight, and the Nips took Hong Kong on Christmas. The enemy is attacking in force in the Philippines and the southwest Pacific. They are taking over Indochina and moving south to Malaya. Surely Singapore will hold out. How can the men on this plane make a difference? These are super kids, smart, ready to get the job done. How many of them will have to die to get this war won? How many will perish as we might today without ever firing a shot? Thousands? Hundreds of thousands of young men and women? A lot Davy, a hell of a lot - more than you ever want to know.*

World War II (1941 –1945)	
Total U.S. Servicemembers (Worldwide)	16,112,566
Battle Deaths	291,557
Other Deaths in Service (Non-Theater)	113,842
Non-mortal Woundings	670,846

Source: Veteran's Administration factsheets

C-46D Commando

Source: National Museum of the United States Air Force

C-46D cargo hold looking forward

Note folded seats and navigator's position forward on right side

Source: National Museum of the United States Air Force

Chapter Twelve

SHARK BAIT – Divert To Hilo, Hawaii

January 1942

After the C-46 Commando crew located the beacon ship and passed the halfway mark to Hawaii, Lt. Comdr. David Pelham finished his ham and cheese sandwich. He washed it down with Coke from the crew's cooler and wiped the residue from his lips. David smiled as he heard the snoring of Captain Mueller above the noise of the twin engines. *Sleep tight, Captain. When you're rested, we'll see if you can return to your Plane Commander status.* Pelham stood up from his sturdy wooden stool in the C-46 cargo hold and walked forward to talk to Lt. (jg) "Tex" Johnson at the navigation table.

"Ben, how goes the Nav fix and winds?"

"I think pretty well sir. Between the rate of bearing drift on the ship's ADF and the latitude sighting with the sextant, we have latitude nailed down to within five to ten miles."

David thought, *damn, I still like this kid. He's a fast thinker and doesn't just deal in absolutes. Instead, he uses estimates and ranges that make sense and are workable in the real world.*

David patted the junior officer on the shoulder. "Good Ben. Please walk me through your longitude calculation."

"Our longitude was west 135 degrees and five minutes when we were directly north of the ship."

"Good work Lieutenant, and what is our Great Circle route to Hilo from there?"

"The circle route starts from wherever we find ourselves. We just start flying it from the Nav fix. I've laid out a series of seventeen waypoints at sixty-mile intervals along the arc and have the pilot flying the first no-wind heading of 265 magnetic."

"Ben, I'm afraid to ask, have you calculated the wind?"

"Sir, the average headwind for the past seven-plus hours was about thirty knots. Using our true headings and airspeed, I've maintained a Dead Reckoning no-wind plot on the maps. When I run a line from that DR plot back to the actual Nav fix, the length of the line represents the wind impact on us for the past seven and a half hours. So we divide that distance of 225 miles by 7.5, resulting in thirty knots of headwind."

Pelham thought, *Davy, you can feel your guts grinding and your gonads retracting - lots of tough decisions ahead.*"

"Lieutenant, we're talking nautical miles and knots on everything, right? And what was the bearing of the wind? Did we get pushed north or south of our course?"

"Yes sir, everything is nautical, and we only got pushed a few miles north. Mostly the wind has been right down our throat from the southwest."

"What do you think Ben, is it going to continue about the same?"

"I think so. I don't know; it might be a few knots less at a lower altitude."

"So here comes the loaded question, youngster. Do we have enough fuel to make it to Hilo?"

"Sorry to be evasive sir, maybe, maybe not. I've only been able to look at the numbers I had projected for Pearl. We wouldn't have made it there."

David thought, *Christ, I'm sweating bullets. What will we do now that we know it's a critical situation? They're all expecting me to have the answers. What would Dad say? If the stakes are high, go for broke - put all your chips into the pot.*

"Johnson, originally you expected a no-wind fuel reserve of 200 to 300 miles of flying time at Oahu, right?"
"Yes sir. With 1,100 miles left to Hilo, we've less than half of our starting fuel load of 1,810 gallons, or about 850 gallons. A two-engine burn rate of 114 gallons per hour gives us 7.4 hours in the air. Take 7.4 times 145 MPH ground speed for 1,081 miles - we fall short of Hilo by nineteen miles."
"So a little less wind, or burning a little less gas, and maybe adding a small amount of speed will get us to land?" David questioned.
"Yes sir, but please remember the pilots need fuel to ditch properly. As the fuel gets low, the gauges are less accurate."
David frowned. "Great, less accurate when you need them the most?"
"Yes sir."
"How much gas do you think they need for a controlled ditching?"
"At least twenty gallons total, ten in each main wing tank."
"Hmm, if we take twenty gallons off the table, that cuts our flight range to an even higher unacceptable number."
"Sir, Mueller put us into this mess, burning extra gas for no good reason, not flying a Great Circle route…"
David cut him off, "Stop it, mister. Right now, it doesn't matter who's at fault. All that matters is what we do about it. We will make up those thirty miles and get everyone safely to Hilo! And whatever we do, the sooner, the better to get the impact we need. Did you have a chance to estimate the effect on flight range by dumping weight?"
"Sir, this is a heavy plane. We probably took off weighing 43,000 to 44,000 pounds. We started with 10,800 pounds of fuel and still have about 5,100 pounds, so we're nearly three tons lighter. As we burn more gas off, the bird gets more efficient and gets more range. Then she flies like a falling twenty-ton safe when it's all gone. I'm saying that a few thousand pounds out of 40,000 may not make much difference."
Pelham snapped, "Lieutenant, don't speculate with me again. I want factual answers or best-guess approximations. I will take you and the Crew Chief forward to discuss this with the pilots. The four of us are going to have a meeting of the minds. In the meantime, we're going down to 5,000 feet and praying for less wind. The pilots told me we were at 44,000 on takeoff, so every 4,400 pounds burned off or dumped is ten percent of our gross weight. So you need to start working those fuel numbers backward and forward and find some more ways to save fuel."
"Yes sir! Please, let's get the fuel flow numbers now and also at the new altitude."
David headed back into the cargo bay, looking for the Crew Chief, and took him aside. "Chief, in about two minutes, we're going forward and meeting with Lieutenant Whitlock and the Army Co-pilot to discuss the severity of our actions in the next few hours."
"Commander, I take it we have a nav fix?"
"Yes we do, a good one within five miles, and the news, she ain't so great. We had an average of thirty knots of wind on the nose instead of the twenty predicted."
"How bad is it, sir?"
"Johnson's back-of-the-envelope estimate has us splashing down short of Hilo. We would've never made it to Pearl. He's refining everything, running the fuel numbers backward and

forward, putting in some fudge numbers for a lighter aircraft. But I doubt that it will be much different after an hour of rejiggering."

Pelham motioned for the Chief to follow him, and they made their way forward ahead of the gas tanks, picked up Johnson on the way into the flight deck, and positioned themselves behind the pilots.

David spoke in a commanding voice. "Here's the straight word, pilots. The computed wind is thirty knots or more, mostly on the nose. It's 1,100 miles from the Nav fix to Hilo. We will run out of fuel short of Hilo and ditch between thirty to fifty miles at sea."

He waited for the impact of his words to sink in; they all looked stunned, including the Chief.

David continued, "Pilots, do you have an opinion on descending to 5,000 feet? Will it help our fuel situation?"

The Army pilot spoke. "Sir, we've been discussing options up here. A higher altitude means less engine efficiency, about a three percent power loss per thousand feet. Essentially, you're trying to cram thinner air down the carburetor, and you get less energy from the pistons firing. On the other hand, higher altitude means cooler air. You lose about three and one-half degrees per 1,000 feet, and the engines work better in cooler air. If we go down in altitude, the engines get more efficient with the denser air, but lose some efficiency because it's warmer. Also, the denser air means more lift from the airfoil, the wing, but more drag, or friction if you like, on the total airframe."

David shook his head. "Captain, you're giving me a headache. Is there an answer in there somewhere?"

Whitlock jumped in. "Like Cappy says, we've been looking at the ins and outs. When we obtained this bird, there were no flight profile charts or proven numbers on flight performance. Rats and I worked on some rules of thumb during a flight up to Seattle from San Diego. We were about half-loaded with personnel, cargo, and fuel. Using the same indicated speed, we tried the bird at different altitudes from 4,000 to 9,000 and logged the fuel flow numbers. With all the variables of thrust, weight, drag, lift, and engine efficiency, it appeared the bird would fly the longest-range profile at 6,500 to 7,000 feet."

"And we're at 8,500 now, right?" David asked.

"Yes sir."

"So if we drop 2,000 feet to 6,500, we may improve the range. And we may have lower winds as a bonus," David said.

"Yes on the range, no telling on the winds," Whitlock replied.

"We have about seven or eight more hours to reach Hilo. Men, we have to make our adjustments now! Also, we need to get a data point on fuel flow when descending." David worked with paper and pencil. "At the end of the flight, if we descend at 500 feet a minute from 6,500 to sea level, it would take us thirteen minutes, and we would travel about 2.6 miles each minute for thirty-seven miles of travel. Seem about right, Pilot?"

Whitlock responded, "What are we trying to prove, Commander?"

"Simply this. You and the Captain will take us down 2,000 feet at 250 feet per minute. Keep track of the fuel flow while maintaining the same airspeed. Then, when we get 150 to 160 miles from Hilo, we'll know what fuel we have and how much we need. We'll then decide on a controlled ditching or going all out, Hilo or bust!"

"Aye, aye, sir, we'll get the numbers," Whit answered.
"Good Whit. You're doing a great job. We're going to make this work. The aircraft is about fifteen percent lighter now, and the chiefs will dump another 3,000 pounds. When we burn off 5,000 pounds of fuel, it has to improve our numbers. All of our computations now have us in dire straits. Does everyone agree?" David asked.

The two pilots, Johnson and the Crew Chief, nodded their heads.

David exclaimed, "Well, that's just bullshit! It's not going to happen. We're going to General Quarters, gentlemen. We will pull out all the stops to fly that extra fifty miles. We have to close that gap, do whatever it takes. Make the plane fly more efficiently and keep the speed up to fight the wind. Do you hear me?"

The flight crew responded in unison, "Yes sir!"

David continued, "This aircraft will have burned nearly 7,000 pounds of fuel in ten hours. We will throw another 3,000 to 4,000 pounds out of the aircraft. That's about one-fourth of the bird's starting weight. Our pilots will get more efficiency at a lower altitude, and the Great Circle route buys us a few miles. When we start downhill at 150 to 160 miles from Hilo, we'll burn much less fuel than one gallon per minute per engine. Pilots assume the headwind is twenty-five knots at 6,500 and fly our maximum range speed. We need to shut down any equipment drawing engine power and not adding to our range. Chief, any recommendations?"

"Sir, we can secure the cabin heaters. They burn gas and use electricity," Watkins answered.

"Good. Gentlemen, my limited aero understanding is that denser air will give us more drag and lift, especially as the aircraft gets lighter. As you say, our engines will be more efficient with a denser air/fuel mixture. Pilot, is 6,500 OK by the rules?"

"Sir, flying west, we're supposed to be at an even altitude plus 500 feet. We're at 8,500 now; our lower choices would be 6,500 or 4,500," answered Whitlock.

"OK, Plane Commander, get ready to take her down to 6,500," David answered.

Whitlock took a deep breath. "Sir, if we get too low, we may miss seeing the Island in bad weather. Unfortunately, Nav has got us on a heading that I think is way too far north."

"Listen, Lieutenant Whitlock; Nav has you on the shortest route to Hilo. He'll be giving you course corrections to port about every sixty miles. You do what he says for heading and speed. That's an order. Decrease our altitude to 6,500 and keep us up on the step. Co-pilot, give Nav the fuel flow now, on the way down, and when established at the new altitude. Chief expect that we will open the cargo doors at 6,500 feet and dump a thousand pounds of weight. Pilots, put both engines on the rear tanks at your discretion so we can toss them as soon as they're empty. Last but not least, everyone figures out more ways to save fuel!"

"The pilots and chief replied in unison, "Yes sir!"

"Whit, isn't there some way you can lean out a piston engine to save gas?"

"Sir, the factory recommends the Auto Lean setting. If we overlean the gas, it will damage the pistons," Whitlock replied.

David was sarcastic. "You mean we might damage the same engines that'll be on the bottom of the Pacific Ocean in a few hours?"

Chief Watkins jumped in. "Sir, you have a point. After we dump cargo, I'll come up here and work with the pilots on manual leaning of the engines. The problem of over-leaning is pre-detonation in the cylinders, but we can take it slow and see it coming in dropping RPM and higher cylinder head temperatures."

"Good man, Chief. We'll clump dump items together to make them heavy and put them out the two doors if they fit. Cut up items too big for the openings for the next round of dumping. Have everyone drink lots of water and dump one of the water tanks. How much weight do we have to dump on the first go-around?"

"About 700 pounds, sir. One of the chiefs has devised an idea to have pushers at the door with ropes back to handlers to keep from losing them as they dump the stuff."

David enthused, "OK, sounds good. Bump the weight up to 1,000 pounds. Work with the pilots and get started when we're at 6,500. Pilots, I will be very interested if you can pull some revs and drop fuel flow after we level off. Have the Chief play with the mixture and keep the same speed after we dump."

Captain Jordon raised his hand. "Sir, another suggestion if I may."

"Go ahead," Pelham replied

"It looks like we have a real emergency, and we should formally declare an aviation emergency on all frequencies. You'd be surprised at the help we might get by doing that."

"I agree; you pilots work with Jinx and Sparks to make the calls after we dump cargo. We're all going to Hawaii to take the fight to the enemy and get our revenge. It won't help the war one bit if we ditch. Each of us should pray for help. It wouldn't hurt to have divine intervention to get us out of this jam. Carry on."

The Chief and Johnson went aft; David stopped to talk to the radio operators. "Jenkins, I need you two to do a couple of things. First, try any way possible to get in touch with the Hilo airport and get their afternoon weather. Next, we will evaluate the radio gear and recommend what we can dump on the next go-around. You could hear what we just discussed with the pilots, right?"

Sparks and Jinx in unison, "Yes sir!"

"OK good. You both have a vital part to play. We need help from other people; we can only reach them by radio. We're ready to declare an emergency."

He stopped briefly at Johnson's Nav table to ask him to use the fuel numbers on this descent and compute the fuel they would burn from 6,500 feet to sea level. It would be critical to their ditch or fly decision when they were closer to Hilo.

David checked on Mueller in the cargo bay, still out like a light and loudly snoring. Then he drew a fresh cup of coffee and sat on the stool thinking, *Final descent is right! A twenty-ton bird with enormous engines and wings hitting the waves at 80 to 100 knots! Christ, what is the best technique for ditching a monster like this? With the waves? Against them? Probably along the troughs, a bounce or two with engines and wings tearing off, she could sink in a few seconds. Work the problem Davy, solve the riddle - put all your chips into the pot.*

Chief Watkins approached him. "Commander, the Bosun Mate Chief, and the Aviation Electrician Chief are worth their weight in gold. I think they should be part of the decision-making."

"Agreed Chief. I have a question for you. Walking around the plane, I see lots of patches riveted on the skin in different places and sizes. I thought this was a new plane; what's with the Band-Aids?"

Watkins looked pained. "They represent some bitter memories, sir. This C-46 was the last plane out of Wake Island on 23 December before we surrendered to the Japs. Captain Mueller kept pushing the launch time back as the Marines loaded more wounded into the bird. We had upwards of sixty men of all ages, including men from the construction crew, fighting and getting wounded. The pilots could see the Japs running across the runway, so we had to go. The last person to board was an Army nurse with a couple more badly wounded men. I felt sorry for our boys. Many of them were bleeding on the deck. We had no way to secure all of them. Rats poured on military power; we barely got airborne in the dark and were heading right for a firefight on the southeast beach. He couldn't climb or turn, so we had to fly right through the Japs, and they gave us a pretty good hosing. Thank God, only one passenger was killed. Both pilots got wounded when some Jap ships opened up on us offshore. Tex had a bullet come by his head so close he could feel it pass by. It bounced off the other bulkhead onto the deck. He carries it with his dog tags for good luck. We gave the boys the best care we could, but another died on the way to Midway."

"Quite a story. How many runs did you make to Wake?"

"We started on the 12th of December and went nearly every other day from Midway. It was safest to land at dusk and leave before dawn. We hauled in 5,000 to 8,000 pounds of everything you can imagine, beans, bullets, guns, aircraft mechanics, and mail. Outbound, we took wounded and older construction workers. Wake had many civilians in their 40s, 50s, and 60s. Rats didn't believe they could survive a Jap POW camp."

"Thanks Chief. I'd like to hear more about this later. Go get the Chiefs, and we can recap where we're at."

"Yes sir. You need to know that this crew bonded forever during those ten days. We're ready to fly through the gates of Hades itself if necessary. I might have given you the wrong impression of Captain Mueller. He's the best pilot I've ever flown with. But he's also young and thinks his luck will overcome adversity."

"Understood, Chief Watkins, thanks."

Pelham told the Chiefs, "Men, there is no way to sugarcoat what we're facing. We have a good Nav fix, and the headwinds are higher than expected - thirty knots instead of the twenty forecasted. We would have never made it to Pearl Harbor. Even though Hilo is 100 miles closer to San Diego, we're still in deep doo-doo. Lieutenant Johnson's rough estimates have us falling short of Hilo by thirty to fifty miles.

"We are now flying a max range profile on the shortest Great Circle route. We're descending to 6,500 feet, hoping for lighter headwinds and a bit more fuel efficiency. Once we've leveled off, we plan to push 1,000 pounds of Navy gear out the cargo doors. If the winds remain the same, we have roughly 1000 miles to fly in about seven hours. Our Navigator will refine those calculations down to a gnat's eyebrow. We need to close the gap and keep us out of the drink. Let's start with questions."

"When you say Plane Commander, is that the original Marine Plane Commander, sir?" asked the Chief Bosun Mate.

"No Chief, it's a Navy Lieutenant. I'm going to have to wake the original PC from his beauty sleep soon, wind him up, and put him back in charge of the flight deck."

"Sir, will we be scuttling the aux oil and gas tanks then?" asked Aviation Chief Norris.

"Yes, for the gas tank supports and cradles on the second dump. The tanks are being drained into both engines as we speak. We'll wait on the oil and the gas tanks until the real Plane Commander gives us the go-ahead."

Crew Chief Watkins raised his hand. "Sir, I recommend that Corporal Jenkins and a crew release the straps on the extra tanks and tip each one as it nears empty so we get every last drop of fuel. When the last one nears empty, we can warn the pilots, so they can switch one engine to another source when it starts to cough."

"Good inputs, just what we need, make it so, Chief. Is there a way for them to communicate?"

"Yes sir, there is a walk-around intercom headset that Jenkins can use," Watkins replied.

"Good. I noticed a portable generator over on the starboard bulkhead; I suppose it uses gasoline, right?"

"Yes sir, but it's a mix of gas and oil, not the best thing for these high-performance engines," Watkins responded.

"Chief Watkins, do you have some of that mixed gas?"

"Yes sir, we have a five-gallon Jerry can in the aft compartment and about half a gallon in the generator."

"Help me out here Chief. If we add the five and a half gallons to one of the 100-gallon tanks, how much damage can it do once it's mixed in?"

"Probably none, sir. It just might work."

"So Chief, if we consider the five-plus gallons of regular gas to be about eighty percent as efficient as Avgas, burning it could give us another three or four minutes of flight time and about nine or ten more miles travel over the ocean, right or wrong?"

"I think you may be right, Commander, but I'm pretty leery of handling open gas containers in the cargo bay."

"Chief, figure out how to add the gas, vent the bay, and ground the handlers. Make sure they have rubber-soled shoes and have a firefighting team at the ready. Toss the gas can and the generator on the next dump."

"Aye, aye, sir."

"Sir, it's pretty damn cold back in the bay, and the heat went off," said the Chief Bosun.

"Thank you, Boats. It should get warmer at 6,500 feet. The heaters take energy from the engines and use precious fuel. So they remain off for now. Have the passengers break out more clothes if necessary. Next?"

"Commander, you say fall short, is that like in falling from the sky and crash landing into the ocean? And if it is, do we have enough flotation gear for everybody?" asked Chief Norris.

"Crew Chief, how about it?"

"Yes Commander, we have forty Mae Wests and four ten-man life rafts."

"Sir, just how far should we go to lighten the plane?" requested Norris.

"Short of throwing the former Plane Commander out, everything is fair game. Add the extra life preservers and two of the four rafts to the priority three list."

Watkins was alarmed. "Sir, are you sure about that?"

Pelham spoke in an agitated voice, "All of you listen. If we get close to making the field, everything goes, the mail goes, the seabags go, and anything we can tear apart without stopping the plane from flying goes. Our shoes and pants go, every man will wear a life preserver, and the extras go. Chief, what type of stuff do you have ready to dump?"

"Well sir, we'll have just over 1,000 pounds, mostly cargo, some aircraft nonessential gear, not counting the aux gas tank cradles. They will take some time," he answered.

"Crew Chief, when we level off at 6,500, I want that 1,000 pounds out the door ASAP."

"Aye, aye, sir."
"Commander, we can dismantle all the seats, the bunk, the Nav table, and maybe some bulkheads and doors," asserted the Bosun.
"Boats, aren't the seat restraints part of the tube and harness apparatus?" David asked.
"Yes, sir, but we can rope the men in sitting on the deck with quick-release knots."
"Crew Chief, what do you think? It's your airplane."
"Well sir, I don't like it, but I don't like the thought of ditching in this beast either. Whatever happens, the powers that be are gonna nitpick our every move."
"You can count on that, so use your best judgment and get the job done. It's Hilo or bust! Dismissed," David ordered.
He saw the chiefs moving about the rear of the cargo bay, talking to the men back there, probably preparing for the first dump as Lt. (jg) Johnson approached him. "Commander, the radio guys have been unable to reach Hilo or get the weather there. Sparks has a pretty wild idea he wants to try out on you."
"It must be all this coffee; I'm game for wild ideas," David answered.

Starboard side forward cargo bay

Radio "stack" left, portable generator center, Nav table to right

Source: Photo inside reconditioned C-46 "China Doll"

Chapter Thirteen

HAM TYPE – Davis, California

January 1942

Police Chief Don Struthers had finished the night shift, made himself breakfast, and glanced at the morning paper when his wife came into the kitchen yawning.

"Well, good morning, sunshine," she said.

Don replied, "Good morning yourself - give us a kiss."

Mrs. Struthers leaned over, complied, and said, "I don't know why you do this night shift stuff. After all, you're the chief."

It was a continuing bone of contention between them. Struthers felt it was effective leadership to take a couple of night shifts a month, and she didn't like the disruption to their routine.

"Yes Margie, can I pour you a cup of coffee?"

"Please and make it to go. The doc is starting surgery at eight this morning; I'll grab something to eat at the hospital."

"What about the kids?"

She gave him a sideways glance. "Believe it or not, they can get up, feed themselves, and get out the door on their own. You get some sleep now, so you can get back on schedule tonight."

"Yes Ma'am. Love you."

She came up to him and wrapped her arms around him, and responded, "Love you, too, snookums. Thanks for the coffee."

Margie left by the side door, and Don could hear her Ford pickup start up and swing onto the street. Struthers went into the garage to lock his police cruiser before going to bed. He put his gun and badge under the front seat and locked the car. Marge did not allow a gun in the house unless he wore it, period. As Struthers left the garage, his gaze shifted to the hobby shop he had built in the front.

He thought, *It wouldn't hurt to spend a few minutes in there and then go to bed. Besides, the warden has gone for the day.* Instead, Marge's admonition, "You get to sleep," caused him to go into the bedroom, kick off his shoes, and lie down on the comforter. Sleep claimed him in a minute.

About four hours later, Don Struthers woke up, entered the kitchen, and poured a cup of coffee. He took the coffee and went out to the garage in his stocking feet. After entering his hobby room, Struthers switched on the radio equipment, turned on the floor heater, and dialed in the general calling frequency. He doubted there would be any traffic as all amateur communications were shut down after Pearl Harbor.

For a few minutes, there was nothing but static. Then there was a weak but readable transmission: "Pan, pan, pan, this is Marine aircraft 78774 en route to Hawaii. We have an emergency. Repeat this is an emergency. Any ham operators in Hawaii or on the West Coast, please reply. Marine 78774, over."

At first, Don was stunned. Was this a jokester? He looked down and noted with a professional smile that he had automatically logged the frequency, time, and call sign. What to do? Would the antenna he had strung between the garage and the house reach that far? He picked up the microphone and pushed the transmit button.

"Marine aircraft 78774, this is Victor Six Xray Mike Xray in Davis, California. Do you read me?"

There was a long pause. Nothing happened. *I knew it, Don* thought. *I should've bought a better rig and probably put up a higher vertical antenna. Damn.* And then he heard…

"V6XMX in Davis, California, we read you medium and clear. We need your help."

Struthers transmitted, "Marine 78774, V6XMX. How can I help you?"

Don heard a new, more mature voice come on. "V6XMX, Marine 774 is about halfway to Pearl Harbor after taking off from San Diego eight hours ago. We have experienced high headwinds of thirty knots, repeat three zero knots. We do not have enough fuel to reach Oahu. So we're diverting to Hilo. I spell, How-Item-Love-Oboe on the Big Island of Hawaii. We have had no luck reaching Hilo or getting the weather forecast for our ETA of four o'clock p.m. local time. Do you copy, over?"

"Marine 774, I copy all. What can I do? Over."

"V6XMX, we request that you call the Hilo tower or airport manager on a landline and relay the field conditions and afternoon weather forecast to us. We will pay any charges, over."

"Navy 774, I have no idea that it will work. There's a war on you know, and my phone is forty feet away in the kitchen. I can try, but no promises, over."

"V6XMX tell them it's a matter of life and death. We may have to ditch at sea if we run out of fuel. We desperately need to contact the Hilo aviation authorities, over."

"Marine 774, roger, I'll give it a try. Do you have any backup frequencies in case this one goes down?"

The aircraft gave Don the frequencies; he jotted them down.

"Marine 774, this is going to take some time. Either way, standby, out."

Don Struthers walked swiftly into the house and found his son dressed and eating a sandwich.

"Morning Pops. On the night shift again?"

"Yes. When are you starting class this morning?"

"I have my first lecture at two o'clock, and I thought I'd go early and get some library time."

"Stick around; things are about to get interesting; I might need you."

"What's up?"

Chief Struthers picked up the phone and dialed O. "Listen, and you'll find out. Operator, this is Police Chief Donald Struthers in Davis, California. This is an emergency."

The operator said, "Is that you, Don? This is Shirley Black."

"Hi Shirley. I received a call on my ham radio from a Marine plane on the way to Hawaii from San Diego. They have an emergency and may not make it to land. They desperately need to get information from the airfield in Hilo, Hawaii, the Big Island, that is H-I-L-O. Can you help me?"

"Well gosh, Don, that's terrible! I've put through a few calls to Oahu, but never to any of the other islands. Hold on."

"Dad, what do you want me to do?" asked Don's son.

"I'm waiting for the operator now, but in case we do get connected, I think you could be on the phone and relay to me on the radio and me back to you."

"Don, are you still on the line?"

"Yes Shirley."

"Don, my supervisor recommends we get the San Francisco long-distance operator on the hook, and maybe she can figure this out."

"Sounds good to me, Shirley."

"OK Don, stay on the phone. You should hear me talking to the other operator. If we get cut off, I'll call you back. You're at home, right?"

"Yes I am, Shirley, and thank you!"

He heard some ringing and buzzing and then, "Long-distance, how can I help you?"

"Operator, this is an exchange operator in Davis, California. I have the local police chief on the line. He's one of those ham radio persons and is talking with a Navy plane with an emergency on the way to Hawaii. These boys are in big trouble and may have to crash into the ocean. Can you connect us to Hawaii?"

The long-distance operator answered, "No operator, but I can put you through to Hawaii long distance."

"Yes, that's just what we need; thank you."

Don heard more buzzing and clicking and then a ring and an answer, "San Francisco overseas long-distance, how can I help you?"

"Good afternoon, Ma'am; this is Shirley Black at the Davis, California local phone exchange. Are you the Hawaii long-distance person?"

"Yes, I'm one of them."

"Wonderful. I have our local police chief on the line; he's one of those ham radio persons. He's talking with a Navy plane on the way to Hawaii that took off from San Diego. Those boys are in big trouble and need to talk to the Hilo airport on the Big Island. Can you help us?"

"Davis operator, I can get you through to Oahu, and maybe they can get you through to Hawaii. I don't think there are any direct lines to the island of Hawaii. Please hold."

Don heard more dialing, clicking, buzzing, ringing, and then, "Aloha, Oahu long-distance, how can I help you?"

"Operator, this is San Francisco long-distance; we have an emergency call on the line for the Hilo, Hawaii airport. Can you connect us?"

"Hello San Francisco. We can give it a try, but Hilo does not always have a good connection. Standby."

There was more buzzing, clicking, dialing, ringing, "Hawaii phone exchange, how can I help you?"

"Hawaii, this is Oahu. I have an emergency call for the Hilo airport from San Francisco."

"This is Hawaii, please stand by, the Hilo operator may answer, or she may not."

Jordy Manelo stood in the Hilo airport tower, sipping his Kona coffee while viewing the field and the surrounding lush green countryside. It was 7:15 in the morning, and they had very little planned during the day. One fire drill for the firemen and a briefing on the construction progress with the contractor at nine o'clock. After that, most of the airport staff were on call at home, with one person left in the tower to scare off any aircraft that came near the field. The phone rang.

"Hilo Tower, Jordy speaking."

"Jordy, this is Mary at the telephone exchange."

Yes, Mary, he thought, *Of course, it has to be Mary. She was the primary operator for the city on weekdays.*
Mary continued, "Jordy, I got a hot potato here. The Hawaii operator got called by the Oahu operator, who got called by the San Francisco operator. There is something about a Navy plane in trouble on the way here from San Francisco."
A new voice broke in. "Hilo operator, Hilo operator, this is San Francisco long-distance. Can you hear me?"
"Yes, San Francisco, I can hear you. This is Mary at the Hilo phone operator's booth. I have Jordy on the line. He runs the airport."
"Thank you Mary. This is San Francisco long-distance. The call is coming from Davis, California. Davis operator, did you hear Mary in Hilo?
"This is Shirley at the Davis exchange. I can hear Mary."
"This is San Francisco long distance. Mr. Jordy at the Hilo airport, can you hear Shirley at the Davis, California exchange?"
"This is Jordy Manelo at the Hilo Airport. I heard Shirley, over."
"Mr. Manelo, this is Shirley in Davis, California. I have an emergency call for you from Don Struthers, the chief of police in Davis, California. Chief Struthers, can you hear Mr. Jordy? Sorry, I mean Mr. Manelo?"
Struthers had a tight grip on his phone. "Yes Shirley, I can hear everybody. This is amazing!"
"Chief Struthers, please go ahead with your call," Shirley directed.
"Mr. Manelo, this is Don Struthers. I'm a ham radio operator. At 12:06 p.m. Pacific time I received an emergency transmission from Marine Aircraft 78774 stating that they left San Diego heading for Pearl Harbor and had thirty-knot headwinds. They are diverting, I repeat, diverting to your Hilo airport and request field conditions and weather for their four p.m. ETA. That is four o'clock your local time. How do you copy, over?"
There was a long pause, then Manelo answered, "This is the Hilo tower. Unfortunately, we are unable to assist Marine 78774. Our airport is closed. It would do them no good to have our weather, over."
Struthers was starting to shake and sweat. "Mr. Manelo, Jordy, you don't understand. They can only make it to your airport and might end up in the ocean. High headwinds are the problem; when will your field be open?"
There was a pause, "Mr. Struthers, Hilo Tower, we won't be open for days or weeks. We're under construction. They can't land here. What kind of aircraft is it, and how many people onboard? Over."
Don thought about his answer. "Jordy, please stand by; I'll relay your information and get those answers. My son will talk on the phone, and I'll call them from my ham station in the garage. Please stand by, over."
Struthers went out to the garage, crossed himself, picked up the mike, and transmitted, "Marine 78744, this is V6XMX in Davis, California, over."
"V6XMX, this is Navy 78774, good to hear from you! Over."
"This is V6XMX. I have Hilo Tower on the phone. Repeat, I'm talking to Hilo on the phone. They report that the field is closed, under construction. I repeat the field is closed, over."
There was a long pause, "V6XMX, this is Marine 774. When will the field open? Over."

"Marine 774, not for days or weeks. It's under construction. They request your type of aircraft and number of personnel, over."
"This is Marine 774, we are a C-46 cargo plane with twelve souls on board, our landing weight will be about 30,000 pounds, and we need at least 2,000 feet of runway. Tell them this is a formal emergency; we must land there, or lives will be lost. We can't go anywhere else. It's touch and go whether we can even make it to Hilo. We need the afternoon wind and weather, even if we just ditch, over."
"This is V6XMX, I copy all, will relay, standby, out."
He raced into the kitchen and took the phone from his son. "This is Don Struthers back on the phone for Jordy."
After a short pause, "Don, this is Jordy Manelo."
"The type of aircraft is a transport C-46, there are twelve souls onboard, their landing weight is 30,000 pounds, and they need 2,000 feet of runway. The only airport they can reach is Hilo if that. Lives will be lost ditching at sea. This is a real emergency, and they must land at Hilo. They require afternoon wind and weather even if they ditch, over."
After a few seconds pause, "Don, Hilo, copy all that. What frequency do you have them on? Over."

Struthers gave him the current contact frequency and two others.
"Don, Hilo Tower, we're going to try and contact them directly, standby, over."
Aboard the Commando aircraft, the Radioman First Class heard it first on the number two radio, "Marine 78774, this is Hilo Tower, how do you read? Over."
He called to Nav on the intercom, "Sir, sir, we have Hilo tower on radio two!"
Lt. (jg) Johnson took the headphones and adjusted the mike. "Hilo tower, this is Marine 78744. We read you loud and clear, over."
"Marine 774, this is Hilo Tower, we are sorry that the field is closed, and you are too heavy to land on the plantation dirt strip on the north end of the main runway. Can you make it over to Kailua, Kona Airport, on the island's west side? Over."
"Negative, Hilo tower; we are east of you and can barely make it to Hilo based on our latest fuel figures. We are dumping cargo and nonessential equipment out of the aircraft. It will be nip and tuck to make the field, over."
"Marine 774, Hilo Tower, you can't land here; the field is under construction. It is closed. You will not get clearance to land here, over."
"Why not, tower?" Johnson demanded. "It's got to be better than crashing into the ocean. You must have some space where we can at least crash land, over."
"Negative Marine 774, construction prevents any landings on the primary northeast and southwest runway 03 slash 21 that usually has 3,200 feet of usable paved runway."
"Hilo tower, what about your east-west runway?" Johnson questioned.
"Marine 774, runway 08 slash 26 only has 1,500 feet of usable landing area; much of it is still coral rock that will not support your aircraft."
Johnson replied, "Hilo Tower, if we start an approach to runway 21, what will we see?"
Jordy was frustrated. "Marine 774, runway 21 has drainage ditches dug across the entire runway about 300 feet from the apron's edge on each end and at the midpoint. They are nearly three feet wide and five feet deep to put in eighteen-inch drainage pipes, over."
Johnson turned around and said, "Commander, did you hear most of that?"

Pelham wore the second set of earphones. "I did Ben, and I drew a little diagram here. If we could land past the first ditch and they made the middle one passable, we would have 2,600 feet of runway to work with, depending on how long we land past the first ditch."
"Sir, how do they make it passable? Johnson questioned.
"I don't know, Ben, that's their problem; maybe they use steel mats, shored up wood, or fill the middle part of the ditch. I've got another of those wild ideas. Tell them to get runway 21 ready for us in seven hours and be firm."
"Hilo Tower, Marine 774."
"Tower, go ahead."
"Hilo tower, Marine 774 will land on runway 21 in seven hours. Request you do whatever it takes to give us a safe taxi over the middle ditch. We will attempt to land just past the first ditch, over!"
Jordy became angry. "Listen, Marine 774, you aren't in charge of this airport, and you are not cleared to land on 21, acknowledge."
Pelham took the mike from Johnson. "Hilo Tower, we are flying a brand-new, expensive, high-performance transport aircraft critical for the Hawaii airlift. We have twelve souls on board and precious cargo to repair the ships and airplanes in Oahu. We are landing on your runway; it's your problem to make it as safe as possible."
"Marine 774, what you are saying is unbelievable. If we have to put fire trucks out to block the runway, we will. And how can you expect us to make these ditches safe in seven hours when it took six days to dig them?"
Johnson took the mike back, "We don't know Tower. It's your problem. Use steel mats, shore up wooden platforms, use earth-moving equipment that's probably there to fill them. Give us a 100-foot shot down the middle; our carrier-qualified pilots will do the rest."
"Nice touch Ben, you threw down the gauntlet, and he's standing there with his thumb up his ass trying to decide what to do," Pelham said. He turned to the radioman. "Sparks, how could I get on the radio and cut out a lot of this aircraft background noise?"
"Sir, if we put the pilot's flight jackets over you, it might help," answered the Radioman.
"Have Jenkins gather them up while I write down what I want you to say," David instructed.
Pelham wrote on a pad of paper. He and Sparks rehearsed the words a couple of times. Finally, Johnson signaled that Hilo was calling.
"Marine 774, Hilo Tower, over."
Pelham directed, "Don't answer them. Let him stew for a while and think about having twelve people crash, burn, and die in full view of their precious airfield."
Manelo, after a long pause, "Marine 78774, Hilo Tower, over."
More time passed, "Marine 78774, Hilo Tower, do you read? Over."
"Hilo Tower, this is Marine 78774. We need the four o'clock afternoon weather at a minimum, acknowledge," Johnson demanded.
"Marine 774, the four o'clock p.m. weather should be just like yesterday, scattered rain squalls, surface winds from the south at ten to fifteen knots, high cirrus at 15,000 to 20,000 feet, field temperature seventy-seven degrees, and altimeter three zero point one two. How do you read over?"
David spoke to Johnson and Sparks, "Don't answer; let him stew some more while we get set up."

Jinx had collected the flight jackets and a canvas tarp. Pelham and the radio operator sat down on the deck with two sets of earphones and a mike. Corporal Jenkins covered them with the jackets and tarp.

"Marine 774, this is Hilo Tower, did you copy the weather? Over."

"Wait, we want him anxious," David said to the radio operator.

"Marine 78774, Hilo Tower, radio check, over."

David turned on a flashlight. "Wait a few more seconds, now Sparks, begin!"

From under the three jackets and tarp, Sparks read, "Hilo Tower, this is Naval Station Pearl Harbor; we read you five by five, over."

"Uh, Hilo Tower, thank you Pearl Harbor," Manelo responded.

"Hilo Tower, NavSta Pearl, standby for the Admiral, over."

"Ummm, Hilo Tower standing by, over."

David spoke in an authoritative voice, "Hilo Tower, this is Commander Pacific Fleet. To whom am I speaking, and what is your title? Over."

"Sir, this is Jordy Manelo, Hilo airport, and tower manager, over."

Johnson was jumping around behind the pilots, also listening to the radio. Unable to contain himself, Tex yelled, "They're buying it, they're buying it!"

"Hilo Tower, this is CinCPac Fleet. Could you be so kind as to spell that last name over?"

"Yes sir, M-A-N-E-L-O, over," Jordy replied.

"Thank you, Hilo Tower. I spell my name N-I-M-I-T-Z, Chester, middle initial W. Do you know who I am?"

"Yes sir, you are Commander of the Pacific Fleet, over."

"Good, my aides tell me we have a plane in distress that can only land on your field. Is that correct?"

"Admiral, they say they can only land here, but we are closed for construction, over."

"Yes, yes, I get it. Either you don't know what to do, or it would be an inconvenience for you to land that airplane and save our boys, is that right?"

"Admiral, it took six days to dig those ditches, and they're giving us seven hours to make them passable. It's impossible, over,"

"Mr. M-A-N-E-L-O, those boys are coming over here to protect you, your family, and your town. Now, where I come from, you can fill a hole up much faster than it takes to dig it. I want that plane to land in one piece and those brave men saved. You are the one who can make it happen, and you will make it happen! That's an order from CinCPac Fleet to you, Mr. Manelo. I want your word that you will get it done, over."

"Admiral, sir, I can't make a promise like that. I don't know where to start..."

Pelham replied, "Listen to yourself, man, I can't, I don't, whine, whine. There is no whining Manelo, haven't you heard? There's a war on. Now, let me give you some motivation. You get this done, or I will see to it personally that you spend the rest of the war sharpening pencils on Midway Island. Now tell me you are going to land that plane! I've got other fish to fry, over," David said with a grin.

Jordy with shock on his face, "Admiral, this is Hilo Tower; we are going to land Marine 774 safely."

"Thatta boy! If I had a thousand more men like you, I'd have those Japs running back to Tokyo in no time. Nimitz out."

A roar of laughter erupted on the Commando's flight deck. The pilots and Johnson were laughing so hard that tears were streaming down their faces. The Army Captain started doing an imitation, "And let me tell you boy, you gonna be sharpening pencils on Midway if you don't land my plane!" Another eruption of laughter and men pounding each other on the back. Johnson added, "Don't you know, Mr. M-A-N-E-L-O, you can fill a hole faster than you can dig one?" More laughter and tears ensued, maybe from relief.

There might be a chance to cheat death and injury - there had to be a chance.

Don Struthers sat drinking coffee and eating a sandwich in Davis, California, while his son stayed on the phone. "Sonny boy, ask them if Hilo is still on the line."

"This is Davis, California. Is Hilo, Hawaii still on the line, over?"

After a couple of seconds, "Davis, California, this is Mary at the Hilo exchange. We are still connected if you can hear me."

"Yes, I hear you."

A man's voice came on. "Chief Struthers, this is Jordy Manelo at Hilo Tower, over."

"Dad, take the phone. Mr. Manelo is back on!"

"Hello, Mr. Manelo. Were you able to raise them? Over."

"Yes Don, they're a large, heavy transport with twelve souls onboard; we're going to land them. It may be a crash landing, over."

"Thank God, and thank you sir, and thank all of you operators. I'm going to church to light a candle and pray for these boys. I hope all of you will do the same. 6VXMX out."

Manelo spoke on the tower phone. "Mary, are you still there?"

"Yes, this is Mary at the Hilo phone office."

"Mary, call both mayors and everyone in town with a truck and a shovel and get them out here!"

"Jordy, I already called both mayors, Williams Construction and the Army Survey office," Mary replied.

"Good, Mary, call everybody. Call the school and church bus drivers. Get the church bells ringing. We need anyone who can use a shovel."

"I'll keep calling all the numbers, Jordy."

"Good girl, gotta go!" Jordy said as he put down the phone.

Chief Struther's son left for school. Don went back to his ham shack and spoke into the mike.

"Marine 78774, this is 6VXMX, over."

After a brief pause, "6VXMX, this is Marine 774, go ahead, over."

"Marine 774, Hilo tower reports that they were able to raise you, and they will land you even if you crash, over."

Lt. (jg) Johnson replied, "6VXMX, thanks for all your help. You may have saved our bacon."

"Marine 774, I hope so. I'm going to church to light a candle and pray for you boys, 6VXMX out."

"6VXMX, Marine 774. Thanks, we can use all the help we can get. Out."

C-46 "Commando" aircraft

Source: National Air and Space Museum

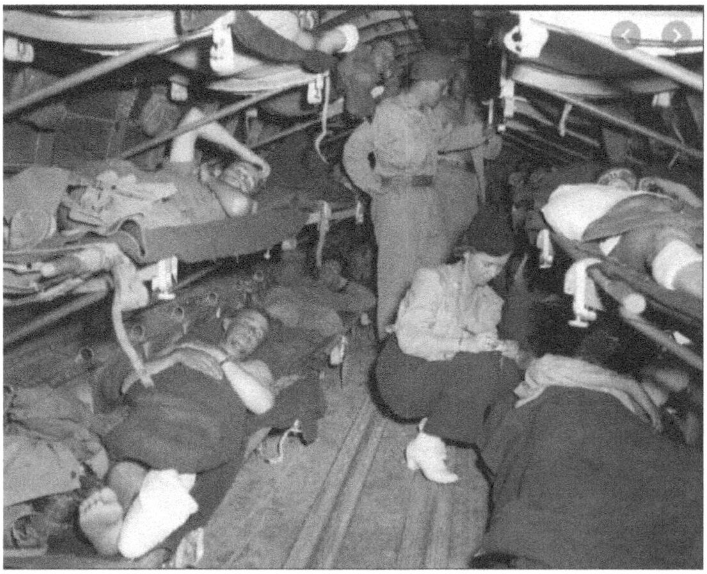

C-46 Stretchers stacked three-high and roomy corridor

Source: Air Force Museum

Chapter Fourteen

I THINK WE CAN. I THINK WE CAN – Flying To Hilo

January 1942

After talking to the Hilo tower, Lt. Comdr. David Pelham left the Commando flight deck and charged his coffee cup. He sat on the wooden stool next to the sleeping Marine Plane Commander. *OK, we're taking action; not sure it's the perfect solution, but when is there one? Johnson is working hard on the fuel figures. We may have a runway if we can get to it. The chiefs are about ready for the next dump. Maybe the headwinds are lower at 6,500. That would be a Godsend. The weather sounds favorable. We should see the island from a long way out. I've got to wake this Marine wildcard up soon. What am I going to tell him?*

David took another swig of coffee and walked forward to Johnson's position. "Ben, I'm going to put Mueller back in the saddle soon. Can you give me a quick study of the plane's fuel system and our current situation, so I can bring him up to speed?"

Johnson drew a diagram on the back of a map. "Commander, the bird has six tanks in the wing, three on each side."

"Six? It seems a bit complicated," David commented.

"I think it has to do with fitting them between the wing supports. The total fuel for those tanks is 1,410 gallons or 8,460 pounds of fuel. The good news is that we didn't fuel up until after dark when the air was cooler. That gives the gas slightly more density and more energy to burn. Also, Rats was able to bully the fuel farm into providing high octane fuel in the main wing tanks, a total of 570 gallons of extra energy."

"What about these tanks back in the cargo bay?"

"Yes sir, we started with four full 100-gallon auxiliary tanks or 2,400 pounds of fuel for longer range. The total is 1,810 gallons or 10,860 pounds of Avgas."

"And what is that in range?"

"Commander, that's where it starts to get a bit tricky. It depends on the bird's starting weight, the engine power settings, and the winds."

"Ben, give this black shoe sailor a break. How far can we fly with no wind and at max range power settings?"

"Sir, the bird is new, and we don't have any manufacturer's manuals, technical charts, or anything - just data we've collected flying her. We think over a long flight with a fully loaded plane; each engine burns .95 gallons a minute or 1.9 per minute for both. Take the 1.9 times 60 minutes for 114 gallons burned per hour. Each minute at cruise speed gets us 2.8 nautical miles in ground travel or a True Air Speed of 168 knots, ignoring the wind."

David figured on paper. "So, if I take 1,810 gallons, divide it by 1.9 gallons per minute, I get 952 flight minutes. Take that times 2.8 miles of travel per minute for a max range of 2,665 nautical miles, right?"

"Yes, sir."

"But if you have a twenty-knot headwind for fifteen hours, it will put you 300 miles behind your desired position. You would have a token reserve for a 2,300-mile flight to Oahu?"

"Yes sir, mostly correct. We had a timed run with half the fuel load and light winds in Northern California. Rats found he could keep her on the step at less than two gallons per minute and produce a ground speed of about three miles per minute. You need a careful launch, climb out, and getting on the step for cruise speed. We figure about 1,400 miles for the first 1,000 gallons and 1,300 for the rest of the fuel."

"Mr. Johnson, twenty-knot headwinds would still put you at Pearl with no reserve!"

"Yes sir."

"What did you do to try and prevent us from being trapped in this airborne disaster?" David demanded.

"Commander, on the two previous flights to Hawaii, we had light winds and flew a lighter bird. I warned Captain Mueller at the preflight brief that if the winds were as forecast, we wouldn't have any reserve at Pearl, and we would not make Oahu if the winds were higher."

"What did he say?"

"Sir, he just waved me off, called me a worrywart, and said we'd be just fine. Then, when we sat with both engines running, waiting for the tower to get manned, I asked him to shut down one engine and return to base to get a fuel truck to top off."

"What happened then?"

"He said, 'Calm down Nav, we'll be fine.' When I found him flying faster than the maximum range speed after we leveled off, I called him on the ICS and recommended 1700 revs and twenty-seven inches MP. He exploded, told me to shut up, and to provide heading information only for the rest of the flight to the co-pilot."

David spoke in a softer tone. "OK, Ben, short of putting a gun to his head, you put your oar in pretty deep and pulled hard. I have nothing but respect for the fighting spirit of the jarheads, but they have a bias to oversimplify a situation. Then they trust in their luck and guts to pull them through. Why not just put more tanks in the cargo hold for more range?"

"Well sir, more tanks mean more weight and less cargo space. Filling them with gas cuts down our carrying capacity drastically. If you put in all eight tanks, you've little capacity left for cargo. Airplanes and I bet, ships are a complex set of trade-offs."

"Yes, thank you, and please forgive my stupid question. Help me with one thing, Ben. I feel in my gut that we should go down to 4,500 feet, dump more gear, and do everything possible to diminish those thirty to fifty miles we're short of Hilo. Whitlock has raised an issue that if we go too low, we'll not see the Island and could miss it."

"I heard that sir, but with the good weather and the fact that Hawaii's volcano, Mauna Loa, is 13,600 feet high, and Mauna Kea is 13,800. I doubt that we can miss them."

"Thanks Ben, just what I need. Isn't there some graph that predicts when we should see the top?"

"Yes Sir, I looked it up. At 6,500 feet, we should see the volcano tops or the vapor cloud at over 200 miles."

David patted Ben on the back, "Once again, you've anticipated my question and speeded up the decision process. Great traits in a Naval officer - keep up the good work!"

"Thank you, sir."

"Ben, work up an estimate of when we need to drop. Ummm, bad choice of words, to descend and land at Hilo. The distance from the field when we should start down."

"Aye, aye, sir."

"One last question, is there anyone on this plane better qualified to control it in a ditching or a tricky landing than our Rip Van Winkle Captain?"

"No Sir. Rats knows the aircraft better than anyone. When we were at Patterson Field in Dayton and the Saint Louis factory, he taught the Army types and even the Curtiss test pilots about quirks with this bird."

"Good inputs Ben. I guess it's time to wake up, Sleeping Beauty. Please go get him water and a sandwich. Carry on."

David walked over to Mueller's makeshift Army cot thinking, *Real fine, the Chief never told me I'd be sleeping on top of 2,000 pounds of explosive Avgas.* He shook Mueller's shoulder and said in an even tone, "Captain Mueller, time to wake up."

Rats, like a lot of military men, immediately woke up. He looked at his surroundings and jumped out of the cot.

Mueller stood at attention. "Sir, what is our situation, sir?"

"Well Captain, first let me explain the rationale for my harsh words. I must admit that I was scared of our situation and my lack of aviation skills to act as SOP. Now we've taken several actions to improve a critical, even a life-threatening situation. We're about two-thirds of the way to Hilo on the Big Island of Hawaii. We need your flying skills in the left seat as Plane Commander and leading your crew. Most importantly, you must ensure that this sailor SOP does not make any fatal errors. Do I make myself clear?"

"Yes sir. As for an explanation, none needed. I'm a Marine; I've had my ass chewed by professionals. Did you say Hilo airport, sir?

"Yes Captain, Hilo is a hundred miles closer to San Diego. We would've never made it to Pearl."

Mueller let out a whistle. "Are you still of a mind to court-martial my Marine ass?"

"Harry, before this flight is over, we both may see the end of our careers."

Johnson showed up with a sandwich and a canteen of water. "Here, oh mighty Plane Commander, you're going to need this!"

"Thanks Tex. No hard feelings, I hope."

"None if you can keep the sharks from chewing off my nuts."

"OK, Commander, I'm awake; I'm ready. Please brief me," Rats said.

"Good, Mr. Mueller. We've descended from 8,500 to 6,500 in hopes of lower headwinds and better fuel economy. The Nav fix we developed from the reference ship gave us headwinds of thirty knots. Johnson has set up the shortest Great Circle route to Hilo and gives course corrections every sixty miles. Mr. Whitlock is in the left seat, and an Army Captain with twin-engine experience is backing him up as co-pilot. Chief Watkins is up there. They're attempting to fly a maximum range profile by manually leaning the engines and shutting down non-essential systems that take engine power. Watkins and two other CPOs have dumped the first 1,000 pounds of cargo and nonessential aircraft parts like the extra water jug and one honeypot. We've tough choices to make right now for the next dump that needs your input. Your bird should be at about 30,000 pounds when we're ready to descend to Hilo. We need you back in the left seat to take stock of our fuel situation. Whit can jump back into the right seat. The Army type, who is sharp, can kibitz as a stand-in Flight Engineer. We need Chief Watkins back here for the next dump. Now hear this, current calculations have us ditching thirty to fifty miles short of Hilo. It all depends on the winds and better fuel economy. They both will be critical for a land or ditch decision when the gas runs low. Any questions?"

"Not yet, sir, maybe later."

"Good, let's go forward and make the crew changes."

They both ducked under the low door opening into the flight cabin. Mueller was munching on his sandwich and taking drinks of water when he noticed Jinx and a First Class Petty Officer manning the radio station. Pelham motioned for Rats to stand behind the Crew Chief and spoke, "Chief, what have you found out about fuel flow and manual engine leaning?"

"Commander, we've kept her flying on the step at 1665 RPM and 26.7 Manifold Pressure. She's giving us about 170 knots of True Air Speed; the fuel flow with manual leaning is down to 113 gallons per hour."

"Good work crew, especially you, Mr. Whitlock. But it may not be good enough. Nav last had us at about 1100 miles from Hilo. If the winds are twenty-five knots on the nose, we just make it to the field. We ditch about thirty miles short of Hilo if they're thirty knots. Anyone disagree?"

All the pilots shook their heads negative, and the mood was somber.

David continued, "OK, crew, we know roughly where we stand. We have to pull out all the stops to use every drop of fuel wisely and fly the maximum range. Captain Mueller is once again our sterling Plane Commander. Whit, you shift to the right seat, and Captain Jordon takes the jump seat and provides sage advice while the Chief works our next dump. After we see the volcano, we can descend to 4,500. You pilots will keep this beast on the step while getting the best fuel flow possible. I understand that the numbers will get better as we get lighter. Chief, when will your men be ready to dump?"

"Sir, the men are ready to dump about 1,100 pounds of cargo and gear, but we still need a decision on the auxiliary fuel tanks and the long-range oil tank."

Mueller was startled. "Whoa, hold it. If you mess with either of those, this bird will be stuck in the Islands."

David said, "The men made sure that we used all of the fuel in our four aux tanks by tipping them as they ran dry. They also dumped in the generator fuel to be mixed with the avgas. I imagine the Genny is going out the door shortly. I see the Chief nodding."

Rats responded, "Sir, this crew went to a lot of trouble procuring this plane for Navy and Marine Corps use. It's been invaluable to both of our services. Please hold off on destroying the gas or oil tanks if possible, please sir."

"Would there be a point that the Captain would be comfortable to top off the engine oil tanks and toss the extended tank?" asked Chief Watkins.

"Chief, maybe when we're closer," Rats replied.

"Chief, if your men are ready to go, let's do the second dump now." David directed.

"Aye, aye, sir."

"Co-pilot, I'm going back to work with Nav; bring Captain Mueller up to speed on our destination and the weather," David said.

"Yes sir, just for the record, I'm still worried about missing the island."

David said, "Not to worry, Whit. You have two 13,000-foot volcanos on Hawaii, and the weather forecast is generally good. At 6,500 feet, you should see signs of the peaks more than 190 miles out."

"Yes sir, thank you, sir."

Pelham strode aft into the front of the cargo bay and drew yet another cup of coffee. Chief Watkins was wearing a headset in the back of the bay and probably talking with the pilots.

David felt the beast nose down slightly and heard the whooshing air from the open rear doors. *There go another thousand pounds,* he thought. *What's next on the list, Davy? Reviewing the fuel numbers with Johnson and doing the next dump ASAP would probably be good.*
Pelham positioned himself behind the navigator. "Nav, let's go through the numbers again. What is your estimate of fuel remaining now?"
"Sir, it looks like we had a bit more than 840 gallons when we last talked. My guestimate now is 805 gallons."
"OK Ben, and it's been what about ten minutes?"
"More like twenty, sir."
"Time flies when you're having fun. If we use my optimistic headwind estimate of twenty-five knots for seven hours, we get pushed back 175 miles in the wrong direction. We still fall short of Hilo and have to ditch, concur?"
"Yes sir, but we should make up some range going downhill on our approach and by getting lighter. If we wait to burn off the main tanks with the higher-octane gas, we should get some help from that."

They both heard the doors close and felt the aircraft tip slightly forward.
"Ben, the pilots are putting out a distress call now. There may be a ship or ships we could ditch next to on the way to Hilo, or one that could run out thirty miles if need be. Time will tell. You and I'll have to find a way to improve these numbers or shift to a controlled ditching as the only plan."
Johnson laughed. "Won't that put the frosting on Mr. M-A-N-E-L-O's cake? He fills in his ditch, and we don't come to the party, ha ha ha. Commander, we could pray for better winds."
"Amen to that. I've given that some quiet effort, but you've given me an idea. Let's figure out what we have to have and get everyone onboard working to that end."
"How so, sir?" Johnson quizzed.
"Let's start with the Lord giving us twenty-knot winds and hope he does even better," David replied.
"Thank you, Jesus!" Johnson exclaimed.
"And then let's figure out what the pilots have to give us to make Hilo with those winds," David said.
"Sir, they have to up the airspeed at the same fuel burn rate or decrease the burn rate and maintain the speed or both."
"Good lad! Let's pressure them for more speed and a lower burn rate!"
Johnson was confused. "But sir, how are they supposed to do that?"
"Ben, the key to Naval leadership is you tell your people what you want them to do, and nine times out of ten, they do it. Most of the time, they just want direction and supply the inspiration."
"Yes sir, but what do you and I do next?"
"We plug in some numbers until we get what we want, then promulgate the requirements. What was the fuel flow when we were going downhill to 6,500?"
"Sir, it dropped from 114 gallons an hour to 105 while we descended at 200 feet per minute."
"And the airspeed?"
"IAS went from 150 to 155 in the descent and 150 when we leveled off with 113 gallons per hour burn. Remember, the air is denser, and TAS is about the same," Johnson replied.

"Ben, it's dangerous, but I've been thinking. If 6,500 is our best range altitude, maybe we should stay up here until we get 100 miles or less from Hilo."
"Yes, sir, but we'll be much lighter by then."
"Meaning what?"
"When Rats and I worked out the best max range settings at 6,900 in altitude, we were about at our current weight. In six or seven hours, and more gear dumped out of the bird, that sweet spot for range could be a bit lower."
"Nav, that's perfect. We get to 150 miles and see the volcano tops. Then we descend to 4,500 at 100 to 200 feet per minute and get the new fuel flow. That'll take us, let's see, ten minutes and move us almost thirty miles along our course. Now we're at 120 miles left to Hilo. At that point, I want your best estimate of the remaining fuel. In ten more minutes at ninety miles from Hilo, we decide to make land, or to ditch. Right now, we need to maintain our speed and drop the burn to 112 gallons per hour."
"One hundred and twelve, Commander? How is that going to happen?"
"It's going to happen because that's what I'm about to require from the pilots."
Johnson shook his head. "Yes sir, good luck, sir."
"Now we have to sell it and get buy-in from everyone onboard. Let's start with the Crew Chief. Would you please go aft and bring him up here?"

In a couple of minutes, Johnson returned with Chief Watkins.

"You wanted to see me, sir?"
"Yes Chief, we've been refining our fuel numbers. If all remains the same, we'll have to ditch."
"Damn, sir, excuse my French, but you're saying we'll lose this plane and no doubt some of these good men, not to mention the crew that's like family to me."
"Well, yes and no; if we can make some changes, we might make it with a slim reserve."
"What kind of changes, Commander?"
"First of all, we're asking the Good Lord and His only son, Jesus, to give us lighter winds as we get closer to the earth. Second, anything you do to lighten the plane has to improve range and speed. Third, the pilots must get more creative; we need the same or more speed with less fuel burned, period. Fourth, we're sending out Mayday messages and must have the best possible ditching plan. What do you think?"
"Sir, if you want more weight out the cargo bay, we can do it. We're working on the aux fuel cradles now, and the last tank back here is almost dry."
"Good Chief, but not good enough. I want 2,000 more pounds out the door in the next two dumps. You Chiefs will have to put your thinking caps on. Everything is disposable except the passengers and crew!"
Watkins looked pained. "What about the mail, the critical parts, the gas tanks, the oil tank, and pump?"
"Chief, it's up to you to leave some items until last. I just want to know that 1,000 pounds will be dumped in the next two rounds, and round three can't happen too soon. If we ditch, it all will be lost anyway. Do you even have a way to cut up the tanks?"
"No sir, we don't have the manpower even if we had the tools."
"Fine, tell me how I sell this to the pilots," David challenged.

"I've thought about that, Commander. I was thinking, good luck with Captain Mueller, telling him you're ruining his baby and that he must do the impossible. Then it hit me. Tell him what you want him to do and dare him to do it like you don't think he has the smarts. He'll kill himself trying to prove you wrong."

"Thanks Chief; I think you've given me the key. Dismissed. Ben, are you keeping up with the course corrections and fuel quantity update from the pilots on the hour, so we can rerun the numbers?"

"Yes sir, I have a timer at fifteen past the hour each hour for our port turns and roger on the fuel numbers in the wings."

"Good, I'm going forward to spread good cheer," David said.

Pelham took some note cards and wrote large numbers on three different pages. Next, he positioned himself behind the co-pilot and overlapped the Army Captain in the jump seat.

"Plane Commander, what is our current status?" David queried.

Mueller answered, "Sir, we're at 6,500 feet, on the step, heading 242 as provided by Nav. Everything looks good."

"If you please, what are our True Air Speed and current fuel flow?"

Whitlock replied, "Sir, TAS is 172, and fuel flow is 113 gallons per hour."

"You pilots have done a great job up here," David said. "All three of you, thank you for your aviation skills and know-how. But, unfortunately, it has not been good enough to keep us out of the drink."

Mueller was exasperated. "Commander, we're doing the best we can; no other crew could get you this level of performance."

"That may be right, Captain Mueller, but if this flight crew doesn't get its act together, we all may be sleeping in some shark's belly tonight. Nav and I have spun some numbers. We know what we need, less headwind provided by the Lord and a faster speed at a lower fuel burn rate provided by the flight crew."

Rats exclaimed, "That's impossible! We're doing our level best for everyone in this bird!"

"How would I know that, Captain? For all I know, you're up here listening to bebop on the radio from San Francisco. You need to get with the program, Mr. Mueller. I have for each of you my requirements. Post these in front of you for all to see. If you can't hit these numbers, people in this plane will die. Figure it out, or be relieved, end of transmission."

David turned around and walked away. Each of them had identical cards:

"FUEL FLOW 111 GPH

TAS 170 - 175: ON THE STEP

NO EXCUSES!"

Mueller smacked his hands on the wheel and started to swear. Whitlock just shook his head. Pappy Jordon coughed and covered a small smile with his hand.

Mueller complained to the other pilots. "Can you believe this guy, not even a pilot, and he's telling us what to do! I've got a good mind to go back and punch him in the nose. It might be worth the court-martial. More speed and less burn, my ass!"

Whit responded, "Rats, I agree with you, but the Commander and Tex are the only ones with the charts and the big picture. He has me scared. I don't think he's a liar, and that means the chances of ditching are much higher than getting to Hilo."

Harry replied, "C'mon now, what was that part about God and the wind? They're just gonna wish the wind away. Give me a break."

"Gentlemen, it wouldn't hurt to say a quick prayer, would it?" Jordon offered.

Whitlock and Mueller shook their heads.

Captain Jordon prayed, "Dear Lord, we're here as your servants in this great war. But we won't be able to serve you if these winds continue to hold us back. We pray for lighter winds in the name of the Son, the Father, and the Holy Ghost, amen!"

Whit and Rats, in unison said, "Amen!"

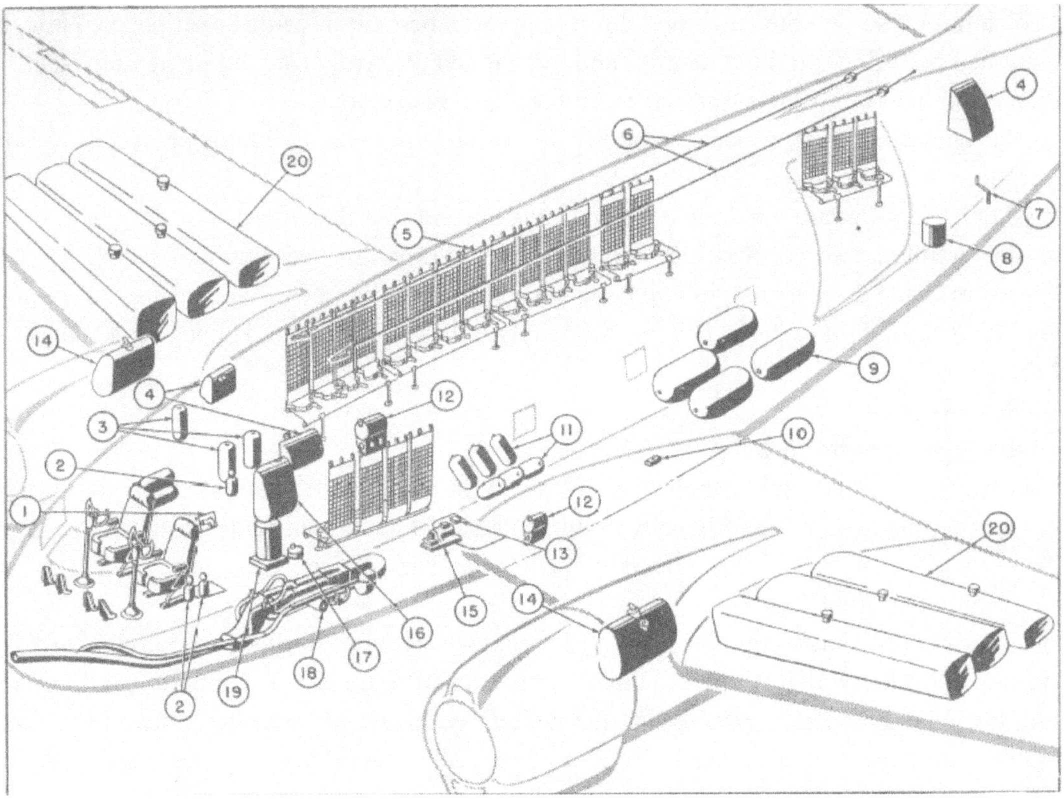

Items #20 are C-46 wing fuel tanks. Items #14 are engine oil tanks

Items 9 are oxygen tanks in a lower bay

Source: C-46 flight manual

Chapter Fifteen

GET IT DONE – Ginger Snap Survival

January 1942

Chief Watkins came forward behind the pilots, saw the paper cards, and whistled. "Yea, verily, you pilots have also been given the impossible to do by the good Commander."

"Chief, if you've any ideas, our ears are wide open," Mueller said.

"I'll think about it. I'm up here to discuss my impossible mission."

"Which would be?" Rats queried.

"Dump 2,000 pounds total in the next two loads most quick. Spare no item including the mail, critical cargo like ship and airplane parts, tools, fire extinguishers, oxygen bottles, parachutes and racks, tube and web seats, personal items like seabags, maybe even our clothes and shoes!"

Whitlock exclaimed, "Good grief, has the Commander gone crazy? How are we going to strap the passengers in if we ditch? We may land at Hilo and not have a plane left!"

Watkins replied, "I think that's his point. What does it matter what remains of the beast if it goes to the bottom of the Pacific? We will tie the passengers in with rope and quick-release knots for ditching or landing. We've got a Bosun onboard; he thinks he can get it done. Gentlemen, if you want coffee or water, or need to take a dump, now's the time. The coffee urn, honey bucket, and last water jugs are all over the side."

"We're going to need water in the rafts," Rats declared.

"We are holding back some in canteens for two rafts."

"Jesus, Chief, you're jettisoning the other two rafts? What about Mae Wests?" Mueller questioned.

"Thirty-five out the door sir, keeping one per man. We can pitch the cradles, but we don't have the time or tools to cut up the tanks."

"Thank the Lord for small favors," said Mueller. "You said tools; don't throw out the hand pump tools."

"Sir, I went over the need for hand pumps for hydraulics, landing gear, engine hand crank, oil crank, chocks, wing and tail locks with the Commander. He replied that our only emergency is too much ocean and not enough fuel - everything else goes."

"The aft tanks are empty, right?" Rats asked.

"Yes sir, our makeshift crew in the bay tilted them to use every drop - which brings up the oil reserve tank. Captain, it's 200 to 300 pounds. When will you feel it is safe to throw it out?"

"Chief, my answer ten minutes ago was never. Now I'm not so sure. How about when we see the Big Island Nav advertised at about 190 miles? Whit, note that we'll call back when we have a visual. Chief, if the engine oil consumption is normal at that point, you'll pump the main engine oil tanks full and dump the aux."

"Aye, aye, sir, sounds like a plan. If you have anything you absolutely must use to get us to Hilo, put it aside. I'm sending a crew up here, and they're stripping everything that can be safely stripped."

Rats complained, "Well, you aren't taking our flight jackets or flight gear bags, are you?"

Watkins replied firmly, "Yes sir, it all goes. You'll be lucky to have your skivvies when we're done. I have my orders."

"I'm not sure it's worth making Hilo; SOP is making the cost too high!" Mueller exclaimed.
Whitlock admonished, "Rats, I don't think you mean that; our first responsibility is to our passengers and then the plane."
"You're right, you're right, damn it!"
Watkins continued, "Sirs, after the stripping crew, I'm sending Chief Norris to check with you on the electronics we can toss, radios, etc. Captain, I have some ideas regarding more speed and less burn."
"Let's hear it, Chief."
"Sir, go over every system that sucks energy out of the engines, de-icing, heating, electrical, hydraulics, oxygen, etc., and secure them if we don't need them to land. Think about selecting the cowl flaps least drag profile for the most engine efficiency. Is it open or closed, or what position for what temps? He was messing with the flaps on a flight I had with a Curtiss test pilot out of St Louie. I asked him why, and he said he wanted to see if he could get better fuel flow…."
Mueller interrupted, "Chief, we all know that flaps provide more lift at a lower speed and produce much more drag; they aren't a way to save fuel."
"My thinking exactly, Captain, but he said he was playing with lowering the step speed. He said that just a teensy-weensy flap might fine-tune the planing, that it was like when some people add a wing on the back of their boats to improve getting the bow down."
"Chief, that's all well and good, but the flaps normally go up and down in increments, as you already know," Rats protested.
"Yes sir, he had me pull the flap fuses right after he actuated the flaps to stop them in different positions. He seemed happiest with two or three seconds. I remember five seconds was too much."
"Whit, what do you think?" Rats asked.
"It's worth a try, but it will be a bitch if they freeze in place. A no-flap landing will have us speeding like a rocket at a field that's bound to be shorter than we need."
Harry looked back at Cappy Jordon. "Army, what do you think?"
"Might be crazy enough to work. I can't think of reducing drag without climbing outside and ripping gear off the fuselage. Based on the basic aircraft design, it might give us a knot or two. Whit is right, though. If they stick and won't go down for landing, we are, as you Navy folk say, screwed, blued, and tattooed."
"Chief, you're a good man. We'll give it a try. Thanks," Rats said.
Whitlock got out of his seat. "I'm going aft. I have a personal dump I've been holding onto for hours."
"Whit, when you come back, be so kind as to bring us some coffee, a couple of canteens of water, sandwiches, and the good Commander," Mueller directed.
Whitlock groaned from stiffness while standing and stretching. "Wilco Rats."

Pelham was back at the Nav station, and Johnson reviewed his approach diagram.
"Commander, If we start down with 150 miles to go to Hilo or less, I think we'll be less than thirty miles out when we reach our twenty-gallon reserve and have to ditch."
David responded, "Doesn't sound like much out of 1,810 gallons, about one percent."
"Yes sir, but I made some critical assumptions to determine that distance. One, that the winds below 4,500 would fall to ten to twenty knots. Two, that we can keep on the step while

reducing the revs per minute. And three, that we'll be able to safely drain the main tanks to ten gallons on each side on our final approach."

"Why would that last one be a problem?" David asked.

"When the gauges get that low, it's a bit of a crapshoot."

"Let's hope not," replied David.

"Sir, if we do a slow descent from 100 miles out and the pilots can drop the burn rate to ninety-five gallons per hour, we might just make the field with eighteen gallons in the tanks. We might be able to land!"

"Of course, we'll land at Hilo, you Doubting Thomas. Good work!"

Chief Watkins was standing nearby and motioned for Pelham to join him.

David grinned. "What is it, Chief?"

"The pilots have something they want to run by you. Do we have some good news, sir?"

"Maybe, we may be drinking our bathwater, but based on some optimistic assumptions, we can just make Hilo."

David walked forward and stood next to Pappy Jordon. "What can I do for you, gents?"

"The Chief remembers that a Curtiss test pilot was playing around with the flaps and might have got a bit more lift from them with very little change in drag," briefed Rats.

Pelham said, "Does that make sense to you?"

"It might. When these Fowler flaps first go out, they create a slot. That could give us a bit more wing surface. The first normal flap-down setting is fifteen percent and has a lot of drag. The Chief says they experimented by selecting the flaps down and pulling the breakers to stop their travel. He said five seconds was too much and two or three about right."

"Then what do we get?"

"We may get a little more lift that would allow us to take off some engine revs, keep the same altitude, and pretty much the same speed."

David was excited. "Let's try it!"

Harry replied, "IAS is now on 147, disengaging the autopilot. Captain Jordon, you select the first flap detent when I call for it. Whit, you'll punch the flap fuses out on my call. All set?"

Both men nodded their heads.

"Standby to lower flaps. Lower flaps now. One thousand one, one thousand two, one thousand three, fuses out."

They all felt the plane rise slightly as if flying over a speed bump.

Mueller looked at his gauges. "Airspeed is down a little. Whit, what do you see?"

Whit looked back along the wing. "It looks like the slot is there, and the flap has some down tilt."

"OK, I think we moved it too far, fuses in and flaps full up. Captain, wait until I say now to move the lever. Whit, wait for the word, now. Standby to lower flaps, lower flaps now, one thousand one, two thousand, now!"

Again, they all felt a little bump up in the aircraft. Rats peered at the gauges.

"Somebody buy the Chief a drink! We just picked up a couple of knots of IAS! Commander, there's something you should know. We're worried that leaving the flaps out in the cold air for several hours might make them stick when needed for landing. On the other hand, the outside air is not bad, forty-nine degrees, the sun is out, and we can warm them up on our approach."

"I'll take the two knots and the risk. What do we do next?" David asked.
Harry gently eased the throttles back. "We put the autopilot back on, take off a few revs, and see what happens. And a few more. Outstanding, we're down nearly thirty revs on each engine and back at 147 IAS, keeping the same speed and burning a little less fuel.
David smiled and laughed. "Way to go, pilots! Good work."
David turned to leave and spoke to the radio operators, "Any responses to those Mayday calls?"
"Nothing strong sir. I think I heard something like Sub Base Pearl acknowledge us, but it was weak, and there was no repeat," Jenkins replied.
David took a couple of strides aft to Johnson's Navigation Station. "Ben, did you get a fuel update from the pilots on the hour?"
"Yes sir."
"Tell me more about the current status of the fuel system."
"Sir, we've used the fuel in the four aux tanks and the two forward wing tanks."
"So, all our usable fuel is in the wings, right?"
"Yes sir, the pilots can select pull rods to feed gas from any tank to either engine."
"And they have tank gauges, do they not? They can tell us how much gas we have?"
"Yes Commander, but remember when we get down to five or ten gallons sloshing around in the tanks, we're flying on a wing and a prayer. If an engine falters, you know the connected tank is going dry."
"Falter how?"
"If you're watching for it, you should see engine temps start to fall or fluctuate, same with RPM and fuel flow. The prop controller will react, and the prop spin will look different."
"The pilots want the Chief to come up and work with them on any additional engine or prop settings that might get us more range," David said.
"That sounds like a good plan, sir. Watkins knows how the engines and props work better than any of us, and after him, probably Whit is most knowledgeable."
"Whit, why Whitlock and not Mueller?" David asked with a surprised look.
"When we went to ground school at Patterson AAF Base, Whit and the Chief attended nearly every class for every system. We had brought a Marine twin Beechcraft with us; the rest of us spent a lot of time flying Air Corps and Curtiss-Wright types around."
"Interesting. Let's get back to the fuel system. You've kept track of our gas usage since we started the engines, right?"
"Yes sir, I keep track of the tanks in use, the fuel flow numbers, and the elapsed time. Once we knew that all the tanks were working correctly, we drained the two wing forward tanks."
"They were empty, as in the engines and props fluctuating, or by the gauges?" Pelham quizzed.
Johnson looked puzzled. "Well, I guess by the gauge."
"So, could more gas be in those tanks we stopped using?"
"Yes sir, I suppose there could be a small amount."
"How do you think we're doing now?"
"The changes have been good. No doubt, we would've run out of gas before making land had you not taken over. We were flying the wrong route at what I thought was the wrong speed and using Avgas for the heaters. Oh, my God, the heaters!" Johnson exclaimed.

"They're off. What about them?"
"I didn't account for them. Stupid, stupid!" Johnson lamented.
"Calm down, Ben. How much do they use?"
"Two gallons an hour."
"For one or both?" David asked.
"Both."
"So four per hour times eight hours or thirty-two gallons. Our eighteen-gallon reserve at the field just became a fourteen-gallon deficit. We're back to falling short by twenty to thirty miles."
Johnson was crestfallen. "Yes sir, sorry sir."
"Don't let it get you down. We're searching for the truth here, and it shall set you free. As much as I hate to beat this problem to death, that's what we will do. Someday when you're commanding a squadron or captain of a ship, I want you to remember this drill. We will work on this situation until we save ourselves, this airplane, and the men onboard. Failing is not an option!"
"Yes sir."
"Ben, please use a 111-per-hour burn rate on the main tanks and tell me our fuel state at 120 miles from Hilo."
"Yes sir. Let's see; it works out to about ninety-five gallons at 120 miles out."
"Outstanding Nav! Ninety-five when we're comfortable with 100. Not bad. Hand me that black marker."

David went forward behind the pilots and asked to see the fuel gauges. Whit pointed out that the forward wing tanks were empty, and the aft wing tanks were nearly empty. David could see that the main tanks were showing almost full. David scooped up the three cards on the instrument panel and control pedestal. He crossed out 111 and wrote, "110 FUEL FLOW NOW!"

Without another word, he walked aft and used the forward head piss tube to get rid of some coffee. Then he went into the cargo bay and sat on his chair. *We might make it yet. The vibration and noise are wearing me down. And the stress, maybe. A couple of lousy judgment calls, and I'm responsible for the deaths of some of these brave young men. I have to rest. I need to sleep.*

David got up and found Chief Watkins. "Chief, I think we've got a solution we can live with if the pilots can get us down to 110 gallons burn per hour before we start on the main tanks. Frankly, I'm beat. I can barely keep up with Johnson's numbers. You and your crew back here are doing great. The pilots have their marching orders. Lt. (jg) Johnson is a true gem. I'm going to take a nap. I need to be rested and thinking straight for the last two hours. Do your third dump without the highest priority items you feel we must save. Go help the pilots get to 110, perhaps by playing with the prop pitch. Wake me in two hours or if we're approaching 300 miles out."
"Yes sir, get a few winks. We'll take care of the ship," Chief Watkins replied.

Chief Aircraft Mechanic Donald J. Watkins took a moment to sit on the deck and contemplate his path to this point. He was born in Princeton, West Virginia, in 1907. His father was a coal miner, a surveyor for roads and railroads, and a self-taught engine mechanic. His mother had her hands full managing a family of six children, four girls, and two boys.

Donald was the youngest boy and the second-to-last child born into the Watkins family. As a boy, he loved the dramatic Mercer County weather. He expected the rain, the rumbling thunderstorms, the warm, humid summers, and the cold, freezing winters. The green fog-shrouded hills, rivers, creeks, and lakes were places of adventure and discovery for the Watkins children. They could count on snow sledding and ice skating every winter, and the blossoms and green leaves of spring.

When Donald Watkins entered high school, his dad had a steady job with the Virginia Railway. Donald and his brother were always tinkering with old cars and engines. They took parts from old wrecks and pieced together a vehicle of sorts that got them to school and back. They helped their dad build additional rooms onto the three-room Watkins home using the new wages as shop foremen. The Watkins home was comfortable but not luxurious. Each family member had a Sunday going-to-church outfit, and the children counted themselves fortunate because they always had at least one pair of shoes and a hot dinner each evening.

Henry Watkins passed his love of hunting and fishing to his boys and taught them how to use tools and work with their hands. He told them always to save a little money for a rainy day and that automobiles were the future of America. Donald felt at home in the high school automotive shop and taught other boys how to rebuild engines, repair clutches, and service transmissions. His shop teacher, Mr. Higgins, allowed Don to use the most expensive tools and gave him perfect grades.

When Don Watkins graduated from High School, his dad said there was an apprentice opening in the Virginia Railroad engine repair shop. Don was also accepted at Concord University in Athens. He was not sure what to do. Watkins went to Mr. Higgins for advice. They discussed his prospects during the booming 1925 economy. Mr. Higgins asked if he wanted to live in Princeton and Mercer County for the rest of his life, but Donald was unsure. He responded that it was a decent place to grow up, and he loved his family, but he was tired of the rain, the fog, and the cold winters. Don Watkins had never been anywhere except a couple of trips with his father on the train to Roanoke, Virginia. He declared that he wanted to see more. He wanted to go to New York City someday and maybe even Europe.

Mr. Higgins assured him that his mechanical skills would always be in demand, and if he wanted to travel, he should consider the military. They drove to Bluefield the next Saturday and talked to the Army and Navy recruiters. Donald said he was most impressed with the Navy during lunch at a Bluefield diner. The Army recruiter had extolled the Army's training and opportunities for advancement. The Navy First Class Petty Officer told his own story. He had come from a small town and traveled on Navy ships to the Caribbean, the Mediterranean, and most of the large ports on the East Coast. Donald took the Navy enlistment paperwork home to discuss it with his parents. His mother cried and said she'd miss him dearly. His father listened intently to his son and said the family would support his choice, whether school at Concord, the railway repair shop, or the Navy.

Donald slept on the decision and had his answer when he woke up. He realized that his family would do just fine without him. He'd be leaving behind four sisters and his brother, who was now working in the railroad shop. None of them planned to leave the Bluefield-Princeton area. His older siblings were heading for marriage and families. So Donald Watkins decided he would join the Navy and see the world.

Source: Author's composite graphic

Chapter Sixteen

ASSES AND ELBOWS – Hilo, Hawaii Airport

January 1942

Jordy Manelo completed his requests for Mary's assistance at the Hilo phone exchange, put down the phone, and abruptly sat down. *My God*, he thought, *What am I going to do? How am I going to do it? No time to waste; what's our situation? It's an emergency. Everyone in town needs help to fill in the ditches and contribute hard labor. We have to fill in at least one trench 100 feet wide, five feet deep, and three feet wide in six hours! Or we could bridge it with something, but what? Steel grates, lumber, thick plywood might work. No, the plane is more massive than most. It could be twenty tons on two tires crossing the midfield opening. I have nothing that would hold that weight, nothing! Why did I promise the Admiral? Why is this happening to Hilo? First, Pearl Harbor. Then the Jap sub shelled our town, and now this. God, we need your help. Help me God, help me, Jesus.*

Jordy stood and looked out on the main runway with its three obscenely gaping open drainage ditches from one side to the other. He sounded the field siren that would be heard all over the airport and for miles around. Immediately, his phone rang.

"Tower, Jordy speaking."

"Jordy its Bill Towers. What's going on?"

William Towers was his Fire Chief and probably had a skeleton crew of three or four men at the field during the construction. Although there was not much for them to do, they didn't get paid unless they came in and logged their time.

"Bill, we have an emergency landing. I need all personnel to report to the bottom of the tower. You need to call in your full crew and every able-bodied man they know. We need trucks and shovels, anything that can move dirt. We have six hours to fill in 100 feet of the center ditch and land a fifteen to twenty-ton transport. It could be tragic!"

"Are you kidding? Fill it in with what, with who?" Towers demanded.

"No time to talk, Bill. Get everyone here, call everyone, their brother, and their neighbors. Tell them we need trucks, shovels, men, and dirt-moving equipment, do it now!"

Jordy hung the phone up, then picked it up again and clicked the cradle buttons up and down.

"This is Mary at the phone exchange; how can I help you?"

"Mary, it's Jordy at the airport. I want you to call my dad. I have a message for him and anyone else you think can help with our emergency. Are you ready to copy?"

"Yes Jordy, go ahead."

"The Hilo airport has a big emergency. We must fill a ditch 100 feet long, five feet deep, and three feet wide in five or six hours to land a Navy transport plane with twelve men aboard. Now, we need trucks, dirt-moving equipment, strong men, and shovels. Everyone report to the airport now!"

"I got it Jordy. I'll call your dad and every number in town!"

"Thanks Mary. Please alert folks in the next two towns, north and south. Beg them to help."

"I'll do it Jordy, but you ain't gonna get nowhere with Wainaku, you know that, right?" she questioned.

"Gotta go, Mary. I've got people forming up at the tower."

He thought, *Dad will know what to do. He started the airport with a bulldozer and a crushed coral and gravel runway back in the twenties when planes were a complete oddity in Hawaii. Now I'm running it; someday, Kiki will be the manager.*

Jordy looked across the field at the small contingent of National Guard fighter planes thinking, *How can they help? Since the shelling, the Army moved in with five or six aircraft, a couple of Quonset huts, and some tents. The shorter runway is OK for them until we finish the construction. No need to call; the siren will get someone from their squadron. What about the construction guys? They dug the ditches and moved on to another job. Another company will lay the pipes; who else? Who else?*

Manelo could see a couple of fire trucks heading toward the tower, some pickup trucks coming in off the coast road, and a jeep departing the National Guard camp. *It's a start,"* he thought. *Leave the siren on for now, so everyone nearby knows it's not a drill. Damn, too bad the National Guard infantry camp is wargaming up at Parker Ranch. We could have got a lot of help there.*

Jordy walked down the four flights of stairs and out into the bright sunlight. Chief Towers arrived in the first fire truck with another one right behind. Two or three pickup trucks were heading his way, and four or five people were already standing expectantly near the tower. They must have been working nearby. The siren was the rotating type. As it came around above him, it was deafening. He thought, *Jeez, I never realized it was so loud. The siren should do something to roust out all the nearby folks. I bet they can hear it at the Kress Building and all over town.*

As the firemen dismounted, Jordy saw that Towers had five men on duty. *Well, they can earn their pay today.*

Towers walked over to him. "Jordy, what's the deal? Doesn't the Navy know the field is closed? Can't they go over to Kaneohe?"

"Bill, I'll fill everyone in a minute," he paused and covered his ears as the siren swept by, "But no. The plane is from the Mainland and is low on fuel. You may have a crash on your hands, here on the field or out in the bay."

Jordy heard a motorcycle pull up behind him, turned, and saw his son Kiki getting off the cycle. He remembered that Kiki was working with the pipe laying company until the University of Hawaii re-opened.

Kiki approached him and said, "What's up, Dad?"

Jordy gave him a quick hug and pointed to the tower. "Son, go up and turn that damn siren off and get back down here. You'll get the full story when that Guard jeep arrives."

"Sure Dad, I'll be right back."

The jeep screeched to a halt in a cloud of dust next to the fire trucks. Jordy could see an officer and an enlisted man, probably a sergeant from the stripes on his sleeve, getting out of the jeep. Jordy had about ten or fifteen people looking to him for direction, so he got up in the back of one of the fire trucks and motioned for everyone to gather around. More cars and trucks were heading for the tower. Then, mercifully, the siren stopped. He thought, *Thank you Kiki, you're a good kid; I bet this is a day you'll never forget. What do I say to these people? How do we start?*

A couple of cars pulled up, and four or five more men joined the throng. Kiki came running out of the tower. Jordy could now see that the Army officer was Captain Armstrong, the detachment commander.

Jordy gestured at the small crowd. "Everyone gather around, come in closer, and listen real good. We don't have time to repeat this situation over and over again. You must get it right the first time. Then we have to get to work. The clock is ticking on whether we can save twelve Marine and Navy boys who have to land here in about six hours."

That got their attention. Everyone was listening intently; the group moved closer to Manelo. He continued, "A large Navy transport plane from the Mainland, San Diego, I think, is diverting here from their planned landing in Oahu because we're closer. They expect to be low on fuel or even crash into the ocean. The aircraft is about halfway here and hit stiff, unexpected headwinds. They may make the field. They may not. Best case, they land here. Worst case, we see them crash in the bay. It's a big plane and needs at least 2,000 feet of runway. We have only 1,400 feet of usable runway with those ditches you see out there. Admiral Nimitz has called me and ordered us to land that airplane – to save those Navy and Marine boys coming here to fight for us."

There was a crowd murmur as it sank in that the top military man in Hawaii was counting on them. Hands started to go up.

Jordy continued, "No questions yet until I'm done. The pilot of that aircraft, Navy 78774, is a Marine aviator. He's asked, no demanded, that we fill in 100 feet of the middle trench. If we do that, sorry, we will give him 2,200 feet of runway when we do that. I've promised Admiral Nimitz that Hilo will get this done! We'll do this by hook or by crook. Questions?"

It seemed like every hand was up, including the Army officer. Jordy picked him first.

"Captain Armstrong, Air National Guard from the fighter squadron on the other runway. Sir, how heavy is this plane?"

"We expect them to be fifteen tons or more," Jordy replied.

"What will you use to fill the trench?"

"Right now, all I can think of is dirt and chunks of concrete that's piled up from the excavation. We don't have any grating or wood to bridge the trench, not in that quantity or in that strength. I'm open to ideas and comments."

"Yes sir, be aware that you must compact that dirt, even put something on it to distribute the load, or the tires will dig in and hit the back edge with terrific force. Also, I recommend you start in the middle and work outward in both directions. A Marine pilot may make do with less than 100 feet if that's all we can get done."

"Thank you, Captain. Good inputs. Maybe we could use the fire trucks to tamp the dirt and put boards on top of the dirt," Jordy replied.

"Jordy, what do you want the emergency crews to do?" asked his Fire Chief.

"Bill, get your crew up to full strength. Augment them if necessary. Prepare for a crash in the water or on the field. Call the Coast Guard station and tell them they're in charge of a crash in the water. Bring in whatever medical personnel we should have and set up a treatment area for burn and trauma victims. Your men can work on the ditch after you make those preparations. Mary and the other ladies at the phone exchange are calling everyone to get more help."

One man put his hand up. "Hey Jordy, how are we going to do this? Fill in the trenches by hand?"

Jordy paused, *Precisely what I'm wondering myself, buddy. Look confident Jordy, find an answer.* A few more vehicles had arrived. Jordy had twenty or thirty people looking up to him for answers.

"Thank you all for responding. How many of you men have worked in construction?"

Probably ten hands went into the air.

"Good; how many of you have been the foreman or in charge of a construction crew?"

Three hands went up.

"You three men go over by the fire hydrant and figure out how to start and who is in charge of what. We've sent out a request for trucks, men, and shovels. My first thought is to shovel the dirt by the side of the field into the truck beds and dump it into the trench. We will try and get earth-moving equipment, thank you!"

A couple of hands went up, including his son's.

Jordy pointed at an older man, "You sir."

"George Bonafin here. I'm a fisherman. I can't handle a shovel but can get some boats out on the bay for a crash. What direction will they be coming from?"

"They will be coming in from the Northeast. Work with the fire chief on what you can do, and get the Coast Guard lifeboat involved if they're in Hilo. Thank you."

He was ignoring Kiki, "You, sir."

"Mr. Jordy, I'm John Williams, a high school teacher. I could run the math on how much dirt has to be moved and track the progress. You will know how you're doing and if you'll make it."

"Great idea sir; please go over and work with the construction foremen. Thank you. Kiki?"

"Dad, the ditch diggers left a skip loader here. I think I could hotwire it and get it going. I know they drained all the gas out.

Jordy was surprised. "You can operate a skip loader?"

"One of the crew let me play around with it some; I think I could learn it," Kiki said.

"All right son, we'll talk about the hot wiring later. Go over and tell the foremen and get gas from the fire crew. Thanks. Captain Armstrong."

"Yes sir."

"Could the Army send a truck to the Guard camp to get help and set up a command post to coordinate people as they come in, tell them what we have to do and why? Direct them to one job or another and coordinate with the foremen and me in the tower?"

Armstrong picked up a walkie-talkie. "Consider it done, sir. We have some other thoughts for later, but they will keep for now."

Jordy addressed the remaining people. "Thank you for coming. If you can man a shovel, please do so. If you can't, work with the Army on what you can do. I'm going back up in the tower to think about what we've missed."

As Jordy climbed the tower stairs, his hands were shaking, his mouth was dry, and he felt he could be sick to his stomach. He thought, *Things are moving, what next? What am I missing? In about six hours, a massive plane is approaching my field to crash land. It's all impossible, impossible.*

Manelo sat at the tower's communication console, put his head down, and wept. After a few minutes, he wiped his eyes, sat up, and vowed to do or die. After all, he and the other Hilo people came from generations of warriors, sailors, and explorers. He thought, *No one can do it if we can't do it.* He willed himself to stand up and look out on the field to see what was happening. Two men were driving a stake into the ground in the middle of the trench. Next to them was the high school teacher working on a pad with a pencil. *Hmmm, not too hard, 100 times five times three or 1,500 square feet of dirt, divide by nine for square yards for 166. How many truckloads? Small trucks hold two or three yards maybe, eighty or a hundred loads?* He looked north down at the dirt pile from most of the center trench and saw the three foremen waving their arms at about eight men with shovels who had started to fill three trucks. *Impossible, Impossible.*

Jordy Manelo walked to the south side of the tower, put his binoculars up to look toward town, and saw his dad's Jeep Willy barreling down the two-lane highway. Further south, the airport manager saw a line of cars and trucks with people sitting in the back-holding shovels - all headed for the airport. Two National Guard two-ton trucks were arriving from the east with five or ten men in the back. *By God,* he thought, *We're going to do this; we have to do this! It's going to be a long, tense day; better eat and get some fluids.*

By the time Jordy's mom bounded up the stairs, he had a sandwich in one hand, a Coke in the other, and his feet up on the desk.

"Hi Mom, how ya doin'?"

Sarah Manelo was a small, wiry woman with an open, loving Hawaiian face and a tough-as-nails heart.

She said, "I'll be doing fine if my oldest, no-account son gets off his backside and gives me a proper hug."

Jordy rose and hugged his mom. "Yes, mama."

Finally, his dad came huffing up the stairs, hugged him, went to the window, and said, "What do we have, Jordy?"

"Dad, a heavy Navy or Marine transport is coming from the Mainland with twelve souls on board. They are experiencing strong headwinds and can only land here. We may see them hit the water right in front of us. The plane weighs fifteen to twenty tons and needs 2000 feet of runway. It has a Marine pilot."

Deacon Manelo looked at the activity at the dirt pile. "How far out, Jordy?"

"They called us with seven hours remaining. It's fast approaching six hours," Jordy replied.

"Thank the Lord they gave us fair warning. What must we do?" Deacon asked.

"Dad, we're filling in 100 feet of the center ditch from the center out fifty feet in each direction, then we plan to tamp the dirt down with fire trucks and maybe place boards on top of the dirt. That will give the Marine 2,200 feet of runway."

"Son, this is your show. You're the airport manager. I'm not gonna take over, but I'll make suggestions that you can take or leave."

"Yes sir, that's what I want," Jordy said.

"Number one, use a lighter truck on the dirt tamping before the fire truck. The last thing you want is a fire truck bogged down in the center of the field."

"Good point, thanks Dad."

"What else you got rollin'?"

"We have the National Guard unit from the other runway setting up a command center to coordinate the jobs and flow of vehicles and people. Bill Towers is calling in all his crew and other medical support to prepare for a water or airfield crash. Three construction foremen are down on the dirt pile directing the ditch filling. A fisherman has gone off to get more boats in the water northeast of the field, and maybe the Coast Guard. You should know that Admiral Nimitz called here on the radio and ordered us to land that plane no matter what and save those twelve Navy men."

Deacon spoke softly, "Navy and Marine men."

"Yes, Navy and Marine." Jordy had forgotten that his dad's brother, Uncle Roger, had been a career Marine and never let them combine the two.

"Sounds like a great start, son. I'm proud of you."

"How you gonna feed all these people and get them enough water?" Jordy's mom asked.

"Gee Mom, I don't know."

"Well, I do. I'll call the churches and get plenty of food and water out here. Jordy, do you know what is coming your way?"

Jordy looked startled. "What do you mean?"

Sarah replied, "Switchboard Mary has got half the town coming out here. People are grabbing every pickup truck in town. The hardware store throws anything that could move dirt out the front door for neighbors to bring out here. The firemen are coming. The cops are coming. The high school is closed, and they're busing the young men here. Dockworkers are on their way on foot. Son, you've got a real challenge coming your way. By the way, where's my Kiki? I thought he was working here."

"He's trying to put gas in a skip loader and hotwire it to speed up the truck loading, Mama," Jordy replied.

Sarah raised her hands to the heavens. "Lord help us, like father like son, like the grandfather - may the saints watch over us."

"He's not trying any longer; come and take a look!" Deacon yelled.

The three saw a yellow skip loader below heading toward the dirt pile from the south tower window.

"Makes you sort of proud, don't it," Deacon declared.

Jordy wiped a tear away from his eye thinking, *We can do this, we will do this.*

An Army enlisted man burst into the tower control center, out of breath. He handed Jordy a portable radio.

"Sir, with the Captain's compliments. Just push this button to talk; volume is on this knob right here."

Jordy put the radio to his mouth, pushed the button, and said, "Command Center, this is tower, over."

Armstrong's voice came back, "Tower, Center, read you loud and clear, over."

"Center, my folks are up here, and tell me half the town is coming. You may have hundreds of people before we're done. The cops are coming; maybe you could use them for routing traffic. Our Fire Chief will take on the firefighters coming from town. My mom is organizing food and water from the churches. And I see my son in a skid loader heading for the dirt pile, which may speed up loading the trucks. Thanks for the radio. Please see if we can get the fire crews on this freq, over."

Armstrong replied, “This is Center, all good news. We’ll be ready, out.”

Manelo returned to his desk, sat down, and finished his sandwich and Coke. He knew what Switchboard Mary was hinting at about Wainaku. Andy Kalwiki lived there and ran the largest construction company on the island's east side. When he didn’t get the ditching contract at the airport, he said Hilo could go to hell before he’d bid them another honest price for quality work.

Jordy decided to go down on the field and encourage the volunteers first-hand. He touched base with the National Guard and his firemen and talked to the foremen directing the truck loading and unloading. About 100 people were working feverishly to load and unload twenty pickup trucks. One foreman complained that emptying the pickup loads into the trench took too long. Jordy rode in one of the vehicles out to the center of the field and had an idea he thought might help.

Back at the dirt piles, he caught the attention of one of the foremen.

“I think one of you needs to be out on the field. They’re trying to shovel the dirt out of the trucks. Maybe instead, they should just push it out, even onto the cement. They’re trying to empty the whole bed with shovels. It might be faster to push out what they can and make faster round trips.”

“Sounds like a plan. I’ll go out there and try it,” the foreman responded.

Jordy went over to the Command Post area. Captain Armstrong motioned for him to come closer.

“Mr. Manelo, I wanted to get some of the info straight in my mind before I propose something to you.”

“Sure, what do you want to know?”

“Well, I think you said they were about 1,100 miles out about an hour ago, and it would take them six or seven hours to get here.”

“Yes Captain, that’s what they told me.”

Armstrong figured on paper, “So they’re closing at about 150 miles per hour. If I wanted to take my P-40s and bounce them at 150 miles, I need to launch when they’re range is 300 miles. That would be, let me see, about 5.3 hours after they called you. Do you know their altitude?”

“No, they didn’t tell me,” Jordy responded.

“Do you know what radio frequencies they’re monitoring?”

“I do. I talked to 774 on one and wrote down two others.”

“That’s what we need. So we’ll probably find the C-46 inbound between 5,000 and 10,000 feet trying to get a visual on the volcano.”

Jordy said, “Your planes could be a great help. If they crash into the ocean, you could mark the spot.”

Armstrong recommended that they inspect the progress on the field. He and Jordy got into his Jeep and drove out to the other side of the trench. Men and women were unloading five pickup trucks. Jordy noted that they were now pushing the dirt out and not trying to shovel every last bit. His stomach sank when he looked down into the pit, and his hands began to tremble. After an hour of work, there was hardly any progress. Armstrong must have agreed as he put his hand on Jordy’s shoulder and gently squeezed him. They went back to the Jeep and were silent for a few seconds.

"This is not working. We have to do better! Captain, any ideas?" Jordy exclaimed.
Armstrong was deep in thought. "Let's think about this aircraft. Nearly all transports are taildraggers. We need to know how far apart the main gear and the main landing wheels are. We first fill in where the main gear will go over the trench and work outward in both directions. We put plywood or something equivalent in the middle for the expendable tail wheel. If it breaks off, it shouldn't be a big problem. We mark the middle with a flag or paint so the Marine pilot can judge the distance. Do you have any specs on the plane?"
"No, give me an estimate, and we'll call them," Jordy responded.
"You said they could be twenty tons or 40,000 pounds. That's a big bird, it probably has a ninety-foot or more wingspan. The gear must be about twenty feet out on both sides from the middle of the aircraft. We can start at forty feet apart and refine it from there."

They drove over to the dirt pile. Jordy stood up in the Jeep and waved two of the foremen over. He told them to stop filling in the middle and start filling twenty feet out from the midpoint in both directions, and plan on covering the middle with plywood or planks. He took Armstrong up in the tower and picked up the radio mic.
"Marine 78744, this is Hilo tower, over." There was a long silence.
"Marine 78744, this is Hilo tower calling, come in please," he transmitted a second time.
"Hilo tower, this is 744, over."
"744, can you confirm you are a tail dragger, over."
"This is 744, that is affirmative, over."
"744, what is the distance between your main landing gear, over."
"744, standby." After half a minute, they heard, "Hilo tower, Marine 744, the distance is forty-nine feet, I repeat four nine feet, over."
"744, copy that, standby." He picked up the walkie-talkie and directed the National Guardsman at the Command Center to tell the foremen to start filling twenty-five feet on both sides of the center point.
"Marine 744, I'm putting on Captain Armstrong from our Air National Guard Unit."
Armstrong said, "744, how does your fuel situation look, over."
"This is 744; it depends on the winds. We will just make it to the field if they're twenty knots. If they're thirty knots or higher, we will have to ditch at about thirty to fifty miles short, over."
"Roger 744, we are deploying some boats out into the bay. I'll check whether we can put some seaplanes further out. What will your altitude be at 150 miles out, over?"
"This is 744; we're at 6,500 now. When we get a visual on the volcanoes, we plan a slow descent to 4,500. Then, at sixty miles out, we plan a slow descent to sea level, over."
"Sounds good, 744. I will intercept you with a flight of four or five Warhawks. We will lead you to home plate from 100 to 150 miles out, over."
"This is 744, thank you, Army. You will be a sight for sore eyes!"
"744, can you give us an estimate of your bearing from Hilo at 150 miles, over." Armstrong requested.
"744, standby." After a thirty-second delay, "Hilo, 744, our Navigator is shooting for the 030 magnetic radial. If you don't see us there, we could be a little bit north."
"Roger that, thank you. Armstrong out."
"744, Hilo tower, we will monitor this frequency for your landing and the other two given to us by the ham operator."

"Thank you, tower. Please write this down for the record. Seven seven four burned over 6,000 pounds of fuel and tossed out about 3,000 pounds of cargo and aircraft gear. We are flying a maximum range profile. Our pilots manually leaned the engines and set prop pitch for more miles. Our landing weight will be about 28,000 to 30,000. This is our last radio; the others went out the door. If we have comms problems, maybe you can have the Warhawks relay, even with hand signals if they have to, over."

"Tower, copy all of that, over."

"This is 744; how goes the trench filling, over."

"We got off to a slow start, but I'm seeing the road clogged with trucks, buses, and people in all directions. Once we get all of them at work, it should go much faster, over."

"744, outstanding. If we run out of runway, we plan to make a deliberate ground loop before the last trench, over."

"Roger that. Tower out for now."

Armstrong looked concerned. "Jordy, what do you think? Will the field be ready?"

"It has to be. I promised Admiral Nimitz. What we need is more equipment. Wait, do you see what I see?"

Armstrong smiled. "I believe I do, hot damn!"

Jordy picked up the binoculars; a mile north of the field, he could see a police car with flashing lights leading a group of flatbed trucks. The first trailer had two skip loaders, the second had two mid-sized bulldozers, and the third had a full-sized angle blade dozer. Behind the equipment were several pickup trucks, each packed with several men holding shovels. A second police car with flashing lights brought up the rear. He could read the logo on the side of the first flatbed, "Wainaku Construction."

"How about that! Even A-W wants in on this job! Thank you, Lord!"

Manelo could see that the caravan was not waiting for the traffic to clear. Instead, they were barreling by the other vehicles heading to the field on the wrong side of the road.

Armstrong shook Jordy's hand. "I'm going to get my boys ready and check the planes. We will be airborne in a couple of hours. I'll talk to you on this primary freq or one of the other two. Good luck!"

Jordy went back down on the field and talked to the foremen. They were now concerned about having too many people and trucks. The construction equipment changed how they planned to fill the trenches. Once the Wainaku equipment was unloaded and ready to work, they recommended switching the pickup trucks and shovel crews to the first ditch. Jordy agreed as he saw a church bus and some pickup trucks arrive with tents, chairs, tables, water cans, and food.

Oh boy, he thought, *There's Big Mama and Little Mama, two of the religious pillars of the Hilo community. Big Mama still believes in the old-style Hawaiian gods like Pele and the four deities. Little Mama violently disagrees with her and practices a traditional Baptist evangelical religion grounded in a vengeful God and a benevolent Jesus. The two of them often disapprove of each other's practices. I hope they get along for the common good. We can use all the help the Gods can offer.*

Curtiss-Wright P-40 Warhawks

Source: Google Images

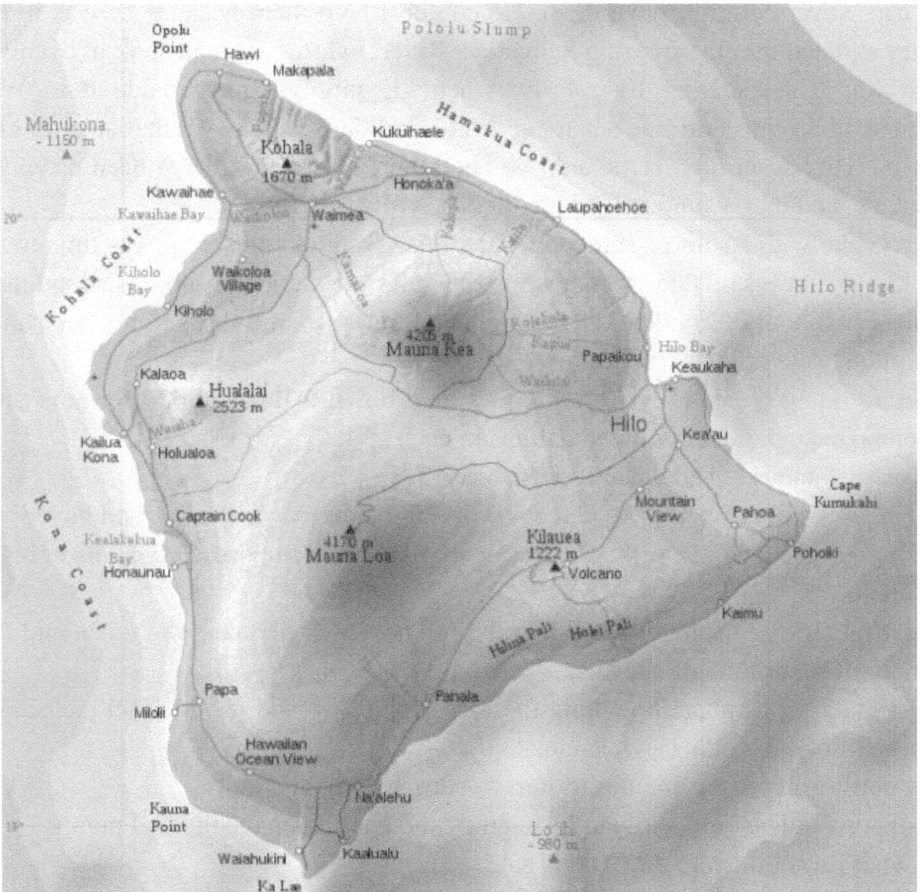

Big Island of Hawaii in 21st century, note Hilo and Hilo Bay on the east coast.

Highest volcanos (Mauna Loa & Mauna Kea) shown in meters above sea level

Source: Wikipedia Commons

Chapter Seventeen

FINAL APPROACH – Hilo Or Splashdown?

January 1942

Before he took his second nap next to the empty auxiliary gas tanks, David Pelham said a short prayer asking the Lord to keep his recurring nightmare away. It didn't work.

As the C-46 aircraft flew toward Hilo, he dreamed of his wife and daughter drowning. His location in the dream varied. Sometimes he was on the riverbank, in the water next to the sinking vehicle, or even in it as they took their last breaths. Occasionally they were attempting to get out of the car, and other times not - just accepting that the ice-cold water would soon end their terrible plight.

Lorraine often fought to open the car door against the water pressure. Other times, she was unconscious and could not hear Mandy's cries for help. He was always powerless to stop the sinking car. Never could he save them. He couldn't help them because he was in England when they skidded into the freezing Saugatuck River, overturned, and sunk in twenty feet of water. It was Thanksgiving Eve, 1940, when his family almost made it to Westport, Connecticut, to be with Lorraine's parents. All the holidays were never the same for him from that day on. He had no joy. He should've been there! He should have been driving on the snow and ice, and they would still be alive!

David awoke with a start as someone patted his shoulder and said it was time to rise and shine. Chief Watkins said they were nearing the 300-mile mark and should be landing at Hilo in less than two hours. David stood and stretched, ran his hand through his hair, and asked how the pilots were doing.

"Sir, they're doing great. They have the fuel flow down to 111 gallons per hour and think we're gonna make the field with ten gallons in each tank."

David was suspicious. "And what does Nav say?"

"Mr. Johnson is still nervous. The plane is working better than any of us had hoped. He says it depends on the real winds. If they're lower, we make it. If they're thirty knots or higher, we have to ditch."

"Chief, let me ask you something. When the pilots switch off the wing tanks and consider them empty, what criteria do they use?"

"Sir, I believe they estimate how long the fuel should last and use the fuel gauge. When it shows near empty, the pilots switch tanks."

"So they don't wait for the engines to start to burp, cough, and quit?"

"No sir, that would be very hard on the engines and could damage the fuel pumps."

"That's what I want them to do," David said.

"Do what, sir?"

"I want them to burn fuel off the forward and rear wing tanks until the engine starts to conk out. Then, when it falters, I want them to rock the plane just like you do with a boat or a car to burn those last drops in the tank."

"Are you sure, Commander? With manual fuel leaning and manual pitch control on the prop, I'm not sure what might happen to a quitting engine."

"Then go back to auto on both until you're sure the four tanks are empty. Then return to manual fine-tuning. Chief, that's an order. Warn everyone that the plane may be rocking intentionally. Get up there and note how long we run on those supposedly empty tanks, and then get some rest."

"Aye, aye, sir."

Chief Watkins went aft to warn the passengers and then back to the flight deck to replace the Army Pilot in the jump seat. Pelham moved forward to talk to the radio men. He could hear the pilots complaining about damaging the engines. Chief Watkins closed the discussion with, "Gentlemen, orders are orders, let's get to it!"

David returned to the Nav station. "Good afternoon, Ben. I've had my beauty sleep. Any good news to report? Are we two hours out?" David queried.

"Yes sir, we're just crossing the estimated 300-mile mark. The pilots are doing great, but the numbers have not changed that much. Whatever the actual winds are will decide our fate. While you were resting, the Hilo tower said there would be a flight of P-40s coming out to lead us in."

"Good. If we have to ditch, the P-40s can report our position. What else?"

"They're sending a Coast Guard rescue boat out about twenty to thirty miles, and there will be other fishing boats, I think, positioned in Hilo Bay."

"Ben, how about the airport trenches?"

"Get this, Commander; the airport manager said he promised Admiral Nimitz the middle trench would be filled in, and he's going to get it done."

David laughed. "It's amazing what a little white lie can do. So let's have a recap. We're about 100 miles from seeing the top of the first volcano, Mauna Kea. Then we fly sixty miles to our 120-mile tipping point. Once there, we need 100 gallons or fifty gallons in each middle wing tank to avoid ditching, agreed?"

The aircraft started rocking; they both grabbed the Nav table.

"Yes sir."

"At 120 miles to go, if it looks more like forty gallons each side, we can ditch in Hilo Bay, or we might just make the field by the skin of our teeth with less headwind," David asserted.

"Yes sir."

"Run me through the sighting of the top of Mauna Kea again. Put us at 6,500 feet and the top at 13,700 feet.

"Yes sir, I have that in my notebook. Hmm, here, if we're at 6,500, we should see it at 210 miles. If we have good visibility, at 4,500 feet, we see it about 190 miles out."

"How do we get together with the P-40 aircraft?"

"The pilots told them we would be at 6,500 on the 030-magnetic radial from the field. They will call us on the emergency frequency."

David grabbed the Nav table as the plane rocked back and forth. "Looks like the pilots are burning off the other tanks."

"Do you think that'll help?" Johnson questioned.

"That's number two, two more to go. It can't hurt to empty those tanks. It might squeeze out a few more miles. To review, if the wind is thirty knots, we'll be thirty to sixty miles short. We'll be five to twenty miles short if it's twenty-five knots. If as low as twenty, we may make the field with a little reserve?"

"Yes sir. We had two more dumps while you were grabbing some shut-eye. The pilots have the burn rate down to 110 gallons per hour."

"Ben, that's good, that's very good. If we see the volcano or its cloud at 190, we could start down if the sea state looks reasonable. We may get better winds on the descent. If we start down at 100 feet per minute, it will take us sixty-five minutes to the deck. I'm going aft to check the cargo bay and talk to the Chief and the pilots up front. I don't suppose there's any coffee?"

"No sir, the pot and all the supplies are long gone. Everyone drank the last of the water and ate all the food. I saved you a Coke and some crackers."

"Many thanks, Ben. After checking with the Chief and the pilots, I'll be ready for that."

The plane started rocking again as he went aft in the cargo bay. He noted that the four auxiliary gas tanks were lashed together, the cradles were gone, the seats were gone, the generator was gone, and the men were sitting or lying on the aluminum deck. The long-range oil tank was gone. As he used the piss tube, he thought, *This is a good crew. Mueller must have decided that the extra oil wouldn't be needed in the last two flight hours. We have to get to the field, even if it's just to crash land. I can't let these men down. I can't. I won't.*

David reached the cockpit to see the pilots select the last tank for complete emptying. "Chief, how many minutes on those first three tanks?"

"About three to four minutes on each one," Watkins replied.

"Does the rocking seem to help?"

"Sir, it seems to get us another twenty to thirty seconds, but that's all."

"Thanks Chief. You've done a great job, and we're close, so close to making the field. Is there anything else you can think we should be doing?"

"No sir. We've held back the mail and some critical parts, but they don't weigh much."

"Good man, Chief! Tex says we've less than two hours until the moment of truth. Hello, pilots; how are you doing? How much of that precious gas do we have left?"

Rats responded in a weak voice, "Top of the afternoon to you, Commander. We're just peachy keen up here. We have the burn down to 110 and can't wait to see the volcano tops in the next hour. As you can see, the vis is outstanding, and the gas gauges tell us we've over 100 gallons on each side."

"Captain, according to Nav, we should see the volcano at 200 miles out. At 160 miles, we need 120 gallons to make the field with ten in reserve on each side. The good news is that it will be near the field if we ditch. We should have the fighters watching over us, and there will be some boats to pick us up. When we reach an hour to go, I want everyone in tip-top shape."

Harry extracted himself from the pilot's seat. "Aye, aye, sir, I need to take a leak and stretch. Whit, you have the aircraft."

David shook Mueller's hand. "Rats, you pilots have done a magnificent job on this flight. You know your bird, and she's performing like a real thoroughbred. When do you think we'll see the fighters?"

"Commander, they've known we were coming for almost five hours. We just reported that we were two hours out. I bet they took off not long after that and will intercept us at about 100 to 150 miles from the field."

"Outstanding. Please think about whether we should descend at 100 feet per minute from 160 miles out when you go aft. Also, Ben has a Coke and some crackers he saved for me. I want to split them with you. You're the most important person on this plane. Only through your skill and knowledge will we safely land at Hilo."

Mueller looked Pelham squarely in the eye. "You ain't so bad yourself, Commander. I'd fly with you anytime, anywhere."

After Rats returned to his pilot's seat, David and the Crew Chief toured the aircraft and talked to every crew member and each passenger. They briefed them on the current situation, shook their hands, thanked them for their efforts, and helped them put on a life vest. By design, the last one they spoke to was Lt. (jg) Benjamin Johnson.

Pelham shook Tex's hand. "Lt. Johnson, flying with you has been a pleasure. No matter the outcome, you've distinguished yourself today and done everything you can to save this magnificent aircraft and the twelve souls on board. Well done!"

"Thank you, sir. Do you think we're going to make it?" Johnson asked nervously.

"It will be close, Ben. Rats is deciding whether we make a long descent from 160 miles out or closer to the field. Whatever he decides, I can live with. He's the man of the hour now."

"Yes sir, if anyone can do it, Rats can. He's pulled us out of some pretty tough spots before."

David took the half-warm Coke and cheese crackers back to his perch on the stool and gulped them down. He put his head back against the flight deck bulkhead and fell asleep. Pelham woke to some whooping and enthusiastic yells from the flight deck. He shook the cobwebs off and went forward to where Whitlock pointed at a distant white cloud on the horizon about ten degrees to starboard. David took the glasses and confirmed a small vapor cloud where they expected the first volcano.

He couldn't help himself; he pounded Tex on the back and exclaimed, "Yes! Yes! Yes!"

There was land out there - a volcanic island under that wispy white cloud nearly 200 miles away.

"Nav, let's assume that's the top of the volcano. What course do you want?" Mueller requested.

"Rats, aim about fifteen degrees left of the volcano. We should be seeing it in a few minutes." Whitlock took the glasses back and was the first to spot the top of the volcano. "I see it! I see it! It's just a dark speck that doesn't move, and it's slowly getting bigger."

They all took a turn with the binoculars; each man confirmed something there, something unmoving and black with a snow cap. David checked his watch and figured they were crossing the 180-mile mark to Hilo and saw the second vapor cloud over Mauna Loa dead ahead.

Pelham could see the primary tank gauges, and his heart sank. They were reading just over fifty gallons on each side - not enough to make the field with twenty gallons to spare. He took a couple of steps backward and slumped to the deck. He felt sick to his stomach and defeated. The Chief exuberantly pounded his back as he went aft and said, "I'm gonna tell the passengers we see land, we see land at last!"

David thought, *It's not fair that we've put in all of this effort, all of this thought, all of these computations, and we still have to ditch. Damn, if we lose the plane; we surely will lose some of these outstanding men. Others will be injured and out of the war. It's not fair that I no longer have a family. It's not fair that my mother died too young and in pain. It's not fair that Hitler is enslaving Europe and laying waste to Russia. It's not fair that Japan pulled a sneak attack on us, sinking our ships and killing our boys on a Sunday morning. Not fair, not fair, not fair!* Now he was angry, thinking, *Whose side is God on?* It hit him like a brick. *God helps those who help themselves.*

David moved behind the pilots, "OK boys, settle down. We look a bit skimpy on fuel. Plane Commander, have you decided what profile to fly?"
"No sir, I think we should rendezvous with the fighters first and then decide. They should know their distance from the field while we're just guessing."
"Very well, how can we help that along? What can we do to help them find us?"
"We're a pretty big target, Commander. They should be able to see us from ten miles out. We could give them an ADF count to help the intercept. Whit, call them on Guard and give them a long count every five minutes."
"What should I call them?" Whitlock asked.
"Hilo fighters should do it," Rats replied.
David picked up the second headset from Jenkins and heard Whitlock's transmission. "Hilo fighters, Hilo fighters, Marine 78744 en route Hilo, about 180 miles out, over."

David's heart sank when there was no reply.

Whitlock on the radio, "Hilo fighters, Hilo fighters, this is Marine 744 transmitting in the blind. We are at angels 6.5, heading 210, about 1-8-0 miles from Hilo; long count follows, 1, 2, 3, 4, 5, 6, 7, 8, 9, 10, 9, 8, 7, 6, 5, 4, 3, 2, 1, Marine 744 out."

Pelham's gut was tight. His heart was pounding. His brain was spinning. They were closing on the last hour of the flight. Something had to give. Something had to happen, but what? What had they forgotten? What was the key to not ditching, not dying, to winning, and not losing? They needed 120 gallons at 160 miles out. He knew they were slightly further out than that, and the gauges were closer to fifty a side than sixty. He went back into the cargo bay and paced up and down, looked at no one, and spoke to no one. The men saw his demeanor and got out of his way. No one talked to him. He paced and paced and paced until, once again, there was yelping and whooping coming from the flight deck.
David quickly moved forward behind the pilots. "What's up?"
"We just heard from the fighters; they're out here at 150 miles from the field waiting for us," Whitlock answered."
"How long until we see them?" David demanded.
Mueller said, "Probably about five to ten minutes or less." He pointed at the first, now easily seen volcano, "Ain't she beautiful. Come to papa, baby! Commander, we get the fighters tacked on, then start our long descent. Whoee, I'm ready for a cold beer!"
"Jenkins, get the Chief up here. He sits in the jump seat for the last hour," David directed.
David put the earphones back on and heard Whitlock give a short count on the ADF, and then there was a medium-strength reply. "774, this is Boomer One, we are at angels nine. If you are heading 210, you should see us soon."
"Roger, Boomer One, we can't wait. It's been a long flight," Whit replied.
David called Johnson up to the cockpit. "Ben, look at the gas gauges. What do you think?"
"Sir, my DR puts us at about 160 miles out. The gas is less than we need, assuming twenty knots of wind. It's not good," Johnson said in a defeated tone.
Rats enthused. "Nonsense, it's all good! We have a visual on the volcanoes, and we have little friends about to join on us. We have this damn fine plane, this amazing flying machine. At Hilo, I'm buying the first round. Nothing could be better!"
Jinx saw them first; the sun flashed off one of the fighter's wings. Then, on the radio, David heard, "774, this is Boomer, tallyho, we're joining. Boomer Five stay up on top."

"Five, roger."
David could now see four aircraft coming right at them and heard, "774, say your speed."
Rats responded, I-A-S 155, say your altimeter, over."
"This is Boomer One. We will use two-niner-niner-two until we get closer to the field; it was two-niner-niner-six when we launched. Joining on your left wing."
David could see a tight diamond formation of four P-40 Warhawks off the left wing and heard, "Boomer Five give us a course to home plate."
"Roger, this is five, 215, over."
"One roger, 774, you copy?"
Rats pushed his transmitter button on the yoke, "Roger, we copy, turning starboard to 215. Are we about 150 miles out? Over."
"Boomer One, affirmative. What is your situation? Over."
David indicated that he'd answer; he took the mike from Jenkins. "Boomer One, this is the senior officer on board; we have declared an official emergency. It is doubtful we will make the field and probably have to ditch ten to twenty miles out, over"
"Damn, 774, sorry to hear that. What are your intentions? Over."
"Boomer, we are starting a long descent at 100 feet per minute at the end of this transmission. We may have enough fuel to get a visual on the field and ditch by one of the boats in the bay. But, when we get to ten gallons a side, we must do a controlled ditching because the gauges may not be accurate, over."
David saw the Chief ease back on the throttles, and the altimeter slowly started to unwind. Then, on the radio, he heard: "Boomer flight, Boomer One, form up in echelon left. Seven seven four, we see you descending. Is it OK if we give you a little lift? Over."
Rats transmitted, "After sixteen hours, you guys and that volcano are the most beautiful things in the world. What do you have in mind? Over."
"Seven seven four, Boomer One, we gonna give you some extra miles, yeeha! It's going to be fun, fun, fun. Number Four in echelon, move over to the right wing tip."
"Four, roger."
"Number Three in echelon, move to the right horizontal stabilizer."
"Number Three, roger."
"Number Two, fall back to the left horizontal stabilizer."
"Number Two, roger."
"Marine 774, Boomer One, are you ready for some fun?"
Harry answered, "Fun away Army. We are burning 95 gallons an hour and have about 100 left. We're ready to try anything!"
"Boomer Four, from one."
"This is Four, go one."
"Boomer Four, on my signal, overlap half your wing with 774."
"This is Four, roger that."
"This is one, overlap now, now, now."
It was unnerving to David. They could see both airplanes before. Now they could only see the top of the cockpits moving very close to the Commando's wingtips.
"774, Boomer One, are you feeling anything?"

"Maybe a little bit. Our rate of descent has slowed a bit; what are the rules for this fun, fun game?" Rats asked.
"This is Boomer One, you don't do anything without telling us first, over."
"Seven seven four, wilco. I'm going take a few revs off to get back on the 100 feet per minute slope, over."
"Roger, tell us when, over."
Rats nodded to the Chief and pushed his radio button, "Pulling some power now, now, now, and we are back at 100 per minute descent."
"Boomer One, copy."
"Boomer Four, take it closer to about three feet from your canopy, now, now, now. Boomer Two and Three, overlap in back, but don't hit each other, over."
"Two roger." "Three roger."
"Boomer Five from One, I'm hoping you can get a snapshot of this at some point, over."
"This is Five. Y'all are looking real good - some sweet flying going on down there."
"All Boomers, we are getting ready to pull a slight amount of power; please acknowledge, over," Rats transmitted.
"Boomer One." "Two roger." "Three copy." "Four roger."
"Pulling a little power, now, now, now."
"Rats, I'm afraid to look. What is our fuel flow now?" Pelham quired.
"Whit, you answer; I'm flying and looking outside only. You and the Chief will have to handle everything else."
"Roger rats. Commander, fuel flow is ninety," said Whitlock.
David was excited. "Ninety? Are you sure? What are the tanks reading?"
"Sir, flow confirmed at ninety. Tanks look like just over forty gallons in each wing; make it eighty-four total," Whitlock said.
Pelham yelled, "Tex, get up here. We need to talk!"
Lt. (jg) Johnson was startled, seeing the Warhawks overlapping the wingtips. "Jesus Christ, what are they trying to do, kill us?"
"They're giving us a lift, private little Army joke. Putting their wings under ours gives us more lift and lets us pull some power. It's helping. We have fuel flow down to ninety and about eighty-five gallons left. What do you think?" David demanded.
"Let's see; we've 120 miles to run at 2.9 miles a minute or forty-one minutes. So we need a lower burn rate or a higher speed to make the field."
David felt the tension and fatigue. "That's it. You've hit the nail on the head. Good job, well done!"
Johnson looked at him, puzzled, as David raised the microphone. "Boomer Five, 774, over."
"Five here, over."
"Five, these other guys are pretty busy. Could you estimate our ground speed and tell us the surface winds when you took off, over."
"This is Five, estimated ground speed is 170 knots. The wind was from the south at five to ten knots when we launched, over."
David transmitted, "Roger your numbers, thanks."
"Ben, refigure using 170 ground speed versus 150," Pelham ordered.

"Yes sir, we have forty minutes to landing, ignoring the five minutes on final approach. Multiply our burn rate of 1.5 gallons per minute times forty minutes for sixty-five gallons needed. If we have eighty-four gallons at the 120-mile mark, we make the field with a nineteen-gallon reserve."

"Like those numbers, Nav!" David exclaimed.

Rats chimed in, "Me too! Crew, let's talk about the landing. We're gonna hold the flaps and gear until the last minute. At a half-mile, we do one-third flaps. At a quarter-mile, we do two-thirds flaps. We go full flaps at feet dry and drop the gear passing through 100 feet. We flare the plane to land at sixty-five knots past the first trench. After the center trench, we stand on the brakes. We ground loop if we're gonna hit the third trench. Any comments?"

They all shook their heads except David. "I have one comment, Captain."

Mueller smiled. "Yes sir, I suppose you want to land the plane?"

Pelham laughed. "Not this trip. I want Whit to be ready to switch either engine to either main tank. We've fully emptied all the other tanks. We can't make a mistake if one engine falters. First, we can rock the plane, and at twenty seconds, Whit must be ready to feed that engine from the other main tank. The two levers must be marked ahead of time. That's my comment."

Rats nodded his head. "We can take care of that, no problem."

On the radio, they heard, "Seven seven four, this is Boomer Five, over."

"Go ahead Five," Rats said.

"Boomer Five has a good lock on Hilo ADF. Recommend you turn right to 220, over."

"Boomer flight, 774, about to turn right to 220, over," Rats transmitted.

"One copy." "Two roger." "Three copy." "Four roger the turn."

"Seven seven four is doing a slow turn right to 220 now, now, now."

After twenty minutes of formation flying, "Boomer Four from One, is your neck starting to kill you like mine?"

"This is Four, that is affirmative."

"Roger Four, move out from under the wing now, now, now. Good. Boomer Four fall back and loosely join on my left, we are switching positions. Boomer Two and Three, you can switch after we are tucked back in, over."

"Four roger." "Two roger." "Three copy."

After another twenty minutes of flight, Boomer One asked how long they wanted the Army lifting service. Mueller replied that down to 500 feet and two miles from the field would be much appreciated. Boomer One instructed his flight that One and Two would turn right and join after detaching; Four and Three would turn left and do the same. Jenkins confirmed that the passengers were roped in for landing.

Several minutes later, Whit called the Hilo tower and reported the field in sight. He requested the altimeter and an update on the field condition. Jordy told them there would be red flags in the middle of the first two ditches. He gave them the altimeter setting and wished them luck. Rats confirmed that the center of both trenches were filled and warned the tower that he'd not be dropping the gear until the last minute. Rats and Whitlock completed the Landing Checklist with flaps and gear remaining.

"Commander, we're passing through 600 feet and have about five minutes to go. You better go aft and get roped in now," Harry directed.

"Nope, I'm watching this landing. It's the most important landing of my short flying career."

"Sir, if we have a bad landing or make a ground loop, you could get banged up really bad," Rats pleaded.

"I'll take responsibility for that. Hold it; what's on the field? Whit, hand me the glasses."

"The fire trucks are clearing. It looks like two people on the field," Rats said in an alarmed voice.

David used the binoculars. "There are two women on each side of the first filled trench."

"Women?" Mueller asked.

"Yes, a small one and a large one," David replied.

Rats got on the radio. "Tower, there are civilians on the field. Please remove them now!"

"Seven seven four, tower. I would if I could. Those women are the community's spiritual leaders. They're praying for you. Unfortunately, it looks like it's become competitive."

Rats was agitated. "Tower, we have to land! They're in the way. We can't go around!"

"Seven seven four, tower, you are cleared to land. Try not to hit the women, over."

"Marine 774, Boomer One, we're detaching now, now, now. Godspeed."

David used the binoculars. "The women are getting down on their knees. They seem to be praying and marking the outside edges of the filled-in trench for us. There's a red flag in the middle of about 100 feet of fill."

Rats yelled, "We can't land, we can't land, we'll have to go around!"

Pelham responded, "Negative Captain, this is your one last chance. The gas is down to ten gallons in each wing. You will land this plane!"

"Half a mile out, selecting one-third flaps," Whitfield advised.

Mueller shouted, "I'll not kill those women. We can't land!"

David replied, "Captain, you can do this. You will land between them. They have faith in you. We all have faith in you. Aim for the small red flag in the middle."

"No, no, what are they doing there?" agonized Mueller.

"Rats listen to me. You're tired. We all are tired. Get a grip and do what you're trained to do. The women are praying for you. Show them what you can do!"

Whitlock calmly stated, "Selecting two-thirds flaps. Rats, you're lined up a little right. Chief, come up a bit on the power."

"Captain Mueller, do you want me to order your Co-pilot to land the plane? Is that what you want?" David threatened.

Rats heaved a big frustrated sigh. "No sir."

"Good. Do your job. Land this plane safely and save the twelve souls onboard!"

Whitlock called, "More power Chief. Going to full flaps, standing by on main gear."

Pelham saw two fire trucks waiting on the outside edges of the approach apron. A flare shot up into the sky from each of them.

"Now what? What the hell?" Rats said.

Whit answered, "They think we're landing wheels up. The lineup is good, standing by the main gear; speed is seventy knots."

As the aircraft went feet dry, Rats directed, "Wait on the gear, wait, wait, landing gear now! Chief, ease the power back to idle."

David heard the main gear thump into place and lock a split second before the plane touched down. Mueller had landed the C-46 on all three wheels just beyond the first trench.

Whitlock warned, "Rats, aim for the flag, aim for the flag!"

"Roger, no brakes until we get past the second trench."
"Speed fifty, she's swerving left!" Whit advised.
"Chief, up throttle on port engine," Rats responded.
Whit said, "Concur, past the middle trench, throttles back, hitting brakes, speed forty."
"Commander, plaster yourself on the starboard bulkhead. If we have to loop her, we'll use the starboard engine," Mueller advised.
Whitlock cautioned, "Speed thirty, it's going to be nip and tuck!"

They could see the open trench coming closer and closer.
"Speed twenty. She's light; it should work. Speed ten, full stop, engines off, parking brake on," Whitlock said with a sigh of relief.

Pelham saw the open trench stretching out in each direction. The cockpit was over the ditch. He wondered how close the wheels were. With the engines finally quiet after seventeen hours, David's body vibrated. His ears were ringing as sounds of men yelling and celebrating in the aft cabin were joined by hundreds of voices on the field. A mass of people, cars, and trucks honking their horns were streaming toward the aircraft. David shook Rats' hand, then Whit's, the Chief's, and Jinx's. Five steps aft, Johnson was sitting motionless with a dazed look and tears in his eyes.
David patted him on the back. "Good job Nav, getting us here safe and sound. You deserve a medal."
"Just happy to be alive, sir. Just happy to be alive," Benjamin Johnson softly replied.
Captain Jordon came up to the Nav table and congratulated them. Jordon looked meaningfully at Pelham. "Commander, ain't no one leaving this aircraft until you leave it. There's a big crowd out there waiting to see you."

David looked down at his white uniform; it was a disgrace. His short-sleeve shirt was dirty and torn. His pants looked more gray than white and had oil smears. His shoes were scuffed, and one lace was broken.
"I can't go outside like this."
"Nonsense, Commander, look around. We're all in a sorry state. It has been a long and difficult flight," stated the Army Captain.

Pelham let the Army Captain lead him to the cargo compartment door. The Commando's passengers applauded and patted him on the back as he walked woodenly past them. He stood in the aft entrance door, and a great roar went up from several hundred people. Most of them were sweat-stained and dirtier than the naval officer they were cheering. It was a sea of brown Hawaiian faces, with hands clapping and bodies jumping up and down.

A young man standing on the top of a pickup truck held his hand out and helped David step out onto the truck cab and down onto a milk crate in the bed. Pelham was dazed but aware enough to notice a truck approaching the plane with the two spiritual women sitting triumphantly on the back tailgate.

After a couple of men helped David off the pickup truck, he stood looking down at the runway asphalt. He had no words. He was spent. The throng quieted down. He knelt, bent down, and kissed the dirty pavement - the workers roared their approval. *Ugggh,* Pelham thought, *Terrible taste, won't ever do that again.*

The young man from the truck helped Pelham to his feet and introduced himself as Kiki, son of the airport manager. Kiki warned him, "You better get ready for some serious hugging. Big Mama and Little Mama are heading our way."

David could see the crowd parting for the two older women. Big Mama was six feet tall and over 200 pounds. Little Mama was five feet tall and shy of 100 pounds. But it was Big Mama who got to him first.

She enveloped him in a crushing full-body hug, "Aloha, welcome to our Island and Hilo. The Hawaiian Gods have brought you here safely to enjoy our love and affection. You will be the guest of honor tonight at our celebration luau."

Then it was Little Mama's turn. She gave him a firm hug and said, "Thank you, Jesus! Thank you, Jesus! Thank you for safely bringing this young man and his comrades to our island. Thank you, God, for hearing our prayers. There will be a combined service of thanks in the Baptist Church in one hour."

She let him go, and he moved through the crowd, shaking hands and thanking the volunteers as the crew and passengers filed off the plane. At the crowd's edge, he saw an Army jeep approaching. It came to a screeching halt next to him, and the passenger spoke.

"Hello sailor, Boomer One here. Would you like to get cleaned up, have a cold beer, and borrow a uniform?"

David smiled wanly at the Army Captain. "Yes, please, but not necessarily in that order."

Pelham got his cold beer, showered, and shaved. He borrowed a summer khaki uniform from the Boomer pilots. Armstrong drove him to the Baptist Church, where he sat up front with the Commando's crew and passengers. The church was packed with standing room only.

The two Mamas put on a combined service, each recognizing their deities and asking for blessings on all the persons present. They asked David to speak; this time, he had some words. He praised the airport personnel, the firemen, and the people out in boats. He commended all the workers and volunteers who had made the field safe for them to land. He recognized the Commando crew for their expertise and courage under fire. Pelham told the locals how the Army P-40 pilots had led them to the field and given them a lifesaving lift through outstanding airmanship. Noting that all his comrades were cleaned up and wearing casual Hawaiian clothes, David complimented the friendly town of Hilo for its caring hospitality.

He finished to a spirited round of applause, sat down, and asked about the Plane Commander's plans.

Rats replied, "Commander, it's time to pull some very serious liberty. This is my first luau, and I can't wait. We'll meet tomorrow at 0900 to assess the aircraft's damage."

The luau was down on the beach near the Marina and docks. The locals apologized for not using Coconut Island because the Army had made it off-limits for civilians. There was plenty of beer, grilled chicken, and a roasted pig dug up and carved after dark. Torches were lit, and out came the ukuleles and young girls dancing. David was drinking his third beer while sitting next to Chief Watkins. He asked the Navy Chief how he ended up on a Marine aircraft crew and got an earful.

"Well sir, I went through the Great Lakes boot camp the summer of 1925. After graduation, I was assigned to the Second Division on a cruiser out of Norfolk, the USS Cincinnati. We fixed the ship's boats. My Division Officer had me work on the engines. When I came aboard, two of the four boats were out of commission because of engine problems. A Second Class and me pulled one of those engines out of a whaleboat and took it apart on the deck. The piston rings were burned, and the internal oil filter was fouled. We changed the oil every month and had no more problems. Good thing, as we sailed down to the Caribbean, and those boats gave us some great liberty runs.

"In '27, we were transferred through the Canal to Shanghai, China. Then back to Newport in 1928. I was striking for Machinist's Mate and made Third Class when we got back in the States. We had two Vought O2U floatplanes onboard, and I got involved with their mechanics. The O-in-C of the floatplane detachment talked me into applying for Aviation Mechanic 'A' School. I got selected and enjoyed my time in Pensacola - met my wife there at the Navy Exchange. We were engaged when I graduated and went on to 'C' school for Aviation Machinists Mates."

"Very interesting," commented David, thinking it might break the Chief's train of thought. There was no such luck – the Chief took a big swig of beer and continued his monologue.

"I got orders to NAS San Diego. Once my fiancé got a job at the Navy Base Exchange, we got hitched. After growing up in West Virginia, three years of shore duty in San Diego was like a dream come true. I was assigned to an aircraft maintenance squadron, made Second Class, and worked on several types of engines and aircraft. In 1932, I had orders to the USS Lexington aircraft carrier as the ship's company. My pregnant wife moved to an apartment in Long Beach since Lady Lex was based in San Pedro. I worked in the engine repair shop and supported all flying squadrons onboard for three years.

"A Chief, a First Class, and me decided which engines would be repaired, replaced, and overhauled. It was a great education, but very hectic. After two years, I was ready for shore duty and wangled some orders to NAS Norfolk as a First Class. Two years later, we had a second child. I got selected for CPO and had orders back to Pensacola as an Aviation Mechanic 'A' School instructor. Two more years of shore duty were not all that bad. Then we were ordered back to NAS San Diego, where I was put in charge of all the multi-engine repairs for transports and seaplanes in early 1940. My sea duty was coming up, and my detailer said there was a mechanic slot with a Marine transport squadron. So I volunteered, that brings us to this crew and flying missions that make a difference – best damn duty I ever had!"

The music was playing, the beer was flowing, a half-moon was rising, and the water lapping on the Hilo Bay shore inside the long breakwater was nearly calm. The Ginger Snap aviators and some C-46 passengers attempted to dance the Hula with several young girls. Rats wore a gaudy Hawaiian shirt, a grass skirt, and had a flower in his hair as he waved his arms and wiggled his hips. He was putting on quite a show to the delight of the locals and his crew.

David had not felt so alive in over a year. He wondered if you needed to have a close brush with death to appreciate being above ground.

His thoughts traced his path from New York to Hilo. *Am I sitting safely on a beach in Hawaii because of fate? Was it God's will that the twelve men on the Commando survived? There were a ton of fervent prayers offered from the seemingly doomed aircraft. How much of life is preordained? How much can you control? So much happened in the last ten years, some good, some bad. I lost my mom to cancer, my wife and daughter to the frigid river, and now I'm going to see men killed and maimed, maybe including myself. War is hell, and then you die.*

Hilo, Hawaii circa 1940s with author's annotations

Source: Hawaii State Archives

Chapter Eighteen

WHAT NEXT? – Honolulu, Hawaii

January 1942

Lt. Comdr. David Pelham stirred in his Army cot and woke to the splendid smell of cooking bacon and fresh coffee. He was in one of four large tents home to the officers and enlisted men of the six-plane National Guard detachment based at Hilo Airport on the Big Island of Hawaii. The night before, David had left early from the welcoming luau thrown by the Hilo community. During the festivities, the story was told and retold how the Marine Corps C-46 Commando aircraft had experienced high headwinds that nearly caused it to ditch at sea.

Pelham thought back over the harrowing seventeen-hour flight - at least the parts he could remember. Finding himself the Senior Officer Present was a mixed blessing. He was not in complete control but bore significant responsibility for the outcome - good or bad. David reviewed the adjustments, and drastic actions the crew and passengers took to make the flight last as long as possible. In the end, the bold flying of the National Guard Warbirds enabled the Commando to land safely at Hilo. The P-40 pilots placed their wings under the transport's wings and horizontal stabilizer during the last 100 miles to the field. It was that extra Army "lift" that saved the day.

At the Hilo Field, there was also a miracle. Its Manager, Jordy Manelo, had about six hours to fill open drainage trenches that crossed the main runway in three places. It took a mighty effort from Hilo's citizens and their neighboring towns to make the field safe for landing.

After their welcome to the Hilo Luau, the crew and passengers on Marine 78744 met at the airport the next morning. Captain Mueller and Lt. Whitlock inspected the landing gear and found no damage. Their C-46 was now in front of the tower. Residents were driving up and having their pictures taken in front of it. Airport personnel had placed a portable ramp at the rear door. The crew needed a new ladder to get on and off the aircraft, since the original was part of the first dump.

The inside of the plane was a mess. The four auxiliary fuel tanks were lashed to aircraft stringers on the starboard side of the cargo bay. The seats and seat belt straps were gone. Pieces and parts of the inside fuselage were missing. For the flight to Honolulu, Chief Watkins decided they would lash folding chairs to the port side and use a thick rope that would go over the passenger's shoulders and laps with a single knot they could pull open. Corporal Jenkins and Chief Watkins carefully checked and re-closed the aft oil and fuel feed lines to forestall an in-flight fire.

The pilots decided the bird was airworthy and went into the tower to plan a flight to Hickam Field. Instead, they ran into Captain Armstrong, who offered to escort them to Oahu with a couple of Warbirds. After hearing about the damage to the inside of the Commando and the missing equipment, he recommended they fly to Marine Air Station Kaneohe instead of the Hickam base. He said that a Marine base would give them better support, and many damaged and burned-out search aircraft at Kaneohe might have some of the C-46's missing parts. A fuel truck arrived, and Corporal Jenkins filled the main wing tanks.

The Commando crew and passengers left for Oahu at 1300. Because of the stripped plane and light fuel load, Rats had the aircraft airborne before reaching the middle trench.

Armstrong had promised them a scenic VFR tour at 1,500 feet. They flew up the northeast coast of Hawaii, between Maui and Kaho'olawe, then between Lahaina on Maui and Lanai, and next along the south coast of Moloka'i. There was a stretch of deep blue water until they rounded Makapu'u Point and flew along the gorgeous eastern Oahu beaches.

Approaching Bellows Army Air Force Base, four P-40 Warbirds joined the three aircraft. Armstrong and his wingman had the lead. The other four fighters split into two flights of two. Marine 774 was escorted by six planes, two in front and two on each wingtip. As Armstrong led them around Mokapu Point, Whitlock received clearance to land on Kaneohe's runway 22. The Ginger Snap pilots descended to 500 feet and completed the landing checklist. The Hilo Warbirds flashed overhead when Rats started his port turn for 22 with full flaps and all gear down. Armstrong turned inverted and flew the length of the runway upside down with his wingman tucked in tight. The flight of four Bellow's fighters followed in a diamond formation behind them. Rats laughed in appreciation of the Army air show as he greased the C-46 onto the wide, unobstructed, long runway at 65 knots.

Lt. Comdr. David Pelham said his goodbyes to each member of the crew and the other passengers. They had been through a severe trial as a team. There were a few tears and a couple of uncharacteristic hugs. It turned out that Jinx and the Chief had squirreled away some of his seabag items. He flagged down a jeep and got a ride over to the Ops Building while surveying his amusing hodgepodge uniform. He had his white cover back, which was good. Someone in Hilo had attempted to clean his short-sleeve white shirt. It was passable, and his shoulder boards announced his rank. His Khaki shorts, compliments of the National Guard, were clean but nonstandard. He wore white tennis shoes with khaki socks because his white buck shoes were stained.

Pelham found the duty Operations Officer and asked how he could get over to Pearl Harbor. The officer said there was a flight to the Ewa Marine airbase once daily. No Hickam or Ford Island hops were allowed because the Navy and Army were still periodically shooting at any planes on sight. David learned there was a bus twice a day to Honolulu that took three hours. The Ops Officer looked closely at his hodge-podge uniform and said, "Say, are you off that transport that crashed at Hilo?"

"There was no crash. They landed safely!" David asserted.

"Maybe so, but rumors are flying here that they pushed some passengers out of the plane to make it to Hilo."

"Now that's a lie! Everyone on the flight made it to Hilo," David retorted.

"So you say. The Navy and Marines are putting together a joint investigation board. Someone is going to fry."

David saw Petty Officer "Tiny" Wright giving him the high sign to come over out of the corner of his eye. He excused himself and followed the big man out of the building.

"Hey Commander, I wangled a lift over to Pearl; wanna ride along?" Wright questioned.

"Would I? Let's get the hell out of here!"

A Marine corporal driver took them on the coast route to Honolulu and Pearl Harbor. Wright said he went to the Kaneohe motor pool and looked at the day's schedule. He had found an afternoon trip over to Pearl to pick up a Marine Colonel and talked the driver into taking them along.

At first, Wright was highly animated and spoke nonstop about their close call on the C-46, the hospitality in Hilo, and the beauty of the young Hawaiian women. Then, as they passed through Honolulu, he became quieter and looked utterly morose as they were halfway from downtown to Naval Station Pearl Harbor.

"What's going on, Petty Officer? Cat got your tongue?" David inquired.

"No sir. I'm a second-generation Navy Diver. I love my rate, but this is the worst part of diving. It can give you nightmares for life."

"What? Repairing the ships? Getting them floating again?"

"No, Mr. Pelham. Retrieving the bodies and parts of bodies after they've been in the water for a month. You crawl around in pitch-black spaces wearing 100 pounds of diving gear and feel a bump. You stop. It bumps again. It could be a life vest, a carton of food, a piece of wood, or a body. If it's a body, you move slowly and try to bag it. With too much force, it comes apart. If It's just a piece of someone, you try to bag it without seeing too much detail. Say it's a head. You try not to look at the face. You bag it and try not to puke in your helmet, 'cause now you know the rest of the torso is in there with you. You need to clear it so the remains can be buried. It's the right thing to do. Are you OK, Commander?"

David spoke weakly, "Please stop talking. I'm not feeling so good."

The car came over a rise where Pelham and Wright could see Pearl Harbor. All conversation stopped. They saw the twisted sunken pieces of Arizona, the overturned massive hull of Oklahoma, and several other battleships and cruisers sunk to the harbor bottom or at odd angles in the oily water. Acrid burning odors and oil fumes were still in the air. As they drove nearer, more damaged ships could be seen at the capital ship anchorages.

Several buildings had blown-out windows and pockmarks from bullets and shrapnel on the base. Low heaps of policed-up shrapnel metal lined the street gutters and littered the lawns. The Corporal driver dropped them at the Uniform Shop, and they thanked him for the lift. Inside, things were in disarray. Parts of uniforms were on tables, not the tidy rows of different sizes on display racks they were used to seeing. Wright asked the Second Class Storekeeper behind the cash register how they could get set up with a new seabag of uniforms. He said if an item looked used, it was free. It probably was recovered, cleaned, and no one had claimed it. There were also full sea bags labeled by rank and size in the storeroom.

Pelham and Petty Officer Wright searched a large room and found dozens of surplus sea bags. After half an hour, they found some suitable uniforms. David's bag said, "Lieutenant USN, probably 6-foot, medium waist size."

He tried on some of the shirts and pants and thought they could work for the time being. The last name and serial numbers were blacked out on the inside of the clothing. He needed to add his information, or the items would never come back from the laundry.

Wright was upset that the shop didn't have any Navy Diver insignia. His seabag had once belonged to an overweight First Class Bosun's Mate. He decided that was close enough. Tiny told David they could always get what they needed in Honolulu on Hotel Street, insignia, tailoring, covers, tattoos, booze, broads, etc. He shook David's hand, wished him good luck, and said the divers hung out at Smith's Union Bar.

Pelham asked the storekeeper how to find one of the destroyers in the Bay. He went down to the main quay, looking for a liberty boat to destroyer row or the orange shuttle boat to East Loch. Once on the dock, David saw several small boats coming and going. One of them had a hand-lettered sign, "USS Phelps DD-360." David went over to talk to the coxswain since he knew Phelps was a sister ship to the Selfridge. "Good evening, sailor. Can you take me out to the Selfridge?"

The sailor looked puzzled. "Yes sir, are you sure you wanna go?"

"Yes, she's my ship. I suppose she's near the Phelps," replied Pelham.

"Sir, I'm supposed to leave on the hour in about ten minutes. Selfridge is on the way to Phelps. Is that OK?"

"Sounds good to me. It's been a long trip to get here," David replied with relief.

"Yes sir, come aboard; let me take your bag. I'm going down to that butt kit on the pier and have a smoke, and then we can shove off."

"Yes, thank you for the ride."

David thought, *It will be great getting back into a fighting ship's routine. I'm hungry. I hope the Wardroom chow is up to snuff. I wish the CO and XO were easy to work with and there weren't too many differences in Pacfleet. With the war on, I should be fleeting up to my own ship in a year or two if I don't fall on my sword as Ops. Not a job for the timid. Something is always going on, something always in the planning hopper, round-the-clock. The Selfridge would be a damn sweet ship to Captain, heavier than most destroyers, fast sailing, and lots of armament.*

The coxswain returned, threw the lines into the small craft, and started the engine. It was just the two of them as they motored along Ford Island at sunset. Pelham could see damaged ships along the Island, burned-out planes, and hangars on the airfield. The sailor pointed out the capsized bottom of Oklahoma and the sunken pieces of Arizona. The oily water had a pungent burnt smell, and several ships were resting on the bottom. Estimates in the stateside newspapers were that thousands of men were killed and wounded. David believed it. The destruction was incredible. He wondered if some of the press rumors were true: that many of the ships' senior officers were ashore, some of the ships never returned fire; next to none of them were able to sortie, and a large percentage of sailors were drunk on Saturday night, or still in town.

Further up the west side of Ford Island, he could see the USS Raleigh listing from bomb damage on the other side of the island and the rolled-over hull of the sunken USS Utah. Abreast the northeast tip of the island, Pelham's heart jumped as he saw a nest of destroyers tied up alongside one another. He counted six "tin cans" - one was his new home. The sailor did something strange. He turned starboard about thirty degrees and no longer pointed the small boat at the destroyer line. David gave the man an inquiring look, and the sailor shrugged his shoulders.

David shouted over the engine noise, "Don't we need to go out to the destroyers?"

"Sir, you asked to go to Selfridge. It won't be long."

David looked out in gathering dusk and saw a large red buoy. Behind it was some sort of derrick or mast about twenty or thirty feet above the waterline. As they drew closer, he could see a second buoy and a smaller flagstaff. The coxswain shut the motor down while closing the distance to the first float.

David was alarmed. "Sailor, what's this?"

"USS Selfridge, sir, what's left of her."

David cried out in anguish, "She's sunk? Selfridge is sunk?"

"Yes sir, when the Jap planes attacked, she was one of the first to fire at them. Phelps was also shooting right away. Selfridge threw off her lines and got underway. She sailed back and forth in East Loch, firing at the torpedo planes and dive bombers. As the story goes, 357 protected Detroit when three dive bombers came in high from the west. She put up a hail of fire that blew up the lead bomber. The other two switched from bombing the cruiser and came at the 'Fighting 357.' They put a bomb through her amidships that penetrated to the keel and blew out her bottom. Another one holed the forward compartments, but it was a dud. The CO and XO were both on the Starboard Bridge Wing and are presumed dead. She immediately settled, so the Bridge crew tried to ground her in shallow water and sounded abandon ship. All but twelve got off."

David thought, *By what twist of fate could it be the number twelve? I just went through hell with twelve men, and now twelve dead men are still on my ship. Is this an ill omen of some kind?*

David pointed below the mainmast. "Twelve men are still down there? You're sure that's the right number?"

"Yes sir, twelve are missing. Some of them might have blown overboard into the bay. We picked up wounded all day on the seventh and bodies for three weeks. We still get one or two every other day that pops to the surface. Why don't you come to the Phelps, sir? We can get you a hot meal and a place to sleep. Maybe our officers will have more details for you."

David spoke quietly with his head down, "Yes, thank you, whatever you say."

Boarding the Phelps, David heard the bosun blow the general call and announce, "Lieutenant Commander, United States Navy, arriving."

David was joined by the Officer of the Deck, a Lt. (jg) who welcomed him onboard, escorted him to the Wardroom, and introduced him to the Executive Officer. Commander Ronny Phillips made David feel at home and had the Wardroom duty steward warm up some dinner for their guest. After some hot coffee and beef stroganoff, Pelham started recovering from losing his ship.

Phillips provided more detail on the day of infamy. He said fleet standing orders required all ships to have one-fourth of their anti-aircraft guns manned and supplied with live ammunition. Once the sailors realized they were under attack, the batteries opened fire, and the sky over Pearl Harbor filled with flak bursts. The frantic din of general quarters klaxons mixed in with explosions of bombs and torpedoes on Battleship Row. Selfridge was outboard duty ship of the destroyer nest. She quickly cut her lines and sailed closer to Ford Island to protect the big boys. Phillips personally witnessed the lead aircraft killed by Selfridge gunners and saw the remaining bombers take their revenge on the ship.

"How could two bombs sink a Porter Class destroyer at GQ?" David asked.

The XO said, "It was only ten or fifteen minutes since the first alarm, and they set partial GQ. There was pandemonium all around, guns firing, calls for more ammunition, commands for the snipes to build steam pressure, calls for slipping the lines, huge explosions on some of the bigger ships, and a hundred things going on at once. We all started in condition XRAY. I'm betting that Selfridge had not completed setting condition ZED when the bombs hit. Later on, we discovered that these bombs were modified to penetrate the big ship's deck armor. They smashed through all her decks, and one blew a big hole in the bottom. The other holed every deck and her bottom but didn't explode. The water came in fast and rose on multiple decks before all the hatches were secured. Dickie Rallins and Billy Williams, the CO and XO, were in the open and probably instantly killed. Selfridge started limping toward McGrew Point as the Bridge crew sounded Abandon Ship. You could tell they were trying to beach her, but she just kept settling. Her gun crews were still firing until they had to start swimming. A brave ship and crew - truly worthy of being called the 'Fighting 357.'"

"She's a good ship. I sailed on her the summer of '37 when she was doing post-shakedown cruises out of Philly. What happens to her now?" David asked.

"She should be back in action a few months after being raised and repaired. Even with the carnage all around Pearl, the word is that most of the ships can be patched up. They'll be back in the fleet in six months or less."

"When do you think they will get to my ship?"

"It looks like the battleships and cruisers come first. Selfridge may be the easiest of all to float, but the priority for a small boy is probably last on the list."

"Why do you say easiest to float?" asked Pelham.

"She just had two bombs; only one exploded, and no secondary explosions. There is likely one large hole to plug and several smaller ones. Pump her out, and she floats on her own."

David whistled. "What a day that must have been. Is it true the Japs completely missed the fuel dumps?"

"Too right. They executed two waves of attacks on our capital ships, mostly the battleships. Once they dropped their bombs and torpedoes, they started strafing ships and people, even swimmers in the water! If they destroyed the fuel tanks with those shells, they could have stopped all of our ships in their tracks. No fuel, no offensive capability. Probably much of the gas and oil would've burned, possibly making all of Pearl Harbor even more of a living hell. Instead, they acted like they were in a shooting gallery and blasted away at anyone they could find. It was the same on the airbases; they were shooting at groups of men and leaving the gas tanks mostly alone. Thank God for small favors and for having the carriers at sea."

"I've been told the Nip admiral was too conservative and should've kept sending more sorties. What do you think?"

"We know he lost several planes here in the harbor and some near the air bases, probably more on the second wave. So you can bet that every gun in Oahu would be manned and ready for another attack, and every fighter that could fly would be ready to take to the air."

"Word is that they destroyed most of the military planes on the island," David said.

"True in retrospect, but what does the enemy admiral know for sure? He knows three carriers are missing that might be hunting him down. He knows that every American submarine and search aircraft is now looking for his battle group. He knows his Airedales have probably exaggerated their bombing results on the airfields and anchored ships. His returning planes may have been shadowed, and Yank bombers could be on the way."

"So you're saying December 7th was devastating but not a complete disaster?" David asserted.

"It was a terrible blow; thousands of Americans were killed and wounded, mostly in Arizona and Oklahoma. Even civilians were killed. Yamamoto failed if he was trying to destroy the Pacific Fleet."

David shook his head. "I just arrived; it doesn't look like failure to me. It looks terrible around here, even a month later. Some press say the Pacific Fleet has been devastated beyond recovery. Many people on the West Coast are in a state of panic."

Phillips responded in a confident voice, "We're far from being destroyed. The operational ships here rotate sixty percent at sea and forty percent at port. So you have a significant number of ships that weren't here. Add in the carriers and the ships in overhaul in the States, more than half the fleet is unharmed. It may not look like it, but eighty percent of the ships you see here are untouched or have light damage. Most of the damage you see is repairable because the Japs failed to destroy the repair facilities. It will take time for the big boys to recover, but ninety percent of the damaged ships will be fighting two months after repairs."

"Thank you," David said with relief. "I'm feeling better. Do you think they will attack again?"

"It's anybody's guess. We've had no end of rumors here. One was they poisoned the water supply on the island. Others had the Japs landing on the north shore, or the south, or the west, or the east – even dropping paratroops. One tale had them marching through downtown Honolulu. Yet another asserted that the Japanese sugar cane workers had cut large arrows in the cane pointing to Pearl and set signal fires before the attack. None of them have proven to be true - not one. If they come again, we'll be ready. They will pay a high price. You can count on that."

"Do you think you can find a bunk for me tonight?"

"No problem, we plan to put you up in the Captain's At-Sea Cabin. But we better get you situated. This ship wakes up early, sets condition YOKE, and mans all the guns before dawn," the XO replied.

David exchanged pleasantries with the Phelps' Captain and turned in for the night in a small bunk adjacent to the Bridge. He felt safe. He was home. The low voices on the bridge at night and the ship's bells didn't disrupt his deep, full night's sleep. He heard the 0530 three bells, got up, used the head, and dressed. The Bridge Messenger guided him down to the Wardroom for coffee. The XO discussed the day's schedule with a handful of Division Officers. David filled out a breakfast order chit and was sipping his second cup of coffee when the Bosun's whistle sounded over the 1MC speakers, followed by, "Reveille, reveille, all hands heave out and trice up, the smoking lamp is lit in all authorized spaces."

The XO came over and sat next to him. "Good morning, sailor. Did you get a good night's sleep?"

"The best. It's great to be at sea again," David responded.

"What's your plan?"

"I'm not giving up on Selfridge. Who would be in charge of raising her?"

"They just put a construction type, Captain Homer Wallin, in charge of salvage. He used to be the Material Officer for the Battle Force. Wallin's a good man, smart, and no-nonsense. If anyone can get these ships floating and fixed, he's the one."

"Thanks; where can I find him?"

"Try over on Ford Island. I think they're working out of a Quonset hut near Oklahoma. Look for the Pacific Bridge corporate logo on vehicles or equipment. They're providing much of the engineering muscle for salvage operations. If you are looking for reassignment, go to the base Admin Building and find a First Class Yeoman named Smith, or "Smitty" as we all call him. He's about 5 feet 4 and weighs 120 pounds soaking wet. He runs the show over there. You want orders. You go to Smitty."

David thanked Commander Phillips and took the first boat ashore. The Coxswain dropped him off on Ford Island. As he walked north through the debris littering the shore, he caught sight of the monstrous hull of the capsized Oklahoma. He saw openings cut in the bottom and wondered how many sailors were trapped inside when she rolled. Up ahead, several civilians got out of a Navy van and gathered around a tall Naval officer in front of a Quonset hut.

David waited until they finished, walked over, and saluted. "Sir, would you happen to be Captain Wallin?"

The Captain shook David's hand. "That's right, Commander, Homer Wallin; what can I do for you?"

"Captain, I just arrived yesterday to find my ship, the USS Selfridge, sitting on the bottom. The boys on Phelps hosted me overnight. I'm not sure what my next move should be. Could you tell me if there is a projected schedule to refloat the Selfridge?"

"Selfridge, huh? What billet? Know anything about salvage?" Wallin queried.

"I was supposed to be Ops Officer. Sorry sir, I know nothing about salvage. I'm more of a strategic and tactical animal, a college history professor in my other life."

Wallin snorted. "OK, I could use some help, but I need engineer types. Let me get the Selfridge folder."

He went into the hut and came back with a red folder. "Yeah, sunk over in the East Lock while they were trying to run her aground. Large hole in the bottom midships, smaller hole forward, and some midship hull damage." He pointed at Oklahoma. "Selfridge is a pretty straightforward problem compared to these big boys."

David replied, "I've got a soft spot for her. She's a darn fine ship. I sailed on her in '37 during shakedowns."

"Right now, it will take months before we get to her. CinCPac has directed us to fix the larger ships with the least damage first, get them back in operation, and off to more repairs stateside if necessary. Then we go after medium-damaged larger ships and small boys with light damage."

David spoke excitedly, "Maybe she's in that category if we can raise her!"

Wallin looked at a page in the folder, "This is not good, damn, this is not good at all. She's not been cleared."

"Captain, how do you mean cleared?"

"She must have slipped through the cracks. This is my fault. I should've carefully reviewed the record on each ship. She's not been cleared of bodies." Wallin thumped the folder. "It says there may be as many as fifteen men still on board."

"How can we get that done?" David asked.

"Only by using experienced divers who are in short supply. It's a grim, thankless job. We owe it to those boys to recover as much of their bodies as possible and give them a respectful funeral. It also removes uncertainty for the families if they're listed as missing. I meet with Nimitz this afternoon for our weekly SITREP. I'll have to bring him up to speed on this screw-up and have a plan of attack. Tell me something, Commander, if I give you some experienced people, can you be responsible for clearing the bodies from your ship and accurately assessing the extent of the damage?"

"Yes sir, absolutely sir!" David responded.

"OK, good man. Go over to Admin at the main base and find Petty Officer "Smitty" Smith. Tell him you're TAD to the Salvage Officer until your ship is salvaged and repaired. Ask him

who can help you. Who should be on your salvage team and has the skills to help you recover the ship?"
"Yes sir. I have a special request," replied David.
"What's that?"
"I met Admiral Nimitz on the train coming out to the West Coast. He directed me to come and see him when I got to Pearl. Would it be possible for me to sit in on your meeting?"
Wallin gave him an irritated look. "Tell me what you want to tell him, and I'll pass it along."
"Sorry sir, it's highly classified. I have to do it in person."
"Pelham, see these four stripes; what's the message for the boss?"
"Sorry Captain, no can do. It's critical information on the operational side."
"I don't have time to argue with you. Report to CinCPac HQ at 1400. I'll meet you there."
David saluted. "Aye, aye, sir."

Sunken USS Arizona with over 1,000 sailors entombed

Source: nps.gov

Chapter Nineteen

FIRST CLASS "SMITTY" SMITH – Pearl Harbor

January 1942

David walked back to the Ford Island small boat pier and bummed a ride to the main base pier nearest the Admin Building. He found Smitty's office and introduced himself.

Petty Officer Smith pulled a file from one of his file cabinets. "Aloha, Commander, we've been expecting you. I have a copy of your orders here; did you bring your service record?"

"No, it is being forwarded to the Pacific Fleet Destroyer Commander through regular channels," David answered.

"So, you're on the Selfridge as Ops O. Do you know what happened to her?"

"Yes, damn shame. I was just talking to Captain Wallin. He said to tell you that I'm TAD to the Salvage Officer. He's placing me in charge of raising and repairing Selfridge, and he thought you might have some ideas on who should be on my team."

"Whooee, you work fast, Commander. I guess that sort of makes you the acting CO of Selfridge, right?"

"Doubtful, but possible if your records say I'm senior."

Smith handed him paper and a pencil. "Please write down your date of rank and lineal number. I'll check the Selfridge roster."

"Isn't that a waste of time? I imagine by now they've been reassigned," David replied.

"Not necessarily, Commander. We have an unusual situation here, more bodies than ships. I've been putting the DD-357 men on working parties as groups, usually with their officers in charge. Captain Rallins and Commander Williams had Selfridge operating like a fine Swiss watch. If she will be salvaged, I'm keeping as much of the crew together as possible."

Smitty looked at a list of names. "Sir, you have a Lt. Comdr. Engineering Officer that is senior to you, but the CO, XO, and Ops billets require an Unrestricted Line Officer like yourself. Congratulations, you have temporary command of a sunken ship."

"This day is taking my breath away. What else?"

"Commander, let's take a smoke break outside."

They got outside, and Smitty led him to a shaded bench with bullet pockmarks on it.

"I should've told you that I don't smoke," David said.

"Neither do I, sir. I wanna propose an agreement to you. If you like it, we're a team. If not, the chips fall where they may."

David was suspicious. "Why would we be on a team? Doing what?"

"Keeping the Selfridge crew together as much as possible, slotting you in as the CO, and making sure no one gets assigned with a higher precedence," Smitty said.

"And what does First Class Smith get out of this bargain?"

"When you go to sea, you take me with you. I've done a bad thing and made myself indispensable for churning out PacFleet officer orders. I'm not gonna fight this war behind a desk!"

"I might have something for you. How much do you weigh?"

Smith replied, "130 pounds."

"Any problems with seasickness or air sickness? How do you handle carnival rides?" asked Pelham.
"Never been sick doing any of them. What do you have in mind?"
"It's an idea I have that needs more work. How do I find the crew? I want to put some of them on my salvage team. Who do you recommend?"

They went back inside, and Smith retrieved the ship's roster. He scanned the names and started giving David a list.
"You want the engineering officer, Lt. Comdr. Ronald Haldane, the senior Lieutenant, Gary Lehman, looks good. He's been on the ship for two years. Have him pick a couple of Ensigns or J.G.s. You have a good group of Chiefs. Chief Bennet could be your senior enlisted on the salvage crew. BMC MacGregor, or 'Mac' as we call him, is the best Bosun on Destroyer Row. See if the Engineer wants to bring a Chief and a group of snipes. Ask your other chiefs to pick six more men apiece, and here," Smith pointed at a name, "This E-6 Yeoman, Petty Officer Brown, can manage your paperwork. Tell Captain Wallin you need a rep from Pacific Bridge Company, one diver on call, and a Navy Yard rep with salvage experience."
Smitty held out his hand. "Skipper Pelham, do we have a deal?"
David shook his hand. "Deal. I have a lot of mods I want to make to Selfridge to make her a better fighter. I'm not an O-5 yet. How do we get by that wicket?"
"Let me think about it, sir. It will come to me. For starters, do a stellar job raising the ship and repairing her. Then convince CinCPac Fleet to give you a battlefield promotion."
"I'm meeting with him this afternoon."
Smitty was surprised. "With Nimitz. What for sir?"
"Captain Wallin is taking me to his weekly meeting. Then I need to confer with the Admiral privately on another matter."
Smith spoke in a low voice. "You know Nimitz? You have pull with him?"
"No, it's not like that. I met the Admiral socially a couple of times. He told me to look him up when I got to Hawaii."
"Outstanding. Please tell him that the Base Personnel Office is listing you as temporary CO of the Selfridge until it's ready for sea. I recommend you tell him about keeping the crew together and putting them on your salvage team. Hold on, Commander, while I make a phone call."

David could hear him talking to Chief Bennet and making plans for the Selfridge Chiefs and David to meet for lunch. Then, Smitty got off the phone.
"Chief Bennet got some burns from the bomb explosion; they have him over in the electrical shop supervising re-winding of all the damaged ships' motors. He thinks he can round up Boats and a couple of Engineering Chiefs and meet us at the base cafeteria at 1230."

At lunch, Pelham was impressed with the four chiefs. They had a good working relationship and seemed to accept him as their temporary CO. Chief Bennet excused himself from the salvage team, saying Bosun could lead the enlisted troops, and he'd be more useful to Selfridge in the Electrical Shop. David asked them how to assess the damage and clear the bodies. Bosun MacGregor replied that most spaces were watertight before the bombs hit. He believed twelve bodies were in the ship, ten E-2 to E-6, one Chief, and one Officer. Ironically, most were part of the forward and midship Damage Control Parties.

David asked how the ship sank if most spaces were in ZED. The Bosun answered that the two shells blasted over foot-wide holes through several decks, and the water rose quickly - probably drowning the Damage Control Party mustered below decks. He added that the ship was heavily loaded with fuel, ammo, and stores. The Engineering Chief weighed in that he got the engines and boilers shut down and all of the snipes out.

Chief Bennet said they needed to float the Selfridge before removing their shipmates. Otherwise, the equipment in the mostly dry compartments would be water-damaged for no good reason if opened in the sunken ship. Pelham asked Bosun MacGregor to pass the word to Lt. Comdr. Haldane and Lt. Lehman that he'd like to meet with them and the Bosun at Captain Wallins' Ford Island Quonset hut the next morning at 0900. The discussion on salvaging Selfridge continued for another hour until David excused himself to go to his meeting.

The meeting with Admiral Nimitz dragged on for over an hour as Captain Wallin covered the status of each ship under repair. When he got to the Selfridge, he begged the Admiral's pardon that as many as fifteen bodies were still onboard.

Wallin turned to David and said, "You're on."

Pelham smiled. "Admiral, good to see you again. I just came from a meeting with Selfridge CPOs. They're sure we need to clear twelve deceased crew members. However, they're adamant that most spaces are sealed and undamaged. I also met with First Class Yeoman "Smitty" Smith today. He's informed me that I'm the senior line officer on Fighting 357. He has entered me into his records as the acting CO until she's fit to get underway. Captain Wallin has asked me to lead the Selfridge salvage effort. I plan to have a team of men from the ship assist me. It will include the Engineering Officer, a senior Navy Lieutenant, two junior officers, and two Chiefs - one from Engineering and Bosun MacGregor. They will lead three teams of eight enlisted. Smitty will take care of TAD orders for the salvage team.

"We'll need assistance from someone knowledgeable about salvage ops, access to one rep from the Pacific Bridge Company, one from the Navy Yard, and a diver on call. At lunch, the men told me we needed to raise the ship first and then clear the bodies to minimize damage to dry spaces. During the next few days, we'll assess the damage. By Monday, we will have a plan for raising her. The week after that, we'll be ready to float her."

Captain Wallin held up his hand. "Hold on there, Commander. You can't raise her until you can go straight into a dry dock. All the main dry docks have larger ships. Floating Dry dock 2, normally your best bet is sunk with wrecked ships in it."

"Yes sir, point well taken. I'll look into that tomorrow. Captain Wallin, I'm new to this salvage world. What are the most important things for me to know?"

Wallin responded, "You must completely understand the damage to your ship. Your plan of attack must be highly detailed and safe. Raising a ship is a dangerous business. You must use a series of precise steps, or you risk sinking her again. Worse yet, she can get away from you and capsize. Plan your removal and cleaning in dry dock to avoid losing any of your men. Expect to empty everything, fuel, ammo, food, water-soaked personal items, and furniture. Expect to clean equipment and spaces with salt water, steam, freshwater, and chemicals as appropriate."

"Yes sir, thank you. Uh, Admiral, would you have a few minutes to discuss some operational matters from my West Coast trip?"

Nimitz smiled. "Nothing I'd like better, Commander, except I have people lined up waiting for me. Come back at 1630 with your running gear on so we can take a brisk walk up to the Aiea Naval Hospital construction site."

"Yes sir, see you at 1630."

David walked across the CinCPac Quarterdeck and down the front stairs. He smiled when he saw a brig crew cleaning up the surrounding area. They were raking up the metal debris, broken glass, pieces of the building, and machine-gun slugs. He thought, *Even with a war on, sailors will be sailors and get thrown into the brig.* He suddenly felt old. Not one of the cleanup crew was even close to his age. *They were just a bunch of kids. We will fight this war with most of our sailors not old enough to drink.*

Pelham walked to the Navy Exchange and bought some exercise clothes and tennis shoes. Next, he signed in at the BOQ. No room was available until he mentioned that he was acting CO of a destroyer. The desk clerk assigned him a small private room that was perfectly adequate. He arrived back at CinCPac HQ at 1620 and noticed that two Marines were doing stretching exercises on the front lawn. They looked out of place wearing khaki running shorts, white T-shirts, and black tennis shoes. They had a .45 pistol on one hip and a walkie-talkie on the other.

Nimitz showed up at precisely 1630 for his afternoon exercise. One Marine went out in front of the two officers, and the other took position behind them. They walked out of the base, across the main highway, and Halawa Road. Nimitz set a fast pace. He told David to wait until they were alone for anything classified. They did discuss moving the Nagy family as David huffed and puffed up the long hill. Nimitz said it took some arm-twisting to set up the move. Military dependents were being sent back to the States; he'd had to field challenging questions about importing a Japanese family. He said the Nagys were scheduled on a Navy supply ship and would arrive in about a month. David thanked him and said they would be worth their weight in gold.

The diamond formation of four men entered the Hospital construction area, returned the gate guards' salutes, and strolled over to a cleared area below the hospital with an odd assortment of tables and chairs. David and the Admiral sat at a table looking down on Pearl Harbor. One Marine took up a position fifty yards to their right, and the other was watching them from a grassy area between them and the construction workers.

"Heck of a view up here, don't you think?" Nimitz commented.

David looked down on Pearl Harbor and Ford Island. "Yes sir, it's gorgeous until you notice that Oklahoma is keel up, Arizona is in two large pieces, other ships are underwater, and I can just make out the main mast of my first command."

"Pelham, don't let the negatives get you down. Most of those damaged ships will return to the war in no time. Congratulations, Captain, I'm sure you'll do a great job salvaging Selfridge. This is one of my favorite places in the whole world. You have a perfect view of the Pearl Harbor complex, including my pride and joy, the Submarine Base. It seems a hundred years ago when Catherine and I were here to build the new sub-support facilities. Back then, in the early twenties, there was one small pier and lots of raw land that we had to clear of rocks and brambles. The Japanese made a big mistake leaving the current submarine base intact."

"Where is your wife now, sir?"

"She's still in D.C., probably will stay there until our youngest daughter finishes this school year and then move back to Berkeley. We have so many friends there. She's getting an apartment and will probably volunteer at a hospital. Enough small talk. What do you have for me?"

"Sir, what I have to say is not technically classified, but it's very volatile." David looked at the Marines, who were just out of earshot. "Do you think anyone can hear us?"

"Not if you keep your voice down. Fire away, Pelham."

"Yes sir, I had three productive days with the USC book author, Professor Michael Nagy. We also visited the farm of his parents. Michael's dad, Yoshi Nagai, was born in Japan. He made it clear that his family's allegiance is to America, took me aside, and asked how far I would go to win the war. Would I make a deal with the devil? I said I expected to put my life on the line and see a lot of combat. He said, would you put your honor on the line and risk your reputation?"

"I don't see where this is going," protested Nimitz.

"Sir, you will in a moment, and feel free to shut me down at any point. Mr. Nagai told me he has an uncle in San Francisco. He believed that his uncle was high up in the Japanese underworld, equivalent to our mobsters and Mafia. The Yakuza run gambling, loan sharking, prostitution, and other money-making criminal activities. Nagai said the Yakuza gangs in the U.S. are connected to gang members in Japan. Therefore, he thought they might be a source of critical inside information for us. He asked me if I wanted to try to contact his uncle, but warned me it would be dangerous. I'd be talking to ruthless criminals."

Nimitz looked interested. "What did you do?"

"I liked the whole Nagy or Nagai family and trusted them. So I decided it was worth a shot. Long story short, there was some cloak-and-dagger stuff to work through. I went to San Fran and met with his uncle, who called himself Kimicho."

"Why would they want to meet with you?" Nimitz challenged.

"My question exactly to Yoshi, sir. He said they were steeped in secrecy, and I was authentic, a Naval officer who wanted to win the war. He said they have a strange code of honor and might be willing to help us. I was impressed with Kimicho. No doubt, he was the boss. He responded to my questions with inside information about the Jap Navy. He knew the interplay of Yamamoto and Chuichi Nagumo commanding the Navy's First Airwing for the attack on Oahu. He gave me some strengths and weaknesses of the Army and Navy. He predicted that Yamamoto would return to take Midway and Hawaii by May or June. I asked why Nagumo did not return and finish us off here."

David gestured at the harbor. "Kimicho explained it as caution on Nagumo's part. He had lost the advantage of surprise and had significant aircraft losses and damage. Also, he was not Yamamoto's first choice for Operation AI. He quoted chapter and verse on what the attack didn't accomplish."

"Tell me the list," Nimitz directed.

"Missed the aircraft carriers."	Nimitz, "Check."
"Missed a million gallons of fuel."	"Check, a much higher number."
"Missed ship and sub-repair facilities."	"Check."
"Missed command centers."	"Check."
"Only destroyed two battleships."	"Three, but Utah was just a target ship."

"Did not cripple the Pacific Fleet, only battleships and some cruisers." "Check."
"Damage to cruisers was not complete." "Three damaged and repairable."
"Did not break the spirit of the American people." "Just the opposite."
"Didn't get American leaders to sue for peace." "Again, the opposite."
"He said Nagumo had six carriers and would attack Midway and Hawaii with four if Yamamoto had his way."
The four-star Admiral was surprised. "How could he possibly know that?"
"On the six that attacked Oahu, I believe he could have named them if I asked. On the four, he said Nagumo would never attack without numerical superiority. The Jap Army has overall control and wants the other carriers in the southwest Pacific."
"Makes sense," Nimitz responded.
"He said that Yamamoto is stuck with Nagumo for the Midway attack since the latter is now a national hero. The IJN is hidebound, and all would lose face if a more capable Fleet Commander is selected."
"What should we expect from Nagumo?"
"Admiral, apparently, he's a logistics and administrative expert. He brought six carriers nearly 4,000 miles to these shores without our foreknowledge. He is an old-school surface officer who knows torpedoes and guns. He's aging and has arthritis. Nagumo is not an aviation expert and probably still believes decisive battles between battleships will exist.
Nimitz snorted. "That ship has sailed. You called it right in your treatise."
"Thank you, sir. Nagumo was initially against the Yamamoto plan for Hawaii. Regarding his failure to send additional strikes on 7 December, several Admirals and Generals warned him to hit hard, hit fast, and then get out of Dodge."
"This Kimicho, he knows about conflicts on the enemy staff? He knows what plans are being prepared?" asked Nimitz.
"He gave me that impression. He said Yamamoto is fighting with the Army, who wants to move southwest, while Yamamoto thinks Midway is key to crushing our Pacific Fleet."
"What does Kimicho want in return for this intel?"
"I'm not sure, Admiral. I dangled cash payouts in front of him, and he didn't seem to bite. I think he'll help us and demand favors or a super big favor someday. There will be damage to us if we renege."
Nimitz replied. "We need intel on their Midway/Hawaii campaign at any price."
"Yes sir, that's how I feel. However, Mr. Nagai would tell us not to forget that this branch of the Yakuza, the Bakuto, are professional criminals. He'd say they can't be trusted and may just as soon make money off both sides in the war."
"What do you think?"
"Kimicho said they make their living off of gin and sin, not drugs, murder, or stealing. He said that the Army and the war are bad for business. The Army, secret police, and local police are shutting down Yakuza operations in Japan. They control the country down to the lowest level; they're forcing all non-disabled men into the military, industry, or agriculture."
Nimitz smiled. "So, recruiting is down for providers of gin and sin?"
"Yes sir. He says his organization wants us to win the war so things return to normal."
"Do you believe him?"

"About as far as I can throw him. It may be worth a shot if we need the intelligence and judge his track record over time. I brought it to you because it's such a hot potato. We would be consorting with known criminals, even paying them for intelligence."

Nimitz looked pensive. "How do we work with this Kimicho and his gang?"

"He said he'd only work with me. I told him I'd be up to my eyeballs in ship's work. Would he accept another person? He said no. I said, what about your grand-nephew, Michael Nagy, the professor? He said that might work. He told me to ask you for a short-wave radio and a set of one-time code pads."

"So that's what put the urgency on moving the Nagy family over here," Nimitz reasoned.

"Yes sir, and both of them can read, speak, and write in Japanese. Beyond that, Professor Nagy has a fine grasp of the Jap psyche and the military culture that can save us many missteps."

"Did you know he has a State Department Secret clearance?" Nimitz questioned.

"No sir, he never said a word about that."

"We started a background check on him, and it popped up. He's been working with State for a few years on our diplomacy with Japan."

"Well, sir, that should get him a fast start over here."

"What about his wife, Rachael, isn't it?" Nimitz asked.

"Yes sir, smart woman, a medical doctor in the USC clinic. She could be an interpreter and evaluate written material, but her heart will be on the medical side."

"There will be no end of medical work here for the next four or five years. Mrs. Nagy will be very welcome. How many people know the content of your meeting with Kimicho?"

"Admiral, just the two of us. Yoshi Nagai knows I tried to set up a meeting, but no more."

"I'm going to have you review all of this with my Intelligence Officer, Lieutenant Commander Ed Layton. He knows all the state secrets here, everything that we know or want to know. Then, after that, the three of us will decide whether to proceed with this and whether to include Professor Nagy."

"Yes sir. I have a request," David replied.

"What's that?

"Please keep me at the Secret level for what I need to know and nothing else."

"Your reason being?" Nimitz queried.

"I'm not completely sure, but I see myself in combat within the next six months. I don't want to know anything that will undermine your plans."

"I see. Lt. Comdr. Pelham, you need to know that one of my objectives is to transfer you to my staff to work on strategy and tactics. On the train, I skimmed your book. Then, while waiting in San Diego for my flight to Hawaii, I read it more thoroughly. I need a man with your background and analytical skills on my planning staff."

"Sir, I'm very flattered, but I'm not interested. I'm assigned to a fighting ship, and that's where I want to be, maybe even as CO."

"Now listen to me, sailor. I know what it's like to have a ship and a crew. There is nothing like it. But you're a Navy man, and you will do what your superior officers believe is the best use of your talents. I'll make you a deal."

"A deal, sir? What kind of a deal?"

"You go raise Selfridge and get her ready to fight. If all goes well, I'll back you as CO at sea for six months. We won't have the men and material we need for serious offensive ops until late in '42, anyway. After six months as a wartime destroyer CO, you'll need some rest. You then transfer to the staff. What do you say?"

"Admiral, how about nine months at sea and a free hand to modify Selfridge with some special equipment?"

"Eight months, and what equipment are we talking about?" Nimitz quizzed.

"Sir, it's not all worked out in my mind yet. We will do it safely, but there is no time to engineer, approve, and test it the traditional Navy way."

"I'll cover for you until the complaints from the Harbor Master and the Navy Yard are too loud. I could eat a starving horse. Let's get back to the base."

Nimitz signaled the Marines that they were about to leave. His car had arrived and was waiting. The sun was setting, and they could see showers of sparks in several locations below them in Pearl Harbor from the round-the-clock salvage efforts. The Admiral offered David a ride back to the base, which he declined. It was all downhill, and he wanted to savor his unexpected fortune.

Nimitz shook his hand and warned, "Remember, Pelham, your ship is not a toy for your ambitions. As Captain, you have a sacred trust to care for your crew, follow your orders, and make a difference for our country."

When David finally hit the rack in his BOQ room, he thought, *What a day! Hiking with an Admiral, acting Commanding Officer of a combat ship, and a good start with the Selfridge crew. Surely the dream wouldn't visit tonight, please God.*

Admiral Chester W. Nimitz, CinCPac and CinCPac Fleet 1941 – 1945

Source: Google images

Chapter Twenty

RAISE HER – West Lock, Pearl Harbor

January 1942

David Pelham did have a long and restful sleep. He awoke, took a shower, shaved, and put on a summer khaki working uniform before breakfasting at the BOQ's fast service café. Pelham arrived at Captain Wallin's makeshift office on Ford Island at 0845 to discuss the Selfridge liaison needs. Wallin said that the Selfridge crew could use a bombed-out tool shed on the island's north shore as their workplace.

When the other officers and Chief MacGregor arrived and introductions were complete, David asked Captain Wallin to advise them on their first objectives in raising Selfridge. Wallin said he was arranging a small diving barge and two divers for the following day to evaluate the damage and remove as many bodies as possible. He didn't foresee any near-term efforts to raise the Selfridge because of its low priority and lack of an available dry dock. David thanked him and said they would have several men available the next day to assist the divers.

The Captain's driver took David and his men over to a dilapidated twenty-foot by fifteen-foot tool shed on the north shore of Ford Island. They decided what action to take to make the space usable. MacGregor pointed out a dock about 100 yards away from where they could stage boats out to the ship. David said he'd see if he could get a boat or boats from Destroyer Row. They agreed to meet at the tool shed at 0700 the next morning. David asked Lt. Lehman to return with a couple of junior officers. He directed Bosun MacGregor to select eight to ten enlisted crew members to work on the shed and man at least one whaleboat.

The Selfridge crew took a boat off Ford Island, and Pelham walked to CinCPac Headquarters. At the reception desk, he presented his card and asked to see Lt. Comdr. Edwin Layton. The civilian secretary on duty picked up her phone and punched several buttons.

"Commander, I have a Lt. Comdr. David Pelham here to see you. Yes sir, I'll tell him. Commander Pelham, please have a seat; he'll be down in a few minutes."

Pelham started reading the latest edition of the Honolulu Star-Bulletin when Layton burst into the lobby, shook his hand, and said to follow him up to the "think tank." Layton took the steps two at a time to the third floor with David following. They entered his office, and he shut the door.

"Top of the day to you, Mr. Pelham. Our boss says you have a hot potato for me. What's up?"

David went over the San Francisco meeting with Kimicho in detail while the Intelligence Officer took notes. Pelham related how he set up the appointment, the Yakuza security procedures, and the discussion he could remember. At first, Layton seemed disinterested except when David mentioned Midway - then he wrote down every word. David finished by saying only Nimitz, Layton, and he were aware of this meeting. The question was whether to brief Michael Nagy about it and use him as a POC when David was not available, which would be most of the time.

Layton responded, "I agree with you that this is a most sensitive matter. Anything we can learn about the Jap plans for attacking Midway and possibly Hawaii should be our highest priority. The boss and I, and probably the Chief of Staff, will have to evaluate this Nagy fellow before agreeing to include him as a go-between. It looks very promising, very promising. If we have to pay money, there are more hoops to jump through, but we can do it. If this Kimicho is correct, we have five months to marshal our forces to beat the Japs at Midway. God help us if they beat us, and Hawaii becomes vulnerable to invasion and occupation. In the meantime, if I want to reach you, how do I do that?"

"You can catch me at the Main Base BOQ or through Captain Wallin out on Ford Island. He's put me in charge of salvaging the Selfridge. Some of my crew will be setting up an office on the northeast tip of Ford Island. Also, DesRonOne and the USS Phelps will probably know my whereabouts."

"Great, good luck with Selfridge. Nice to meet you, Pelham. I'm looking forward to working with you," said Layton while he ushered David out of his office.

Pelham had lunch at the base cafeteria and took a boat out to Phelps in the destroyer nest. The Phelps CO offered one of their whaleboats, and Commodore Early said his gig was available when he didn't need it. Pelham got a ride in the whaleboat back to Ford Island and made his daily report to Captain Wallin before going to the Main Base and meeting with Yeoman Smith.

The next morning at 0700, David found some of his men busy cleaning out the storage shed. Others were repairing holes in the roof with galvanized sheets. A handful of men were unloading furniture from a four-by-four, one-and-a-half-ton truck. Others were loading trash and the shed's contents into a second vehicle.

Bosun MacGregor approached David and saluted. "Good morning, Captain. My apologies; we had to get an early start to borrow the trucks. Lt. Comdr. Haldane and Lt. Lehman are out on the ship working with a two-man dive team to secure their support barge to the ship. Chief Romano from Engineering is over at the Navy Yard, trying to Shanghai a salvage expert to help us this morning. Once he gets back, we can signal the Phelps to send us a liberty launch to go out to the ship. I plan to leave a small crew here."

"Outstanding, Bosun, you're getting more done than I could ever imagine. How many men do we have?"

"Four officers, counting yourself, two chiefs, and ten enlisted. There will be more tomorrow when we know what we want to do."

David pointed at the shed. "What's the plan for this space?'

"We found some office furniture, a couple of couches, and some chairs in the Hickam dump. We plan to build an eight-by-twenty covered porch out here in front so all troops can get out of the rain. We scored a couple of blackboards in the Navy salvage dump to set up a small briefing and meeting area inside. Right now, we have only two walkie-talkies, one out at the ship, and the other is with Chief Romano at the Navy Yard."

"What is the plan today? Are they going to start diving this morning?"

"No sir, they're just setting up the gear. The Engineer wants you and the salvage rep from the Navy Yard onboard the barge to decide what we should do today."

"Good. Do you know if the divers are ready to clear any of the bodies?"

"I expect so, Captain. This isn't their first rodeo," MacGregor responded.

Pelham could see two men in dungarees at a nearby pier wiping down an Admiral's Gig.

"Bosun, tell me, do we have a flag officer in the vicinity?"

"No sir, that's your personal boat sent over by Captain Early, er, Commodore Early. We have it on loan until 1500."

"Wonderful. It sounds like Phelps is going all out to help us."
"Yes sir. Being our sister ship, I think they want us back on top of the water."
"Outstanding. Pick who you want at the ship for the initial meeting, and let's get underway."
"Aye, aye, sir, we'll join you shortly."

Pelham walked over to the dock and greeted the coxswain and his line handler. He boarded the gig, as did the Bosun and two men from his ship. They shoved off for the half-mile trip out to Selfridge. As they approached Selfridge, David noticed something different from his first visit. The top of the forward searchlight was awash.

"Bosun, what are the tides at Pearl?"
"About a three-foot maximum range, skipper. Low tide in about an hour."

They moored the gig against the divers' barge and disembarked onto the twenty-by-thirty-foot working area. David returned salutes to Selfridge's Engineering Officer, Lt. Comdr. Haldane, and Lt. Lehman. The two divers were preparing portable generators, air pumps, and masks. David made a point to shake each man's hand and learn their names.

"Engineer, what are your recommendations?" Pelham requested.
"Captain, we've discussed some of the problems with the divers. Maybe we can review those and all make a decision when Chief Romano gets here with the Yard rep."
"Sounds good. What's our situation?"
"Sir, we may have some luck as Lt. Lehman, Boats, and I think it was two bombs that sank us. One went through all the forward decks near frame forty-five. It put a hole in the bottom but didn't explode. The other bomb hit just starboard and aft of the forward stack around frame ninety-five. We think it went all the way through the ship, through the bottom, and exploded when it hit the mud. The seawater came in so fast that the lower compartments immediately flooded. A fourteen to sixteen-inch shell hole that's twelve feet below sea level lets in several thousand gallons per minute. As each compartment fills, the water is forced higher and higher, probably too fast for damage control to find the holes and plug them."
"How far along were you in setting ZED?"
"Pretty far along, most hatches, doors, and fittings were probably closed. The bridge called away modified ZED to get the guns fully manned, and then ZED about five or ten minutes later."
"What happened to the CO and XO?"
"We know they were on the Starboard Fly Bridge when the first bomb hit. We think that the first plane kept coming and strafed them before pulling out. Their bodies are still missing. We have a corpsman who checks new bodies each day."
"How did she go down?"
"I was in Engineering, but by all accounts, after the bow flooded, she just slowly settled. Someone called the bridge and told them the mess decks were flooding. The OOD headed for the beach and ordered abandon ship when the weather deck was awash. The Porter class is known to be top-heavy and prone to capsizing. It was a good call as most of the men were able to find egress routes before the freeboard went completely under."
"So, you think she may be in good shape when we raise her?" David asked, hopefully.
"Too soon to say, but there is a good chance that the engineering plant is operable. We closed down all the boilers, engines, and pumps before we went up the escape trunk," Haldane replied.

There was some static on the walkie-talkie; they could hear Chief Romano say he was approaching the Naval Station East Loch pier. David jumped in the gig and directed the coxswain to take him over and pick up the Chief. Chief Romano introduced him to Danny Price, a salvage type from the Navy Yard, who was in working clothes. On the way back to the ship, David asked Price how to raise the Selfridge.

Price spoke above the noise of the gig's engine. "First, Commander, we have to assess the damage. If the Chief is right, it may be very straightforward. The bombs passed through all the decks, and one exploded under the ship. The water rose so fast that there were no persistent fires. The Chief thinks most of the ship was in ZED, another good factor. We find out from the divers today what they think the bomb damage is and the amount of flooding. Then they give us their best guess on how to raise her."

"Then, what after that?"

"We develop a salvage plan that has safe, careful steps to re-float your ship, and identifies the resources needed."

"What are the risks?"

"First, she rises and sinks again on us in a more difficult angle. Or worse, she rises too fast, develops a list, and the loose water capsizes her."

"Loose water? Is that the same as free water?" asked David.

"Yes, compartments full of seawater are mostly neutral in the salvage effort, but the ones that are partially full can be a real problem. As the ship takes a small list, the loose water shifts and accelerates the list. Too much free water on the wrong side of the ship, and over she goes."

Pelham asked, "The flooded spaces could have slow leaks along electrical runs, ventilation pipes, leaking valves, and such?"

"Exactly. As Selfridge rises, it's a tricky business to dewater the ship without inducing a list to port or starboard."

"So what is the general approach?"

"We meet with the divers, and they map out a plan to assess the damage. That will probably take an hour. They get all their gear working, and we support them for about a two to three-hour dive. First, they will assess the hull where the bombs hit. Then they will go inside to determine the extent of flooding and the flooding boundaries. With no reported fire or oil slick, that data will go a long way toward defining our salvage plan."

"So we can expect minimal damage?" David asked.

"Hard to say. The bottom could be damaged, and the shafts and screws may be bent or smashed. An explosion underneath the hull can damage plates, bulkheads, equipment, and mountings throughout the ship. Only close inspection and testing will tell the story," replied Price.

They reached the diving barge and conferred with the divers. Based on what they had heard from the Selfridge crew, the experienced divers thought the job was pretty cut and dried. They asked if anyone had the ship's plans. The Engineer replied that they were on their sunken ship. David sent Lt. Lehman in the gig over to the Phelps to see if they could borrow their plans and arrange a liberty boat to bring the rest of the working party from Ford Island. The divers briefed Bosun MacGregor and Chief Romano on how they wanted their lines handled. They cranked up two generators, two air pumps, and went over the side. The line handlers paid out the safety line and air hoses as they descended to the ship's main deck and then to the harbor's bottom.

David and Haldane used underwater viewing buckets to watch the divers. The divers worked their way up on the main deck and started rapping on it with crowbars to determine the amount of flooding. They opened a hatch near where the first bomb penetrated, and David could see the two lights disappear into the hold as they took more and more lines.

Lt. Lehman returned with the gig, the Phelps plans, and a bonus. The Phelps' mess deck crew had made them twenty box lunches. They also sent water, iced tea, and a tub of Cokes. David had everyone, except the line handlers, chow down and get some drinks. The day was warm with a high overcast, a light breeze, and high humidity. Chief MacGregor found some canvas on the gig and rigged a sail shade for the officers and Mr. Price on the barge. After an hour, a liberty boat from Phelps arrived with the rest of Selfridge's working party. David set aside lunches for the divers and air hose handlers. He had the small boat crews and the newly arrived enlisted group divide the rest amongst themselves.

After another hour in the water, the divers returned to the barge. Shivering, they both put on long-sleeve dungaree shirts and lit cigarettes. David took over the box lunches and Cokes for the divers and handlers. "Well, how does she look?"

The First-Class Diver thanked David for the drinks and said, "Let us catch our breath and finish lunch. Have everyone ready who needs to know the results. I don't want to have to repeat myself. We have another dive in a couple of hours."

David had the Engineering Officer, Lt. Lehman, the Bosun, Price, and Chief Romano join them under the shade sail. He questioned MacGregor, "Bosun, what do you think? Could we use this equipment and do some of this on our own?"

"Skipper, I've used it during a tour on a sub tender. Things are pretty tightly controlled here in the harbor. You could get yourself in a lot of trouble."

Danny Price admonished, "I second that. You'll have a couple of Admirals and Captains up in arms if you try an end run."

The divers came over to them; the First Class motioned the Second Class to speak first. "I went down the port side and didn't see any damage along the hull. The ship has settled several inches into the mud. It may be tough to break her free. We saw a clean hole through the main deck forward and amidships when we got back on deck. We expected to see all of the decks below flooded."

The First Class Diver chimed in, "We sounded some hatches and compartments under the midship main deck. The ones around the shell holes are flooded. The ones amidships over to the port side are clear or partially flooded. We went down a deck and found the space flooded with another hole. Down another deck, probably the mess deck, was flooded with another bomb hole and some bodies."

"Excuse me, how many bodies?" David asked

"We could see five with our lights. There could be more."

David thought, *five down and seven to go.* "How much damage so far?"

"Not much sir. The lowest deck was a small compartment in the C-Strake area that must have filled very rapidly and had another shell hole. There was an access hatch to what seemed to be a tank below that, maybe a water tank, because we did not see any oil. There's definitely a hole in the bottom, but we could only see dirt and gunk where it is. It's hard to judge its size."

"How wide was the dirt and silt?" asked Price.

"About ten feet by six or seven feet. We didn't see any water movement. The hole in the bottom is plugged with muck and mud."

"And the condition of the starboard hull amidships?" Price questioned.

"Sorry sir. Yes, there is some vertical damage up the side of the hull, and it might be slightly buckled. There is no damage to the starboard hull forward. Up forward, we found a couple more bodies down two decks and were unable to see the hole in the bottom of the ship."

"Anything else to add?" David said.

"No sir, except normal procedure would be to bag and clear the bodies as soon as possible. We will defer to you and your crew, as it has been very traumatic for some sailors to see their shipmates carried out."

David replied, "We'll clear the bodies now. Anything less is disrespectful to their sacrifice. Lt. Lehman and Chief Romano take most of the enlisted back in the gig and work on the shed. Bosun, we'll place the bodies in the other Phelps' boat and deliver them to the Naval Hospital for processing. We can talk about our next salvage steps during the trip and then take Mr. Price back to the Navy Yard. Any questions?"

They all shook their heads. The Commodore's gig departed, and the divers prepared to return to the Selfridge's mess deck.

"Mr. Price, how long do you think it will take to have a detailed plan?" queried Pelham.

"About a week or two," Price answered.

"And how long to raise her?"

"A destroyer is going to have low priority. We're working on a lot of capital ships. Give it two or three months."

David exploded. "Two or three months! That's unsatisfactory. It won't work! We have operational obligations that require us to be ready to fight by the end of next month."

"Commander, I can't see it going any other way. Everything is in short supply. You need a couple of larger barges with high-capacity air pumps, more electrical power, and water suction. You need large pontoons attached to each side of the main deck to lift the ship or flood to fight her listing. You need tugs to push her into a dry dock. And here's the real bottleneck; you need a dry dock ready and waiting."

"What's the dry dock situation?" David asked impatiently.

"All of the large docks are taken by battleships or cruisers. On a good day, they would put Selfridge in floating Dry dock YFD-2. You now have USS Shaw in there with most of her bow blown off, and tug Sotoyomo completely sunk. They had to flood the dock, and she's still down and damaged. We plan to give Shaw a false bow and send her to the West Coast in February."

David answered sharply, "That's too long. We need to refloat Selfridge next week and start repairs the week after. If there is no dry dock available, is there a shipbuilding way where we can pull her up on the shore or room on a seaplane ramp?"

"I can't think of anything at Pearl that would work for you. Don't bite my head off, the Salvage Officer, the CO of the Navy Yard, and the Naval District Commandant set the priorities based on the guidelines from CinCPac Fleet," said Price.

"You're right, Danny. You've been a great help to us. Thank you. Let's get these bodies cleared, take you back to the Yard, and I'll go to YFD-2 personally to assess the damage."

It surprised all of them how quickly the divers could bag the bodies and put them one by one in a body-sized scoop net. The Selfridge men raised their shipmates carefully and placed the gray body bags in the Phelps' liberty launch. They all embarked on the small boat and headed for the Navy Yard main dock, with that done. Two ambulances met them and took possession of the bodies. The divers and Danny Price disembarked at the Navy Yard HQ pier.

Price gave Pelham and MacGregor directions to the dry dock. They found Floating Dry dock 2 resting on the bottom and listing to one side. Bombs and strafing had severely damaged it. There was no pier or functional dock in sight, so David had the coxswain beach the liberty boat. He and the Bosun thanked the boat crew, slid off the bow into two feet of water, and waded ashore. David recommended that Haldane take the boat back to the Selfridge working pier on Ford Island and release its crew to return to Phelps.

Pelham and Boats MacGregor made their way through a tangle of old, rusting ship parts up to the quay wall and over it to the lower side of the slanted dry dock. They went up several ladders to the top deck and could hear metal grinding and the snap, snap, snap of welding. The aft two-thirds of the destroyer USS Shaw was floating in the dry dock well. To their left was the sunken twisted heap of what had been the forward third of USS Shaw and a submerged tugboat.

David walked over to a man on his knees, grinding the sides of a shrapnel hole, and yelled, "Hey! Excuse me, Hey! Can you hear me?"

The man stopped grinding and asked for a hand up.

David pulled him upright. "Excuse me, you are?"

The tired-looking, grimy, unshaven man answered, "Lt. Ron Kordowski, sir." He took his glove off and started to extend his right hand, but looked down at the grease, oil, and blood on it and thought otherwise with a shake of his tired head.

David could hear some other grinding, pounding, and arc-welding inside the dry dock walls. "Lt. Kordowski, what's the status of YFD-2?"

"Yes sir, she's full of holes, not seaworthy. My crew and I started working on her a few days ago. They had us working on the big ships last month."

"Lieutenant, I plan to bring a Porter Class destroyer here in two weeks. What has to happen before I can do that?"

"Commander, we need to get her shipshape again, plug the holes, fix the pumps, and repair the electrical system. Then we can right her and refloat her by pumping the water out of the well. After that, we can put pontoons on the tug and the Shaw wreckage to float them out. Once we clear the wreckage, we can float the larger part of Shaw out and put her on a pier. At our current rate, we may get there in a month or two."

"Lieutenant, we're from the USS Selfridge crew. I'm the acting CO and salvage coordinator. Boats here is in charge of the deck crews. You may know that the Japs sunk DD-357 out in the East Loch. We cleared some bodies today and determined she should be straightforward to raise. We plan to do that next week or the week after. We need this dry dock to patch her up and repair the flooded compartments. I plan to have ten men here to help you tomorrow. Double that the next day, and double that the day after. Please work with Bosun MacGregor to figure out what they can do to repair your ship. Boats, also determine what additional tools they'll need."

Kordowski declared, "Commander, first and foremost, we need base electrical power back on. These portable generators help, but they never will get the dock back in full operation."

"Good, that's what I need to know. I'm going to go see the Yard CO and change some priorities. Boats, get the cook's tour, and decide what we need and who we should put here in the next few days. I should be back in an hour."

MacGregor smiled and saluted. "Aye, aye, sir."

Pelham climbed down from the non-floating, listing dock. He flagged a work truck down and got dropped off at the Navy Yard Commander's office. He introduced himself to a civilian secretary outside the CO's office and said he needed to see Admiral Furlong.

"Sir, the Admiral is very busy. Do you have an appointment?"

"No, but he'll see me. I'm the Commanding Officer of the USS Selfridge. Admiral Nimitz sent me to talk to the Admiral."

"Standby sir." She picked up a phone and punched a button. "Admiral, I have a Navy O-4 here who wants to talk to you. He says he's CO of the USS Selfridge, and Admiral Nimitz sent him. Yes sir, yes sir." She returned the phone to its cradle. "The Admiral is wrapping up with the Yard Engineering Officer. Then he can give you a few minutes."

"Thank you, thank you so much. Please tell me, who is the Yard Engineer?" David requested.

"That is Captain Manson, a very busy man," she answered.

"Thank you."

He took a seat in the outer office and thought about what he'd say to the Admiral. In a few minutes, the door opened as the Admiral ushered out a Navy Captain with some parting remarks. David got up and approached them.

David held out his hand, "Captain Manson, David Pelham, acting CO of the USS Selfridge DD-357. You need to hear what I'm going to say to the Admiral. Kill two birds with one stone, so to speak. It will only take a couple of minutes. Admiral Furlong, I know you're both super busy. I believe I can help."

Once inside the Admiral's office and seated, David continued, "I'm sure you are aware that the Selfridge sank in the East Lock. I'm on Captain Wallin's salvage team as the officer in charge of raising her. By a twist of fate, I'm currently the senior line officer on the Fighting 357, which makes me the acting CO. You may not know that we evaluated the damage with divers this morning, and she's in pretty good shape. There is a small bottom hole to plug at frame forty-five and a larger one starboard amidships. Plug a handful of others on the above decks, pump out some water, and she floats. Most of the spaces are in condition ZED. We removed five bodies of the twelve expected to be on the ship."

"Commander, that's all new information and good news. What's this about Nimitz sending you to see me?" Furlong questioned.

"Sir, I guess I used a little poetic license there. Admiral Nimitz expects me to have the Selfridge raised and ready to fight by the end of February. This morning, we had expert help and advice from Mr. Price from your Yard. When I said we would be raising Selfridge next week, he advised me that we needed a waiting dry dock berth before that would make sense. I've just come from Floating Dry dock 2, where I met Lt. Kordowski on his hands and knees repairing his ship."

Captain Manson interjected, "Whoa, hold on there, skipper. I don't know where you're going with this, but let me tell you our priorities were decided on 8 December."

"Yes Captain, that's exactly why I'm here. Priorities change when the tactical situation changes. I need YFD-2 up and running next week or the week after. I need the tug wreckage and the Shaw pieces out of there. I need shore power restored to the dry dock tonight. I'm putting ten men at Lt. Kordowski's disposal tomorrow, twenty the next day, and forty the next. We will plug the holes, fix the pumps, and whatever else she needs."

Admiral Furlong was irritated. "Lieutenant Commander Pelham, although I admire your enthusiasm, I don't see how you expect us to upset our plans just because you want us to. This meeting is bordering on impertinence. You are late to the party, and telling us your demands is not okay with this officer."

"Admiral, request permission to speak freely."

Furlong paused with a glance at Manson. "Granted."

"Sir, it's true that I'm late to the party. Because of that, I have to work twice as hard. The CO and XO of Selfridge were missing overboard and presumed dead. No one has been defending her. No one has been her advocate. The crew knew the damage was limited. Did you two gentlemen know that until today?

"The crew also knew that their shipmates were entombed in her. No one cleared any of the bodies until today. Captain Wallin admitted to Admiral Nimitz yesterday that Selfridge had fallen through the cracks, and no divers had worked on her. I'm not trying to be impertinent. I'm trying to work as an advocate for Selfridge and her crew, using the latest facts to get you onboard. I need the dry dock next week or the week after. I board YFD-2 and discover a Lieutenant on his knees, manning a grinder to try and bring his ship back into the war. I find a Mahan Class destroyer in pieces with more priority than a Porter Class ship with minimal damage.

"Selfridge is a lead destroyer that can be in that dry dock and repaired within four to five weeks. Now, the USS Shaw is an impressive ship. I'd be proud to be her skipper. That's not my current fate. I'm in charge of a bigger ship, with a larger crew and twice the firepower of the Shaw. Admiral Nimitz and I have a critical TOP SECRET mission for the Selfridge at mid-year. I don't have time to be extra polite. I'm a bull in the China shop. I'll get this done even if I have to run her up on the beach and dig a tunnel with my bare hands to repair the two holes in her bottom. Thank you for listening to me."

Furlong stared at Pelham. "Commander, you've made some good points, even if I didn't appreciate the tone of the delivery. We agree it's time to get YFD-2 back in operation. Captain Manson and I will discuss your proposal and get back to you. How do we reach you?"

"I'm sleeping in the Main Base BOQ. During the day, I'm working with my men out of a shed on the north end of Ford Island and coordinate daily with Captain Wallin."

Furlong wrote on a pad. "Got it, dismissed."

David returned to the dry dock and collected Bosun MacGregor. They bought some sandwiches and Cokes from a roach coach and wolfed them down. The Bosun flagged down a passing van and talked the driver into taking them back to a shuttle dock for Ford Island. Again on Ford Island, David found Captain Wallin alone in his makeshift office and briefed him on the day's events and the nature of his discussion with Admiral Furlong.

Pelham laid out his planned approach, 1. Repair YFD-2. 2. Remove ships from YFD-2. 3. Float Selfridge. 4. Place Selfridge in YFD-2 on blocks. 5. Dewater spaces and remove remaining bodies. 6. Unload ammo and nonfunctional flammable materials. 7. Unload foodstuffs, trash, and damaged furniture. 8. Perform required salvage functions and clean all spaces. 9. Repair all combat damage. 10. Restore the ship's vital services.

Wallin looked at him and seemed to agree. "There are a lot of actions to complete, getting from step 1 to step 10. Whatever you do, do it safely, and keep me posted."

Bosun MacGregor briefed David on the YFD-2 needs and repair strategy as they walked from Wallin's office to the Selfridge shed. Boats estimated that the dry dock could be floating and operable in a week or two with the right parts and enough manpower. As they approached the shed area, David noticed the remarkable change in his ship's work area. The old parts and trash were gone; there was even a crushed rock pathway down to the pier. David introduced himself to Yeoman First Class Brown and quickly dictated the day's progress for the ship's log. Brown took notes and promised to type up David's remarks before returning to the barracks.

During the next week, Pelham's activities fell into a predictable schedule. Chief Romano picked him up in the morning, and they went to Floating Dry Dock 2, or the Selfridge shed. By Monday, he had forty men from Selfridge and the Destroyer Row duty sections working on the dry dock. David received word from Admiral Furlong that he'd have priority in the dry dock if it was operational in less than two weeks.

David met with Yeoman Smith and determined that as many as 130 Selfridge officers and men could be available by the end of two weeks. He decided to have a Selfridge all-hands muster, in working uniforms, at the Main Base enlisted barracks at 0700 the next morning.

Smitty went to the barracks to post a notice and put out the word. He agreed to call the chiefs and officers who were on temporary assignments. David took a boat to the Phelps and met with Commodore Early and the CO/XO of the Phelps for a working lunch. He laid out his plan to them and asked for their help in raising Selfridge. At 0700 the following day, he was standing in front of over 100 Selfridge crewmen in his service dress whites. An officer or a Chief was in front of each of the six groups. Lt. Lehman acted as his XO. Lehman called the men to attention and ordered the divisions to report. David heard:

"First Division, all present or accounted for!"

"Second Division, all present or accounted for!"

"O" Division, two men missing!"

"E" Division, three men missing!"

"C" Division, two men missing!"

"S" Division, all present and accounted for!"

Lt. Lehman returned the salutes of each division leader, did an about-face, saluted David, and sounded off, "Captain, all assigned men present and accounted for except seven men missing!"

David returned his salute. "Very well! Selfridge, parade rest!"

They moved their left foot out and clasped their hands behind their backs in perfect sync. David knew that part of command and leadership is to look and act the part. His uniform was spotless, from his buffed white buck shoes to his officer's hat with the gold-braided chin strap above the bill and the eagle flying above the crossed anchors emblem. He looked intently at each division leader and then at each man in the division. This took some time. He wanted each man to see him as their Commanding Officer, to feel no more uncertainty. As his gaze

reached the "S" Division, he had to smile. Yeoman First Class Smith was the first man in the first row. "Petty Officer Smith, welcome to the Selfridge!" David boomed.
"Thank you, Skipper. Proud to be here!"
"Selfridge crew, at ease!" David ordered
Pelham, with Lt. Lehman in tow, took a few steps forward. "Men of the Fighting 357, I'm Lieutenant Commander David A. Pelham, USNR, your acting Commanding Officer. I have been an East Coast sailor for ten years. So, imagine my surprise when I got orders to Selfridge in Hawaii. Let me tell you, the weather is much better here." He paused for the laughter to die down. "I have had the pleasure of meeting and working with many of you during the past week. What you have accomplished in the past few days is remarkable and impressive. Captain Wallin, the Pacific Fleet Salvage Officer, and Vice Admiral Furlong, CO of the Navy Yard, have noted your outstanding performance. In fact, we've done so well; they're challenging us to do more." He swept his gaze back and forth at the crew, letting them build some anticipation of his following words. "Many of you have been working on YFD-2. You see me there every morning because we need to get her back in operation a week from this Saturday. Then we clear the tug and the Shaw out of dry dock because we will be floating your USS Selfridge into YFD-2 the next week!"
A burst of yells broke out, and men pounded each other; Pelham waited for them to settle down. "With two weeks of preparation, a team of divers has told me they can float our ship in two days of concentrated effort. Navy divers say the damage is light. As you know, most of the spaces were in ZED. There are two manageable holes in the bottom and smaller ones above decks. We did not have a fire, and our fuel tanks are intact - all good news. Many of you will be there when we bring her up.
"Our Chiefs will decide who will be on the ship and who should be at the Floating Dock. After all, we all know who really runs the Navy. It's the Chiefs!"
That got a laugh from all the Chiefs and the men who worked tirelessly under their leadership.
Pelham continued, "The divers and a Selfridge working party will start next week, placing equipment, pontoons, and bigger barges. We will be out there at the crack of dawn each day to assist in any way we can. USS Phelps and the destroyer squadron Commodore are committed to helping us raise their sister ship. Now listen, men, I'm told this is a very tricky business. We could get her up and lose her. The divers and the Navy Yard reps will tell us what to do. Once we have her safely floating, the divers will collect our shipmates' bodies. Two tugs will be waiting to push Selfridge gently into dry dock over the forms and cradles that many of you will put in place. Not one of us will step on the ship until our shipmate's bodies have been recovered."
David moved back to his original position with Lt. Lehman facing him and ordered: "Selfridge, attention! Selfridge, parade rest!"
He looked over at Smitty. "Petty Officer Smith, what will we do next week?"
"Raise her, sir!"
"Damn right! All you Selfridge men, you can drop the 'sir' at the end." David pumped his fist in the air. "What are we going to do!?"
"Raise her."
David yelled, "I can't hear you, people. What are we going to do?"

The formation shouted, "Raise her!"
David yelled louder, "They can't hear you on Ford Island. What are we going to do!?"
The Selfridge sailors boomed, "Raise her! Raise her! Raise her!"

David waited. They were getting their home back. It was a big deal.

"Selfridge, attention! Executive Officer take charge and dismiss the divisions."
Lt. Lehman saluted. "Aye, aye, sir!" He did an about-face. "Divisions, carry out the plan of the day. Dismissed!"

The division leaders saluted the acting Executive Officer, and Lehman returned their salute.

Ship locations in Pearl Harbor on 7 December 1941, note USS Selfridge in upper middle row of destroyers
Source: Pearl Harbor Visitor's Center

Chapter Twenty-one

DRY DOCK YFD-2 – Pearl Harbor Naval Base

January 1942

In the next few days, Chief Romano and David followed their usual routine. They checked on the progress with Floating Dry dock YFD-2, provided update reports to Captain Wallin, had discussions with Yeoman Smith in his office about ship's staffing, and took the Commodore's gig out to Selfridge to view salvage preparations. Each day there was progress; to David, each step in their plan seemed to take forever. YFD-2 would be ready in a week thanks to Captain Manson reinstating electrical power and the daily efforts of destroyer men augmenting the dry dock crew. Captain Wallin said two large barges were scheduled for raising Selfridge. After a week of concentrated effort, it seemed possible to complete the first phases of the salvage in another week.

The following Friday morning, Chief Romano and David started at the dry dock. David climbed up to the top deck to get the daily progress report from Lt. Kordowski. The next stop was Ford Island to brief Capt. Wallin and drop in at the Selfridge shed. Then they made their daily trip out to Selfridge using the Commodore's gig. They had lunch at the cafeteria on the Main Base and decided to drop in on Chief Bennet at the Yard Electrical Shop. He showed them around and explained how the shop refurbished motors, instruments, and sensitive gauges.

David and Romano returned to the dry dock late in the afternoon. Lt. Kordowski was on the pier, taking a smoke break. David went over to see if there was anything new to put in the ship's log. Kordowski said progress was outstanding, and they would test the pumps in two days.

He pointed at an Army coupe parked near the dry dock. "By the way, Captain, someone in that car came looking for you. I think they're asleep now."

Intrigued, David went to the car and found an Army nurse fast asleep in the driver's seat. He leaned into the open passenger window.

"Ah-hem, hello there. Were you looking for me?"

The nurse awoke. "Yes, if you're Commander David Pelham, I am."

"Guilty as charged, but only a Lieutenant Commander, like a Major in your Army."

"Don't patronize me, sailor; I know all the ranks. Believe me, they all look the same when we work them over naked on an operating table."

David noticed that she was above average attractive, and the rank insignia on her khaki shirt collar, "What can I do for you, Captain?"

She got stiffly out of the car. "Whew; I've been sitting there too long. To answer your question, a mutual friend told me you would make a good fiancé."

"What? Is this some kind of joke?"

He wasn't sure he wanted it to be a joke now that he had a good look at her. The nurse O-3 wore dark brown shoes, a trim khaki skirt, and a short-sleeved khaki shirt. She had dark brown hair with a reddish tint, a curvy figure, and daring blue-green eyes.

The nurse put out her hand. "Captain Beatrice White, Army Nurse Corps. Could we walk a bit? My muscles are stiff from that car seat."

David smiled. "Sure. If I'm to be your fiancé, we should get to know one another. Let's walk over to the snack bar. I'll buy you a Coke."

Betty put on her fore and aft hat. "That's just ducky. What are you doing with the dry dock?"

"If all goes well, I'll have my ship in there by the end of next week. So who's this mutual friend you think we have?"

"None other than Captain Rats Mueller, USMC," she said as they walked.

"Rats? You know Harry?"

"Yup, your old buddy, Rats. He flew me out of Wake Island before it fell. We were the last plane to leave with a ton of wounded and older men. Last night I ran into him in a bar in Honolulu. I was getting in my cups after a twenty-four-hour surgery training shift. I told him about my problem, and he recommended you. He said you would be ideal."

David walked along beside her. "Me? I'd be ideal, in what way?"

Betty stopped and looked him directly in the eye. "He said you were reliable as an old shoe; you were someone I could trust to keep my secret."

"Your secret being what?"

"That's simple. I don't really want a real fiancé, just a pretend one. One that's conveniently gone most of the time, like a Naval officer. And Rats said you have a yen for combat, you probably will be killed. Even better for me, a dead fiancé."

"I don't get it. Why do you want a pretend or dead fiancé?" David asked as he handed her a Coke.

Betty showed a flash of anger in her eyes. "Because if I'm betrothed to a respectable ranking officer, the men I work with will have to keep their hands to themselves and swallow their nasty comments. It's getting so I don't want to go to the hospital anymore. The married men are the worst, with their come-on stories about how they can make me happy. All of them, officers mostly, think it's funny to pat you on the ass or ha, ha, accidentally touch your breasts. I'm just sick of it! I'm thinking of carrying a scalpel in my pocket and slashing some fingers and hands."

"Can't you complain?" David asked.

"Oh, some women have tried. Unfortunately, it just seems to get them more grief."

Their slow walk had brought them back to her car.

"Miss White, do you have time to go for a boat ride to see my new command, the USS Selfridge?"

Betty smiled. "I'd love a boat ride. Let's go!"

David told Chief Romano to take the ship's car and make it available to Yeoman Brown. He climbed into her car, and she drove to the central pier, where they boarded the Commodore's gig. David directed the two-man gig crew to shove off and take them out to Selfridge. He pointed out the destroyer squadron and sister ship Phelps on the way across the bay. David told her that the gig belonged to a real Navy Captain or Colonel in her world. He explained that while Captain Early commanded a Destroyer Squadron, his rank was Commodore, a one-star Admiral.

Once they reached the ship, the sailors tied the boat to the divers' barge. David helped her down from the gig and pointed at the masts. "Well, there she is, a fine fighting ship, a heavy destroyer, just a little waterlogged at the moment."

She looked at him first with alarm and then with pity. "This is your ship? You're the Captain?"

"Acting Captain, until we get her raised and repaired. The CO and XO went missing in the attack."

Betty took his hand, "Damn this war! You should see the boys we worked on after the bombing." Tears streamed down her cheeks. "They were so brave. They were so hurt and mangled, so terribly burned. They were just kids. So many died; it was horrible."

David spoke softly, "How did you end up on Wake Island?"

Betty had a faraway look. "We had caught up on most of the emergency surgery here at Tripler. Then we heard about the fighting on Wake. I worked with an orthopedic surgeon with great hands who could diagnose, operate, and fix a patient faster than most people get their vitals. He said we had to go to Wake and help the men there. He asked me to assemble an emergency surgical kit and bring it to the Hickam tower because a plane was flying to Midway and Wake. He wanted me to come to Midway and pilfer some of their medical supplies while the plane refueled."

"That's when you met Captain Mueller, right?"

"I met them all, Rats, Whit, Tex, the Chief, and Fitz."

"Damn good crew!" David exclaimed. "But I don't think I know this, Fitz."

"They're the best. Corporal Fitzpatrick is one of the Plane Captains, I think. He and Chief Watkins treated me like royalty. I'll never forget that C-46 crew. They all want to see you again and buy you some drinks. Whatever you did, you're like Superman to them."

"Enough about me; why did you go to Wake?"

"We got to Midway, and I talked their dispensary out of some medical supplies. Harry and my surgeon got an update that there were many wounded to evacuate, only three Corpsmen on the island, and two doctors. My surgeon asked me to nurse the wounded on the return trip. How could I say no? We landed at Wake just after dark. Rats said we had to take off before dawn, or the plane would be a sitting duck in daylight. My surgeon and I tried to find sickbay, as you call it. The hospital took some direct bomb hits that wounded several medical personnel. Sickbay was no more than a couple of huts dug into the coral with wounded men on the floor and on the tables.

"There was no decent sanitation. We wouldn't have operated on a dog in that environment, but there was no choice. We cleaned one of the tables, laid out our gear on a clean towel on another one, and started to operate. Three or four men held Coleman lanterns up for us. We had to cut a couple of legs and arms off right away. We put those men on tables that someone had cleaned with bleach. The military doc asked me to help some patients on another small island, so I missed the flight back. Everyone, especially Rats, took a lot of grief for that. No women were allowed on Wake to prevent them from being captured by the Japs. My doctor friend and I found plenty to do in the next couple of days, especially when there was a lull in the bombing. Captain Mueller and his C-46 "Rat Pack" crew returned for more wounded and older men on the construction crew. They gave me no choice. I had to leave the island.

"A few days later, we went back again after sunset. Carrier planes had caused a lot of damage and strafed many of our Marines. My surgeon and I worked in the command bunker, and two Corpsmen started helping us. They would put a man up on a table, I'd prep him, and the surgeon went from one man to the next, sometimes changing gloves, sometimes bathing them in bleach. Then, if I had an extra minute or two, I checked the boys on the floor with cuts and sewed them up as best I could. Then it got really nasty, nasty, nasty."

"How so?" Pelham asked.
Betty looked stricken and cried. "Rats found us in the bunker and said we were flying in two hours. We had to start loading the plane - my surgeon said I'd have to triage the men. We had been at it for four or five hours. He was still cutting men open and sewing wounds in their guts, legs, chests, and everywhere. The other two doctors worked in the medical revetments over at the airfield. I asked the Corpsmen to help me choose. They just shook their heads and sobbed. Finally, they said that they would load the plane and that I was the officer; I'd have to select the passengers. They knew all the men like brothers. They couldn't pick one over another.

"Harry was turning pale, watching the surgeon hold up a man's intestines looking for a bleeder. I took him outside and asked him how long it took to Midway. He got a grip on his nausea and said two hours to load and seven to eight hours of flight time. I asked him how many men we could take. He said forty. I told him we would load the plane with men who looked like they could survive ten hours. We tagged each man with their treatment if they already had morphine, and what they needed. Two were already dead. I picked out fifteen to be loaded and five to be left behind."

Her voice broke; she was shaking. David guided her over to an overturned wooden box and had her sit down.
Betty continued. "I put some quick stitches in some more boys, about five, and bandaged them. I marked them with an L for loading. My surgeon marked the man with the stomach bleed with an S for Stay. He had not found the bleed. Two or three others would die if he spent too much time with one man. I told the medics to bring some of the worst cases from the other doctors to the plane."
"Tell me about that last flight," David directed.
Betty sighed. "Before I left, they brought in a man with a terrible leg wound above the knee. His thigh bone was smashed to pieces. I knew it had to come off for him to live. The Corpsman and I got him on the table and prepped for my surgeon, who said he had to take a break or he was going to pass out. He sat down on the floor, and someone got him water. He said I'd have to take the leg off above the knee. He said I'd seen him do it a hundred times and should know how by now. So I grabbed one of the Corpsmen and said I'd do the cutting if he'd do the sawing. He agreed.
"The man, the patient, was listening to us wide-eyed. I took one of his hands in mine and told him to save his life; we had to take the leg - that I could give him something for the pain. I told him it would get infected if he stayed on the Island, and he'd die if we didn't amputate. He was older and must have been a civilian with long hair and mutton chops. He was crying. I was crying. He said he had a wife and three children. He'd do anything to get back to them. I told him I'd put him on the plane and get him to Hawaii."
She looked at David with a long, mournful look. "We took his leg. It was nip and tuck in the plane. I had to give him three transfusions. Now, when I'm having a bad day, I go and visit him at Tripler. When he sees me coming, he yells, 'Everybody, here comes my lady doctor! Hey Doc, what did you do with my leg? It seems to be missing!' I hold his hand, and we both have a good cry together."
David questioned, "Did you fill the plane?"

"As the battle really started to flare up, we still had ten seats, no, make that ten places to sit on the floor. There were 1,000 contractor civilians on that island, a thousand! Many of them were fighting the Japs somehow, helping to fire guns or acting as riflemen. Rats decided we would take the ten oldest ones. The word went out, and we had ten or twelve men in their 50s and 60s trying to get on the plane. We took them all. Rats told me no way they would survive in Jap POW camps. We got out of there. I worked a couple of hours and slept for half an hour on the flight, and started again. Rats and Whit were both injured by shrapnel as we took off. I had to sew them both up.

"The crew and the older men gave blood that I used on the wounded. Two boys died on the way to Midway. At Midway, the crew, I, and the older men lay in the shade under the plane. We put wadded-up clothes or towels under our heads and slept on the concrete. Medics from Midway and other volunteers came on the plane to move the wounded men. We slept like that for three hours. Rats was called in front of the brass, got his backside chewed, and threatened with a court-martial. He and Whit had a doctor evaluate their injuries and ended up in the hospital overnight. Then we flew to Oahu."

"Let's take a break. I want to talk to you about my ship."

David went over to the gig and picked up a couple of towels. They knelt on the barge's edge and used a viewing bucket to look down on his ship. He pointed out the significant features and answered Betty's questions. They discussed the necessary steps to raise the destroyer and put her back in operation. Betty listened intently. The Navy was new to her. She knew a lot about the Army because she was an Army brat. David looked at her face as she looked down at his ship and saw the faint streaks of red hair, the blue-green eyes, and her cute upturned nose.

"Ginger Snap," he said.

"What? I beg your pardon."

"You look like the girl painted on the Rat Pack's plane, Ginger Snap."

"Well, I'm not! That's Marybell; she's my younger sister. We all call her Ginger."

David was astonished. "Harry Mueller has a girlfriend, your sister, and you end up on his plane. What a small world!"

"Yes, it is. And now, Rats is the one who tells me about you, my potential make-believe intended. He forgot to say you're very easy on the eyes. You might as well know they're married, Harry and Ginger. He could be your brother-in-law someday."

David laughed. "Captain Rats Mueller, a relative. What a strange, small world."

"Yes, you could do worse."

"Ha, Mueller has far less chance of surviving the war than I. You saw him in action. He'll go where the bullets are flying and trust his luck to pull him out in one piece. How's your sister going to handle that?"

"They've decided to take what they can get when they can. Ginger has moved to San Diego. They secretly got married in Ohio in case she got pregnant. She asked him; it seems to run in the family. How about you? Girlfriend? Married? Family? I can't be engaged to a married man."

David paused and looked sad. "No, I had a family, a beautiful wife, and a nearly six-year-old daughter. They died in a car accident over a year ago. It was all my fault."

"Oh my God, you poor man. You were driving?"

“No, I was in England on orders. They tried to drive in the ice and snow on Thanksgiving Eve, skidded off the road into a freezing Connecticut river, and drowned.” He sobbed. “If I’d been there, they would still be alive.”

“David, we can’t control life and death. We can only soldier on with the amount of time we have. My surgeon and Harrold Mueller have taught me that. So give 110 percent to live fully and thank the Lord for small favors.”

David got control of himself. “Your surgeon, was he more than a colleague?”

“Yes, he was my friend, and my lover. He was married to a woman in Philadelphia who refused to move to Hawaii. I didn’t care. For a year, he was mine, all mine.”

“Where is he now?”

“When I boarded that last flight out of Wake, the Japs were overrunning the island. We were up to our knees in injured men. I helped him for several hours until he said it was time for me to go. There was no hug, no goodbye kiss, just a quick glance up while he was suturing a boy’s eye socket, a tired nod, and a wink. The Japs have him now. It was Ginger Snap’s fifth flight to Wake. All of us were ground down to nubbins except Harry. He seemed to have a reservoir of energy that was limitless. The engines were turning. I was the last person on the plane. I went up the ladder, the Chief pulled it in, and slammed the door shut. The plane was full. You could not move around the cabin without stepping on somebody. Bradley Whitlock, the Co-pilot, told me later that they could see Jap troops running across the runway. He said that Rats barely made it into the air. Brad said the engines were at full power, and only something called the ground effect kept us up for a mile or more. That’s when the two of them took some shrapnel. On the way to Midway, I tended to the wounded and did a headcount.”

“How many were there?” David asked.

“Counting myself and the crew, we had sixty-one souls onboard. They got off Wake with the shirts on their backs, nothing more. Two of them died on me. When we got to Midway, I got down on my knees and thanked God for being alive and free.”

“I think we should be heading back. Can you give me a lift down to Hotel Street?” he requested.

Betty gave him a coy look. “Why, Commander, what could you possibly want on Hotel Street?”

“There’s a bar there, Smith’s Union Bar, where the Navy divers hang out. I need some diver advice on my re-float plans.”

“They won’t be there this early. How about we go down on the beach at Fort Derussy, have some Mai Tai’s by the water, followed by dinner at the O Club.”

They drove into Honolulu and were sipping Mai Tai’s with chunks of pineapple and little umbrellas on the Army Officer’s Club veranda when David said he’d like to swim. Betty went over to the beach office and wangled swimsuits for them from the lost and found bag. When Betty came out of the ladies' shower room, she wore a two-piece swimsuit; David gave her a wolf whistle. He complimented her and said if he was going to be anyone’s pretend fiancé, it would be her. Betty bet him that their mock engagement would be a done deal by the end of dinner – her mind was made up. She also told him he was too white and skinny; she was buying dinner to fatten him. They dove off the beach platform into the warm Pacific water at Wakaki and swam out to the float. She let David sun himself on both sides, just short of a sunburn and said it was time for dinner. They had a leisurely meal and discussed their childhood, school, and college experiences.

After dinner, Betty drove him over to Hotel Street and parked in front of a rundown jewelry store. He found the divers' bar, and some of them were letting off steam with a few beers in the back bar. They had heard about the Selfridge dive earlier in the week and told David to plan on floating her the next weekend - that it would be a piece of cake. They knew First Class Diver Ricky Wright and the C-46 flight from San Diego. They said floating David's ship would be payback for saving Tiny's neck. David thanked them and bought a round of beers. He walked back to the car and found Betty waving to him from inside the jewelry store. She showed him a ring with a tiny diamond chip for fifty dollars. He told her she should not get engaged to any guy that cheap. He found a half-carat ring with some minor flaws for $450. David had the shopkeeper set it aside. Betty drove him back to the Navy Base BOQ and kissed him goodnight.

The following week was a blur of activity that saw the dry dock righted and emptied of derelict ships. On Tuesday morning, two liberty boats were tied up to the temporary diving barge. David and two of his officers, three Selfridge Chiefs, and ten enlisted men were in the launches or on the small barge they had moved forward of the Selfridge's submerged bow. Then, about 0900, two tugs pulling large barges arrived, stowed their tow lines, and nudged the barges into place on each side of the ship.

The barge sailors dropped anchors fore and aft to hold them in place. Two crewmen and a couple of divers manned each barge. They spent the next three or four hours preparing the pontoons and other equipment for use during the coming weekend. Finally, one of the tugs came close to the small diving barge, and a Chief told David it would take a few days to get all the equipment in place and set up. He said the tugs would return midday on Saturday and tow Selfridge to the dry dock if all went well.

David and Bosun MacGregor drove downtown to Smith's Union Bar on Thursday night. Petty Officer First Class Ricky Wright was there, and several other divers. They had diagrams of their plans for Saturday. They said the pontoons were in place on each side of the ship, and other equipment was ready. Wright said the first task Saturday morning would be to plug the shell holes in the decks above the ship's bottom. They turned over cardboard templates of the holes and asked the destroyer men to bring the necessary plugs.

Wright recommended raising the hull, bow first, as the next task. He said it was tricky to get her free of the bottom muck and dangerous if the whole ship broke loose. They expected to plug the two holes in the bottom by 1000 and cautioned David that progress would seem very slow at the start. They estimated that he could put the ship under tow by 1300. David bought a round of beers and left the Bosun to learn any additional gouge.

Early Saturday morning at the Selfridge site, there were more Selfridge crewmen, additional divers, and several barge equipment operators. As the generators and pumps were tested, David could see Petty Officer Wright talking to the divers on the port side barge. David and Bosun MacGregor took one of the boats over to that barge and unloaded the plug materials that the divers had requested. David shook Wright's hand and asked how things were going above the din of the generators and air pumps.

"So far, so good, Commander. The Salvage Chief will have us start pumping air into the forward pontoons in an hour. Two crews are ready to patch the shell holes and put air hoses into the watered spaces. We can start lifting the forward pontoons soon after that. If they don't break the bow loose, we slowly pump air into the other two midship pontoons. Next, we start pumping air into the lower flooded spaces after we plug the deck holes."

David saw two diver teams from each barge enter the water with long metal pipes connected to air hoses and said, "What are they going to do?"

"They're gonna start blasting away the muck under the bow. As Selfridge lifts, the muckers will work their way aft until we can see the forward hole in the bottom. We have to move the ship from negative buoyancy to neutral and then just slightly positive. We know we're in business if we can get the bow to lift an inch or two. Then it's steady as you go. We raise her an inch at a time. Then we plug the forward hole, inspect the larger midships hole, and patch it with mattresses and a large, heavy canvas bandage. We raise her until the flooded bridge is half out of the water. We open it up and bag a couple of bodies."

David thought, *That'll be two down and five to go*. "Then what?"

"We raise her until the bridge is dry and start pumping the spaces below it. That'll be the most critical point in refloating her."

By 0800, the divers signaled to start pumping air into the two forty-foot pontoons to lift the Selfridge's bow. About an hour later, Wright said, "The forward pontoons are nearly full of air and no movement yet."

"How can you tell?" David asked.

"Look at the ribbons we tied on both masts at sea level. They're still in the water. So now we start pumping the second set of pontoons. If most of the spaces are dry and in ZED, that should get us some action," Wright replied.

Ten minutes later, the Salvage Chief stopped the second set of pontoon air pumps and ordered the muck blasters to take a break. David thought he saw a slight movement of the forward mast, or was it just some waves passing by?

Wright pointed at the ribbons. "I think we're getting some movement, just a tad, and that's what we want. The boys blasting the muck helped her start upward. Shit, that's a thankless job. You can't see the ship for all the loose crap in the water. You have to stand with one hand on the ship and the other shoving the hose under it."

They saw the Salvage Chief signal to the other barge and hold one finger up while pointing to the amidships pontoons. He made a pumping motion with his hand to start one minute of air into each float. David looked over at a crewman on the other barge, who looked at his watch and then shut his air pump off. The Bosun grabbed his arm and pointed at the mainmast. The ribbons now hung slightly above the water. She was moving! She was rising!

Wright said, "Now we've got to treat her like a virgin on her wedding night. Nice and slow does it, no sudden moves. Captain Pelham, you can help by going back on the forward barge and signaling any list. If there is no list, give me a thumbs-up. If it's more than two degrees, hold your hand up, showing us the direction of the list."

David got in a boat and returned to the smaller divers' barge forward of Selfridge's submerged bow. He couldn't see any list on the mast, so he gave the diver a thumbs-up. There came a cheer from the men on the barges and the small boats. Everyone could see the mast slowly rising about an inch or two each minute. When it was about ten feet above the waterline, the movement stopped. David could see some new divers going into the water with mattresses and a large canvas sling with multiple lines tied to each end. He got a viewing bucket and kneeled to see four divers pushing the mattresses down the ship's starboard side and two more swimming under the vessel while pulling the makeshift canvas bandage. In a few minutes, the lines from each side of the sling were cinched down on the ship's railing. Two more divers worked a smaller patch and canvas into place until they plugged the shell hole in the bow.

Suddenly, he could see nothing. The de-muckers were back at work, and visibility was zero. He stood back up and saw the Chief signaling for another minute of pontoon air pumping. David checked the main mast list; it was about a degree to starboard. It was rising steadily, inch by inch. The upward momentum stopped when the superstructure was half out of the water. Divers swam over and slowly opened the port door on the Bridge. In a couple of minutes, they floated two black bags out of the opening and into a waiting mesh basket lowered by the port barge's crane.

A couple more hand signals from the Chief, and the ship rose and stopped with the water level below the bridge's entrance. David saw divers hauling air hoses into the lower spaces below the pilothouse. He thought, *I get it. They will go space-by-space, emptying seawater as that compartment reaches the surface.* The ship, his ship, again slowly started to rise. Wright waved at him and pointed to the mast. It had about two degrees of starboard list, so David held his hand up vertically and leaned it to that side. Wright made sure his chief got the word. The Salvage Chief signaled to the other barge, and David assumed they pumped more air into the starboard pontoons. After a few minutes, Wright pointed to Pelham again, who gave him a thumbs-up; the mast was straight up and down.

Slowly, ever so slowly, the ship's forward guns and lower superstructure started to rise above the water. Selfridge was dirty and had a green slime of marine growth everywhere. To David, she was the most beautiful sight in the world. The forward guns were fully exposed as the bow broke above the water. David knew there were 380 feet of fighting ship to go. He had to hold himself in to stop jumping up and down joyfully. The main deck was now awash from bow to stern.

Wright was checking with him every few minutes. The list was a degree to port. He gave Wright a thumbs-up and then leaned his hand slightly to the left. Wright waved, smiled, and returned the thumbs-up. There were some more hand signals, and the mast straightened. The main deck was now above the waterline. The salvage barges slacked their anchor lines and tied up to the Selfridge. Multiple air and water pump hoses were pulled over to her deck, and divers swarmed over the ship, tapping hatches and doors with hammers and placing the suction pumps.

David saw two divers signal for the mesh basket and position two more body bags. *Four down and three to go,* he thought. Now he could see the pumps on both barges sending multiple streams of water into the bay. The ship was slowly rising, and the barge crew was detaching the bow pontoons and lashing them back against the barges. Finally, the crane swung back over Selfridge and removed the final three bodies. David could see that Wright was motioning for him to come aboard. He and Engineer Haldane got in a small boat with Lt. Lehman, went to the destroyer's starboard side, and climbed up a Jacob's Ladder. Once they were all on deck, Wright asked his Salvage Chief to provide David, the Engineer, the Bosun, and Lt. Lehman his damage report.

"You gentlemen live in a shoe. Most of the spaces are dry - more than ninety percent. Nearly all of the seawater is toward the bow and just forward of amidships, where the two shells penetrated. There is a good chance that your power plants are intact. We have the bow and midship sections half pumped out. We've searched all the spaces that aren't flooded and didn't find more bodies. The Yardbirds are putting some portable generators and pumps onboard so you can keep pumping right into the dry dock. After the midship pontoons are detached, you'll float with a slight starboard list. That should only improve as we pump the amidships mostly dry, which should take another hour or two. We check the adjoining spaces for flooding and bodies as the water goes down. So far, so good."

Pelham looked at his watch. "Chief, you and the divers have done an outstanding job today. You made it look easy, but I now realize the week's preparation got the ship floating safely. I expect to thank each of you at the Selfridge beer bust tomorrow afternoon. How do we work with the tugs due in twenty minutes?"

"Sir, one will put a tow line on your bow and the other on your stern. When you're ready, the bow tug will tow you over to the dry dock. He'll have you release the tow line and position himself on the wind-leeward side of the bow. The aft tug will back down to slow you to a stop in front of the dry dock while the other one lines up your bow for entry. The dry dock crew will receive your bowlines and move you forward from one capstan to the next. I'm sending two divers to make sure you settle correctly on the blocks."

David shook the Chief's and Wright's hands. "Petty Officer Wright, Tiny, we can't thank you and your divers enough."

Bosun MacGregor took David aside and handed him binoculars. "Skipper, you might want to take a look at that car on the point."

Pelham saw a brown Army car with a woman standing next to it. He focused the glasses and looked at Betty as she looked at him through binoculars. She threw him a kiss, raised her hand in an A-OK sign, and held up a sign that read, "FORT D. 1900 SHARP!" He laughed, gave her a thumbs-up, and turned back to see his men smiling in approval. More Selfridge crewmen were coming aboard from the liberty boats borrowed from Phelps.

"Mr. Lehman, divide the men up and have them set fire and rising water watches throughout the ship. Look lively. We have destroyer row, and the squadron Commodore is watching. We aim to be the top ship in DesRonOne!"

The tugs arrived on time and secured bow and stern lines to Selfridge. Lines were cast off from the three barges, and the lead tug started towing while the following one maintained slack in the thick aft rope. They slowly glided past the somber wrecks of the Oklahoma and Arizona. David could see the dry dock floating without a list from a quarter-mile out. As they got closer, he could see a hundred or more men on the dry dock and the quay walls on each side. The tugs did their job without incident; Selfridge was secured to the dry dock with DD-357 sailors attending each of her lines. Lt. Kordowski waved happily at David as he had air pumped into YFD-2's flotation tanks. As the dry dock rose, Selfridge slowly lowered until firmly on the blocks. Crewmen on the top deck of the dry dock used a hoist to swing a gangplank over to the ship, and the Bosun's deck crew secured it. The first person over the plank was Mr. Price and two assistants. David and Lt. Lehman climbed down the slippery bridge ladders to meet them. Price had the Bosun close the gangway.

"Captain. You're in dry dock now and subject to a whole new set of rules," Price directed.

"Tell us what we need to do," David said.
"You'll need a skeleton crew onboard for security and fire watch. No one, and I mean no one, can move back aboard this ship. There are no services - no electricity, steam, water, sewer, ventilation, nothing. All the heads must be secured. All hands must use portable heads on the pier, with no exceptions. It's up to you how you let your men onboard to claim personal items. Just know that they will want to roam the ship and could get into trouble or be injured. We're going to check for toxic air today. Be aware that spaces with batteries, food, or other chemicals may contain deadly gases. We'll mark those spaces with a large red X. The ship should be in condition XRAY tonight. You and your crew must not activate equipment or start working on the ship. Yard personnel will start at 0800 to install ventilation and clear all the spaces. Tomorrow, have ten men ready to be trained on properly dewatering and cleaning flooded spaces."

After Price left and several Yard workers came onboard, David met with the XO to discuss the ship's watch bill and monitor crewmen who wanted to board Selfridge. They decided that after the yard ventilation was in place, watch standers would escort any man who wanted to retrieve personal items. Lt. Lehman set up an officer and chief watch bill rotating every four hours. David's work day ended as he roamed the ship with Boats MacGregor, getting a firsthand impression of the necessary work to get Selfridge back at sea. This included discussions with his Supply Officer, Lt. Baxter, and Chief Gonzalez about the All Hands Party to ensure they would have plenty of food, soft drinks, and beer. While crossing the gangplank to the dry dock, David saw a large structure pushed by the tugs against the end of the dry dock. *Will wonders ever cease? I know what that is. It's a floating barracks that'll probably hold half the crew.* David walked back onto the ship and told Lt. Lehman and the Bosun to figure out who should be in the Barracks.

Then Captain Pelham departed high and dry, Selfridge using the gangplank, went down the dry dock ladders, and out of YFD-2. Yeoman Brown was waiting with a car next to the quay wall to take him to the BOQ. David looked back at the floating structure and his ship's masts above her. He thought, *Maybe I should pinch myself. This salvage is going too smoothly.*

YFD-2 Dry dock (raised) many years before the Pearl Harbor attack
Source: Google Image

Chapter Twenty-two

WILD IDEAS – CinCPac Fleet Headquarters

February 1942

Pelham slid into the backseat of Selfridge's Navy car, and Yeoman Brown drove him to the BOQ. David shaved, changed into civvies, and packed an overnight bag. Next, they went to the jewelry store on Hotel Street in downtown Honolulu where David paid for the pretend engagement ring. After being dropped off at Fort Derussy at 1800, he took a long swim in the ocean and sunned himself on the Officer's Club veranda when a shadow fell over his face.

Betty spoke with mock indignation. "I might have known, here I'm all dressed to the nines, and my escort is trolling for dames at the beach."

David smiled. "I'm trying to get less white, as directed by my favorite medical advisor. As far as trolling, it looks like it worked."

"Well, Captain Pelham, this dame is hooked, at least for the night. We have a dinner reservation at 1900, window seats for the sunset, no less," Betty replied, kissing him on the cheek.

"Fancy a quick swim, Betty?" he quipped.

She pouted. "Take a closer look, sailor. This swank, sophisticated, and alluring feminine appearance took two hours of hairdo, nail polish, makeup, and new clothes. Now get dressed so you can buy this delectable gal a drink."

David jumped up and saluted. "Yes Ma'am! Right away, Ma'am!"

They ordered drinks in the dining room and looked each other over with approval. Then, David started talking about raising the ship and activities that she couldn't see from shore. He shifted to all the action items he and the crew had to get done before Selfridge could sail again. Abruptly he realized he had been talking nonstop for ten or fifteen minutes.

"Sorry Betty, I got carried away. How was your day?"

"Fair to middlin'. I spent two hours this morning watching some Navy men raise a ship. When I saw you go on board, I was pridefully busting my buttons. I was laughing and crying at the same time 'cause how important it was to you and your crew. Did you get the bodies out?"

David snapped his fingers. "Yes, all of them. You've reminded me that I need to see about their burial ceremony. I can take it up with Smitty tomorrow."

Betty sipped on her drink. "Who is Smitty?"

"That would be Yeoman Petty Officer First Class Smith. He's on the shore staff and pushes the best men toward the Selfridge. In exchange, we'll take him to sea so he can fight the Japs."

"Is that even legal? Can't you get into trouble?" she warned.

"Is the war legal? Sometimes you have to bend the rules to get the job done. I'm starving. Let's order and eat."

Their table viewed the waves rolling in on Waikiki Beach and pink clouds framing Diamond Head as the sun set. Betty ordered wine; they enjoyed the leisurely dinner and conversation. After finishing dessert, David led her outside on the shady side of the Officer's Club veranda, where they could see and hear the waves breaking on the beach in the fading twilight.

He took both of her hands in his. "I have something for you. Will you be my pretend fiancé? Will you write to me every day and send me cookies? Can I count on presents for Christmas and my birthday?"

Betty smiled with tears in her eyes. "Are you sure, my Captain? We might get busted for fraternization."

"I don't think so. We're only one rank apart, members of different services, and from different commands. There is no coercion to prosecute except the alluring way you look tonight. There is only one thing that could make you look more attractive."

Betty squeezed his hands. "And what is that?"

David took her left hand and slipped the ring on her finger. She held her hand up so he could see the sparkling diamond. She put her arms around him and gave him a soft kiss. Then she looked at him, looked at the ring, and gave him a much longer kiss.

Betty looked into David's eyes, "I have something for you, my make-believe suitor. Here, take this."

"What is it?"

"It's an extra key for my room at the Royal Hawaiian. A room equipped with bourbon, ice, and spring water to celebrate the raising of USS Selfridge."

David held her closer. "You, Miss White, are trouble."

"So are you, Captain Pelham. Now you kiss me!"

Early the following day, she dropped him at the Navy Base BOQ. He changed into service dress khakis and decided to walk to the dry dock. He reviewed his most enjoyable evening with Miss Betty White along the way and remembered to arrange a meeting with Lt. Comdr. Layton. When Pelham reached the upper deck of the dry dock, he saw a makeshift quarterdeck at the gangway leading to his ship. A Bosun's Mate blew his whistle and announced with a handheld megaphone: "Selfridge arriving."

David strode down the gangplank as the quarterdeck watch rang the ship's bell four times. He noticed men swarming around Selfridge's upper decks, blasting off marine growth with high-pressure hoses, and others working on the ship's bottom on the dry dock floor. Large ventilation tubes, water, steam, and electric lines ran through openings in the main weather deck in several places to service the ship's lower levels. He could hear pumps running and see water going over the side.

Pelham was surprised that Daniel Price and his assistant were waiting to meet with him Sunday morning. They declared all the spaces safe for inspection and told him that only fifteen percent of the ship compartments had any damage. They reminded him that all of the ship's storage tanks would probably have to be emptied and cleaned due to saltwater contamination through the air vents. Price again made David acknowledge that no one could live aboard the ship or use the heads. Once Pelham agreed to that, they were gone.

Captain Pelham double-checked that Supply was ready for the 1300 beer bust and went up three ladders to inspect the Bridge. It was a mess. There were broken windows, bullet holes, and green growth on the deck, making it hard to walk without falling. The seawater destroyed the Bridge logbooks and navigation manuals. His At-Sea Cabin was filthy, and it needed a new bunk mattress. He found his acting XO, and they discussed how the ship's watch standers could attend the all-hands party. They decided to relieve the afternoon watch an hour early to enjoy the last two hours of the beer bust.

David roamed the nearly empty ship for a couple more hours and saw Yeoman Brown waiting for him down on the pier. Brown usually had the ship's paperwork for his review while they were driving in the car. As David slid into the backseat, he saw that Brown had some new correspondence for him to read and sign.

"Would the Captain like a lift to the ship's party?"

"Yeoman Brown, that sounds great. Where is it?"

"Over near the Marine barracks at Quick Field."

"OK, let's see if any other men need a lift."

They picked up four shipmates and drove over to the field. Selfridge's Supply Division had three large Bar B Qs cooking dogs and burgers. There were tubs of beer and soft drinks on ice. Several picnic tables had condiments, potato salad, and chips. Someone had rigged two sail cloths from a couple of trees to seven-foot posts for shade. There were about sixty men there, and a baseball game was in progress. David grabbed a beer and started meeting his crew. He asked them their names, rate or specialty, hometown, and where they were when the ship sank. After thirty minutes, two vans and a large truck pulled up with twenty divers. Pelham saw Tiny Wright and went over to talk to him.

"Petty Officer Wright and company, welcome. Please get yourselves a beer or Coke and dig into the food. Say, it looks like you could easily field a baseball team. How about a five-inning game against the Selfridge?"

"Captain, you're on as soon as we get a couple of beers and some food," responded Wright.

David found Chief Romano sitting at one of the picnic tables, eating a burger. "Chief, let our people know that they are playing ball for another half hour. Then, put together a Selfridge ball team to take on the divers with me on it for one inning at first base."

Several vehicles were arriving with more of the Selfridge crew. David went over to meet as many of the men as he could. He met electricians, machinists, pharmacists, gunners, seamen, quartermasters, firemen, coxswains, torpedomen, radiomen, cooks, carpenters, stewards, and more. The variety of talents, training, and depth of a destroyer crew to support each other amazed their new skipper.

Pelham found Smitty in the crowd and told him to report to the ship at noon the next day with the Selfridge billet structure, the crew list, and his plan for fully manning the ship in one month. A Navy sedan arrived with five officers in uniform that David didn't recognize. They were performing temporary duties on the base or with ships under repair. He brought them up to speed on the DD-357's condition. David told them to report onboard in one or two weeks and see Smitty if they needed orders.

A car arrived with Lt. Comdr. Haldane, Lt. Lehman, and Chief MacGregor. David went over to them and received an update on the ship, the planned watch bill, and the criteria for assigning men to the floating barracks. Lt. Lehman complained that Mr. Price and the yard workers seemed to think the ship was theirs for a month or more. Lehman said there was minimal damage, and they should be able to power up the engineering plant and move onto the ship in a few days. Haldane weighed in that the engineering spaces were intact, with only a few inches of water on the deck.

David responded, "Tomorrow, we learn what the Yard thinks we must do. We'll push them on the repairs while quietly cleaning up the ship and taking over the intact compartments. Let them clear the dangerous gases, test the void spaces, and empty the fuel tanks. Then, we'll leave it up to the Bosun to blast off the sea slime. We need to clean the bottom, but don't paint it yet. Plan on moving some of the men back on the ship next Sunday if we can get new bedding."

MacGregor smiled and sounded off, "Aye, aye, sir!"

"Mr. Lehman, I'll return to the ship by 1130 in the morning. Then, you and I will meet with Smitty to discuss ship billets and manning at 1200. Who else should attend?"

Lehman said, "Probably the Admin Officer, the Admin Chief, and Yeoman Brown."

"Good, make it so. The rest of you are dismissed. Go have a well-earned beer or two. Mr. Lehman, stay with me a moment."

After the others left, Pelham said, "Smitty recommended you as acting XO. Are you all right with that? How long have you been on the ship?"

"Sir, I've been on Selfridge for two years and two months. She's the finest of ships, and I'd be proud to be your XO."

"Excellent. If we do a good job of repairing the ship and putting her back into fighting shape, we both may get battlefield promotions from Nimitz," David stated.

Lehman smiled. "What, sir? Do you have some compromising pictures of the Admiral?"

"Nothing that exotic. I met the Admiral on the Super Chief train out to the West Coast; he tasked me with researching Japan. I plan to meet with his Intel Officer tomorrow morning."

"You told Mr. Price we're going on a secret mission. Can you elaborate on that?"

"No XO, I took some poetic license there. It's something forming in my brain that's not yet clear."

Pelham played an inning of the Navy divers versus Selfridge baseball game and did his best to meet the rest of the crew. After the watch changed on the ship, he greeted the ship's off-going watch standers as they arrived at the field. The ship's party closed down at 1700. David had Brown drive him to the BOQ and directed him to return with the car at 0900. Selfridge's Captain was exhausted from talking nonstop for four hours. He took a nap before eating a late supper.

Yeoman Brown was waiting for his Captain at 0845 the following day and drove him to CinCPac HQ. David checked in with the receptionist and picked up a magazine.

In a few minutes, Lt. Comdr. Layton met him in the lobby. "Good morning, Captain Pelham. We meet again." Once they were in his office, he said, "By the way, I read your book, and so did my partner in crime, Comdr. Joe Rochefort. I see why the boss wants you on the staff. New subject, he wants us to test this fellow in San Francisco, give him some tough questions to answer."

"I have something that would test him. It's a half-baked operation that could use his help to succeed." David replied.

"OK, shoot."

"It is all about chess, whether you and the Admiral are willing to trade a pawn for a queen."

"Who is the pawn? Who is the queen?"

"USS Selfridge and a Japanese aircraft carrier, respectively," David responded.

"Congratulations on raising Selfridge. A fine ship. I had destroyer duty myself. It's very rewarding. You want the old man to sacrifice a destroyer to sink or damage a Jap carrier; is that the caper?"

"Yes, in a way, that's the gist of it. We would be taking a high risk of losing the destroyer and gambling on a low probability of sinking the enemy carrier. I have much of it worked out in my mind, but nothing on paper yet. It would require some very clever cover and deception at the outset. There's one piece missing that could improve the destroyer's survival odds while increasing the carrier's destruction odds," asserted Pelham.

"What's that?"

"I need a Japanese signalman with current fleet experience who will fight against Japan."

Layton whistled. "Whew! You might as well ask for the British crown jewels or an autographed picture from Hirohito."

"Yes, it's asking for the impossible. On the other hand, if the Yakuza deliver, it would be beyond impressive."

"Agreed; tell me more about your plan," Layton directed.

"It's preliminary, so please don't shoot the messenger."

"Pelham, the boss, is taking heavy rolls from D.C., namely Admiral King and FDR. They want us to strike, to attack Japanese forces now, whether it makes sense or not. So, any idea you give me is worth a look, no matter how crazy. Look what Yamamoto did to us with his outrageous idea to raid Pearl Harbor."

David replied, "The end game is to put Selfridge next to an enemy carrier near Wake Island in the dead of night."

"How near to the flattop?"

"A quarter mile or less."

"Jesus, if you could do that, even your five-inchers could do some real damage!" Layton exclaimed.

"Commander, if we get that close, we'll sink her. We will put so many damn holes in her in five minutes that she'll list violently to starboard and rollover. I plan to add a six or seven-inch gun that will gut her. In five minutes. It will put ten large shells into her set to explode amidships."

"Where are you going to get a bigger gun?"

"From the Army or the Marines. We'll take off the damaged aft torpedo launcher and put in a bigger gun to be manned by the same service."

"The Yard BUSHIP boys are not going to go for that!" The Intel Officer retorted.

"Too right. That and several other changes will drive them crazy. I'll need the Admiral's protection; maybe a designation as a research ship would help," David replied.

"Tell me more about the attack approach. How will you get past the destroyer screen? There will probably be a cruiser on each side of the carrier. How do you avoid getting creamed by them? The Japs are damn good shots at night."

"If they buy our deception scheme, it will be carriers, not one carrier. So we convince them we're only sending one carrier, so they will send two."

Layton was impatient. "OK, OK, how will you get into your firing position? Even if there is no moon, you'll have a bow wave and a phosphorous wake they can see miles away."

"I'm working on that. Some scientists at USC and the San Diego Scripps Oceanography lab are working on dampening those little plankton buggers."

"What about the escorts? How do you handle them?" Layton demanded.

"We create a diversion. We put small boats out there with weird lights. The Japs are very superstitious. So we give them something that looks supernatural. Something that spooks them and makes them prone to making mistakes."

"And your turncoat signalman, he gives you a few more minutes of closure?"

"That's the idea. If we're detected, he signals about suspected subs ahead. We look like we're screening the starboard carrier and its cruiser. He also tells us what ship lighting to have or not have as if we're a Jap destroyer."

"Damn, Pelham, you're starting to pique my interest. What is your intent with the carrier?"

"To put her under in five minutes," David asserted.

Layton was incredulous. "A destroyer is going to sink a carrier in five minutes?"

"Her flight deck will be burning, and aircraft are exploding, maybe on the hangar deck too, while we pump her full of holes."

Layton put his feet up on the desk. "Lordy! It is getting thick in here! How do you plan to do that?"

"Light and heavy mortars and small arms fire into the airplanes."

"You're serious? I suppose those weapons will come from the Army or Marines also?" Layton demanded.

"Yes sir, the Jap flight decks are wood on top of light armor. The light mortars will destroy the aircraft, and the heavy mortars with a slight fuse delay will open holes down into the hangar deck and cremate it."

"What do your five-inchers do? How many shots do they get off?"

"We have eight five-inch guns. They will be firing at the waterline and a few feet above it. We'll get off over 200 shells that hit home in four minutes. Each hit will put a one to two-foot hole in the ship's starboard side."

Layton shook his head. "I'm starting to feel sorry for this Jap carrier crew. Here they're in the middle of the night, probably cruising at All Ahead Slow, a majority of them asleep, and Selfridge stalks them like an Apache warrior. Then, all hell breaks loose with no warning. Let's say she's listing and sinking. What do you do then?'

"We become the bogeyman."

Layton's mouth was open in disbelief. "The bogeyman?"

"Yes sir. The Japs aren't afraid to die. They're not scared of American ships, but they're afraid of the unknown, the unexplainable. So if they see their carrier go down in five minutes and then something that scares the bejesus out of them, they will run."

Layton put his head down on his desk. "Hold on for a minute. Let me think. Let me think this through." Then, after a long pause, he looked at David. "You're near Wake, right? You're taking on two Jap carriers. How will you handle patrol aircraft from them and the island?"

"We'll shoot them down."

Layton snapped his fingers, "Shoot them down, just like that," he said with doubt.

"Yes sir. If it's daylight, they will think we're a Jap ship. We will entice them into a close fly-by and shoot them out of the sky. Our signalman will make them think we're one of their ships if it's night. We'll have to run if the aircraft marks our position and stays outside our deadly fire. By the time another plane comes looking, we'll be fifty to one hundred miles away."

Layton looked shocked. "In an hour or two, you'll be 100 miles away? What is the top speed of the Porter Class? That's your class, right?"
"The stated top speed is thirty-seven knots or forty-two miles per hour. Selfridge will be capable of forty-two knots or forty-eight miles an hour."
"How can you say that? Do you have a magic wand?" Layton said sarcastically.
"No sir. In dry dock, we've started cleaning the bottom and all underwater structures. We're taking all the layers of paint off until she's as clean as a baby's bottom. Then we rough her up with heavy grit and apply a new chemical from Dupont. It has copper to hinder marine parasite growth and is very smooth on the surface. So the ship will, in essence, be very slippery. I have some changes for Engineering also."
"Pelham, you're giving me a headache. There is a lot you aren't saying, am I correct?"
"Yes sir, no sense in you getting court-martialed along with me."
"Tell me one more thing that would make me your champion with the staff," Layton challenged.
"Yes sir, as you know, the biggest problem for a small boy is the short visual horizon - only a few miles to detect subs, and maybe double that for an aircraft carrier."
Layton pounced. "Exactly. Tell me how you'll solve that problem!"
"With the aft torpedo launcher gone, we put the larger gun on the port side and a gyro glider on the starboard."
"A gyro what?"
"It's essentially a type of kite with a circular set of vanes and limited flight controls. The circular vanes provide lift. A pilot can steer it up or down, left or right. The company that makes it thinks we can get lift-off for a 140-pound pilot at twelve to fifteen knots."
"How high can he go? How does he talk to you?"
"At least 300 feet, maybe more. The pilot uses a sound-powered phone to talk to a talker on the 2MC circuit."
"Pelham, who is going to pay for all of this? And when do you propose to be ready?"
"Pelham Oil, or I'll pay. I'm thinking mid-May so the Japs can see it as a prelude to their Midway attack."
"Captain Pelham, you've given me a lot to think over. Let's pivot to our special friend in San Francisco."
"Let's," replied David.
"I think you need to see him and ask for your signalman and answers to three tough questions. I don't know what they are yet. They won't be easy. You can promise him $5,000 in the currency of his choice if he delivers. When will you be ready to leave?"
"I'd like to see the ship in a steady repair routine, say about a week," David replied.
"Done. We'll have our questions ready. Let's talk about Michael Nagy."
"Yes sir."
"He reads, writes, and speaks Japanese?"
"Yes sir, and so does his wife, Rachael."
"It seems he has a diplomatic perspective; does he have a tactical military perspective?" Layton asked.
"Not necessarily. Nagy's real strength is projecting the Jap's reaction to our moves and keeping us from making decisions that defeat our own purposes. He understands their social

norms and psychological reactions better than we do. So he's quick to swat down any Pollyanna notions that the Nips. No matter how hopeless their circumstances, they won't surrender or abandon the Emperor and sue for peace."

"I can't wait to meet him and try my rusty Japanese with him. I studied the language for two years in the country. Then I went back for a tour as Assistant Naval Attaché. I don't rate them as inscrutable, but what they're thinking is never clear."

Pelham said, "I think Nagy can help with that and in liaison with our friend in Frisco."

"Good. How can the staff help make your crazy-like-a-fox plan come to fruition?"

"We need to telegraph to the Japs that there will only be one carrier there. There should be some strife, some staff infighting, and poor security, like draft plans in shit cans. Some loud arguments overheard by cleaning or food service personnel would help. Careless doodles, sailors boasting in bars, and Marines thinking we would attack Wake could help. We need an exercise that looks like a rehearsal for attacking Wake in April. Even the troops should think that. We need the Japs to believe they know the exact date we will attack. I'm not sure how we do that. It's more in your bailiwick. They will appear armed to the teeth two days before that date."

"I said I was done with the questions; sorry, here is another one. Say you sink one carrier, and the other runs away. How do you deal with the aircraft at Wake? They're going to hunt you down and sink you."

"First of all, if we bagged a major aircraft carrier and lost a destroyer, that's a fair trade for our side. Second, we're going to start the attack phase by shelling Wake. We will destroy their planes and knock out the runway for at least a day."

"Pelham, you have a lot of slick answers and extreme bravado. You need more facts and figures. Come back in five days at noon. We'll have sandwiches with the Admiral and discuss your west coast trip, not this somewhat crazy, but possibly worthwhile operation."

David stood up. "Yes sir, thank you, sir."

"One more question before you go."

"Yes sir."

"Why do you want to take on this risk for your ship, your men, and yourself? What's your motivation?"

"Just look out the window, Commander. The Japs have to pay for what they did here. I want revenge, and I'm going to get it."

"OK Pelham, keep refining your plan. It might fit into the grand scheme of things or not."

As Lt. Comdr. David Pelham descended the three ladders to the first deck of the CinCPac Fleet building, he thought, *Yes, they will pay for attacking Pearl Harbor, Guam, Wake, Hong Kong, and the P.I. And the Nips will pay for other parts of my life that didn't go the way I wanted. Someone has to pay.*

Pelham returned to the ship and again found Price and his assistant waiting for him. They reported that five teams of six men were working on the damaged compartments. Each man had a cleaning outfit that included rain gear and knee-high boots. Price said all free saltwater would be pumped out by late afternoon. Next, ruined materials and trash would be cleared from the flooded compartments, and a freshwater washdown would commence.

Selfridge personnel were trained in using Tectyl on fittings, pumps, and motors. They used Consol fluid on metal surfaces. Fresh warm water mixed with steam was used to wash off the green slime. Delicate instruments removed from the Bridge and engineering spaces were turned in for testing, refurbishment, or replacement by the Yard. Price said most electric motors in watered compartments could be cleaned with fresh water, treated with Tectyl, and tested again when dry.

The bomb hole in the bottom amidships was irregular, about seven feet by five feet. There was some damage to nearby ship stringers and bulkheads. Repairing these and the shell holes on each deck would take about a week to two weeks. None of the ammunition spaces were damaged, so the ammo could stay on board. Men wearing portable breathing gear removed all foodstuffs during the morning. Fire watches were set, and fire hoses from the dry dock were deployed about the ship. The engineering plant appeared intact and undamaged. The rudder, screws, and shafts seemed to be unharmed.

Pelham responded, "Gentlemen, this is all great news. I'm about to have a meeting regarding our personnel needs. How many more men can we provide you to speed things up?" David questioned.

Price replied, "We have about sixty to seventy now. As we train them, you could add ten or twenty per day until we have about 100. That's not counting the men you have cleaning the ship's outside. Machinists, metalsmiths, carpenters, and ship fitters can work directly with our damage repair crews. Oh yes, your aft torpedo tubes were damaged."

"We're removing those anyway, and the main deck torpedo storage lockers. So we'll sail with four torpedoes in the forward launcher only," Pelham declared.

Mr. Price gave him a curious look. "Have you cleared this with the BuShips configuration control folks?"

"No, I cleared it with Admiral Nimitz. Selfridge is gearing up for a special mission. We're going to make several changes; consider us as a testing and research ship, and stay out of our way," David warned.

Price shook his head. "OK Captain, it's your funeral. You report to the Shipyard CO while you're in dry dock. He reports to Commandant Fourteenth Naval District, Admiral Bloch."

"Mr. Price, Danny, don't get me wrong. We appreciate everything you're doing. However, we need to know what has to happen for us to start moving men back in a week from now. What tests do we have to run? Do we use our power plant or run from shore power? The full crew will be aboard three weeks from now, and we'll be running under our own power. We float her, test for leaks, and start sea trials four weeks from now. My men will work with your people to get this done. I don't have time to argue about it now."

Pelham abruptly walked away from Price and the other yard type. He found Smitty, his XO, the Admin Officer Lt. Forest, Lt. Lehman, Yeoman Brown, Bosun MacGregor, and Chief Sorensen in a makeshift office on the dry dock top deck. They had a beat-up desk, several chairs, and a sailcloth strung above for shade. David introduced himself to Lt. Forest and Chief Sorenson. He questioned Smitty.

"Petty Officer Smith, could you tell me the ship's manning on 7 December?"

"Yes sir, fourteen officers and 174 enlisted."

"That would be the peacetime manning, correct?"

"Yes sir, wartime manning is a maximum of 290 bodies, officer and enlisted."

"Where are we at now?"

"Sir, your Bureau listing that I update every month shows eight officers and 132 enlisted."

"For wartime, it sounds like we could set a goal of fifteen officers and 220 enlisted. Do you think that's possible?" Pelham queried.
"For the officers, yes sir. I can have the five TAD officers back next week. We had 174 enlisted, lost twelve, and then had thirty-one depart on Permanent Change of Station orders. Of the 132, we have about 112 back on the ship's muster. I have about thirty bodies stashed on TAD jobs that can be back next week. I'm picking up roughly ten new men a week in the needed rates and one junior officer every two weeks."
David figured using a pad. "So by next week, when I start moving into the ship, we can plan on about twelve officers and over 142 enlisted?"
"Yes sir, that seems doable."
"Excellent, then three weeks after that, we can have thirteen officers and expect thirty more enlisted, nearly back to pre-war strength. Chief Sorensen, what are the sleeping and messing requirements for 220 enlisted versus 174?"
"Not really good, Captain. If Divisions want to share quarters, some may have to hot bunk. We have permanent racks for 190 enlisted and eighteen officers. Taking the officers first, you have the Captain's Cabin and the Captain's At-Sea bunk next to the bridge. The XO and the Engineer rate single staterooms. We have four double cabins for officer department heads and a bunkroom that can hold up to eight Ensigns and Midshipmen. Once we go past 190 enlisted, we have to hot bunk or sling hammocks all over the ship. As far as meals, more crew means more food and fluids needed, and more men mess cooking, more cleaning up, and dirty laundry."
"I think I should pare back the enlisted number from 220 to 180. We will do some new weapon experiments and have extra bodies onboard for sea trials. There will be another increment in May; as many as thirty additional men may be necessary."
Bosun MacGregor spoke up. "Sir, do we have a cruise schedule?"
"Boats, I have some goals to run by the Commodore. I want us to be on the ship's power in three weeks, floating in four weeks, and at sea no later than the fifth week for fourteen days. My plan would rotate two weeks cruising with two weeks in port. After our April cruise, we should be ready to sortie as assigned."
"Sounds like a good plan, skipper. It gives us time to slot the new men into their jobs and get the crew combat-ready."
"XO, work up a training schedule for the first two-week cruise period. Emphasize Anti-Submarine Warfare, gunnery, and damage control. Lt. Forest, who was our Ops O that I was supposed to replace?"
"Sir, that's Lt. Donald Wolfe. He's been selected for 0-4."
"XO, how about you? Does Lt. Wolfe outrank you?"
"No sir, we're in the same year group; I'm ahead of him about six months. And yes, I got selected, too."
"Roger that. Work with Smitty to produce a chart showing me the leadership structure for the divisions and the operational jobs. Show me who is in each billet from Second Class up, or what we need. For now, the XO will be XO, and Lt. Wolfe will be Ops. Lt. Forest, what are our critical personnel deficiencies?"

"Sir, we need an Engineering Chief, a Storekeeper Chief, and a Gunnery Chief in the worst way. We also have eight to ten First Class billets that need filling, thirty Seaman billets, two L-T's, and a couple of 0-4s."

"Skip the O-4s for now. We need to go out and recruit the rates we need," David asserted.

They were interrupted by a truck arriving at the quay wall and honking its horn. David asked the Bosun to take some men down and unload several pallets of five-gallon containers with the DuPont logo. The discussion returned to personnel, unfilled billets, and watch standing bills.

David smiled as he thought about how hard it had been to explain the ship's organization to Betty. He told her that the ship's complement or crew consisted of several specialties or rates. Each vessel was assigned a specific number of men with these skills and varying levels of experience. Each man belonged to a Division based on his specialty, training, and expertise. The six divisions assigned their men to one of three Sections. The three sections stood watch and worked around the clock according to the watch system. David and the XO had decided that the three-section watch system would commence when the ship was nearly fully manned with twelve or more officers and 160 enlisted. Until then, they would work twelve hours on and twelve hours off. Officers and enlisted were assigned an additional four-hour fire watch and quarterdeck watch.

After discussing sources of more ship's crew, Pelham went down to the quay wall where Yeoman Brown was waiting for him with the ship's sedan.

David slid into the back seat. "Yeoman Brown, please take me to the Marine Corps Barracks."

At the Barracks, David had Brown wait while he went in and located the Duty Sergeant. He asked the Sergeant how to find information about Marine Corps sniper teams or an actual unit in Hawaii.

"Sir, I heard rumors that there might be one up at MCAS Ewa or over in Kaneohe."

"Thanks. How about if I want to learn about Marine mortars and artillery? Where would I go?"

"Of course, you can find the Artillery Regiments at Camp Eliot north of San Diego or at Quantico on the East Coast. Here you would need to find a Gunnery Sergeant with a artillery tour."

"Sounds like good advice to me. Do the Marines have a favorite bar downtown like the Navy divers?"

"Yes sir, you can find them in this area at the Pearl Tavern or the NCO club nearby at Ewa Airfield. The Army has closed down most of the downtown bars."

David shook the Marine's hand. "Sergeant, thanks for all of your help."

Brown drove David off the Navy base and over to Ewa, where they parked in front of the NCO Club. David took off his cover and entered the small building. A sergeant behind the bar saw him first and sounded off, "Attention on deck!" There were a couple of tables of Marines drinking beers and Cokes. They all shot to attention as he entered the room.

"As you were, men. I'm just looking for some information." He headed for the bar. "I bet the sergeant can help me. The next round is on the Navy!"

"What are you drinking, sir?" asked the bartender.

"Make it a Coke and put it on my bill.

"Yes sir. What can I do for you?"

"I'm acting CO of one of the destroyers in Pearl. I heard there might be a sniper team here or at Kaneohe. I have some questions about how one of those teams might work on my ship."

"Sir, there's a team passing through on TAD orders. None of them are here in the bar. I can tell you how to contact them. Is there anything else?"

"Yes, I'm interested in the different sizes of Army and Marine mortars and how to use them on my ship."

"That one is easy. The big guy with the crew cut silver hair at the first table is Gunnery Sergeant Giles. He knows every weapon in the Corps and the Army, too." He motioned at the table. "Sergeant Giles, front and center!" he shouted.

Giles came over, shook David's hand, finished his beer, and swept up a new one that the bartender automatically slid down the bar.

"What can I do for you, Commander?"

"Sergeant, I'm David Pelham. I have command of a destroyer over in Pearl. I've thought that sniper squad and mortar weapons might be effective in certain situations."

"Can you name a couple of them for me?"

"I can if we go outside. My plan is not for the general public," David responded.

Once outside, Giles said the sniper grunts might be pitching horseshoes over by the gym. So they went over to the side of the gym; Giles called one of the men over.

"Lieutenant Commander Pelham, meet Gunny Hortsbock, sniper team leader," Giles stated.

David shook hands with a skinny, short Marine wearing running shorts and a white T-shirt. He was not much taller or heavier than Smitty. Giles saw David's surprised look and quickly made him aware that the Gunny was one of the best shots in the Corps. David discussed his plan with them, and they both agreed that sniper fire and mortars could cause the desired confusion and lethal results. He decided he'd need two snipers, each paired with a spotter, and four mortar teams led by a Marine.

David asked Hortsbock and Giles, "If you were hard up for men, where would you go to find replacements?"

Giles answered, "I don't know about the Navy, but we would see if we could Shanghai men out of the brig and off of light duty at the hospitals."

Hortsbock nodded in agreement. David thanked them and got back into the Navy gray sedan.

"Petty Officer Brown, take me to the main Navy Brig."

"The brig sir, you're sure?" Brown responded.

"Yep, we're going to do some radical recruiting."

After they parked at the brig, he had Brown accompany him; they found the Marine Captain in charge.

David said, "Captain, we're from the USS Selfridge, which we just raised off the bottom of the bay and put into dry dock. We plan to sail her in one month, and we're shorthanded. Would you be so kind as to show us a list of your current prisoners?"

"To what end, sir?"

"We have some serious deficiencies. If you have some men who can fill those billets, I might try to get them transferred to my brig to finish their sentences and enjoy some challenging sea duty. How many do you currently have locked up?"

“Fifty-two here and thirty-seven on the base clean-up crew. Here is our current roster for both. It has the man’s name, rate, unit, serial number, offense, and sentence.”
“Do you mind if I place a mark next to the ones that interest me?” David asked.
“No sir, help yourself,” the Marine answered.

David put a checkmark next to some names and a double checkmark on men from Arizona, Oklahoma, and severely damaged ships. He handed the list to Petty Officer Brown.
“Captain, how would we go about it if we wanted to interview these men?”
“The clean-up crew is back in their barracks by 1700, and they chow down at 1745. Then they’re on their honor to stay in the barracks. The men in the brig get three daily exercise periods. They have three squares a day for thirty minutes. Other than that, they’re locked up in cells or working in this building. Reveille is at 0600 and lights out at 2200.”
David thought for a moment and said, “Maybe we could interview men in both groups, in their barracks or their cells, with one of your men present. Can you help us out? We need to get our ship manned so we can go kill some Japs.”
“It is a bit unusual, but as long as you don’t touch them or give them anything, it can’t hurt to talk to them.”
“Excellent. Some of my crew will return tonight and coordinate with your men.”

David returned to the car and directed Yeoman Brown to drive to the Naval Hospital. Brown looked in the rearview mirror and quietly asked, “Captain, do you think those brig inmates are good prospects? That they can fit in on Selfridge?”
“Yes Brown, if we interview them and weed out the thieves and crooks. Half the boys in the brig will be there for getting drunk and fighting. We just need to give them some good discipline and fight the Japs. Any of us could end up in the brig for a dozen reasons. I bet there are a few times it could have happened to you."
"No sir."
“Never, Petty Officer Brown?”
“No sir. I’m strict Southern Baptist, don’t drink, don’t smoke, and I leave the fighting to Jesus. But, Sir, are you sure it’s legal for us to transfer these men we want?”
“All I know is there is a war on, and we need more men. So I’m willing to stick my neck out to get them.”
Brown shook his head. “Yes sir, you’re sure doing that.”

Marine Corps sniper rifle
Source: NRA Rifleman publication

Chapter Twenty-three

AGGRESSIVE RECRUITING – Pearl Harbor

January 1942

Pelham knew Yeoman Brown was trying to warn him that his recruiting ideas were unconventional. He was skating on thin ice, and he knew it. *What the hell is the brass going to do? Stick me on an expendable destroyer and send me to war? No, they've already done that. I'm just making it work.*

They reached the hospital entrance as it started to rain. "Brown, you're probably right. But I will do what I must and let the chips fall where they may. Wait for me."

David went in alone and walked the halls until he found a Navy Chief Hospitalman. He asked him if any engineering, gunnery, or supply chiefs were on light duty or nearing recovery and release from the hospital. The medical Chief told him that a CPO Storekeeper was taking inventory of hospital supplies while he recovered from some burns and shrapnel wounds. Pelham followed directions and found Chief Donald Gonzalez counting boxes of medical supplies in a third-deck storeroom. He had bandages around both legs and gauze wrapped around his forearms.

"Excuse me, Chief, are you the supply type that I heard about?"

"Well sir, that depends on what you heard."

"I heard that you were nearly recovered and ready for duty. I have a big challenge for you."

"What would that be, Commander?"

"I'm Lt. Comdr. David Pelham, acting CO of a Porter Class destroyer, the USS Selfridge. We recently refloated her and have her in dry dock. I want to start moving my men onboard in one week and feeding them shortly after. We've removed all the food from the ship and have to restock her from the ground up. I need a man with your experience to procure everything we need and organize the mess decks and ship's stores. Then this supply magician will have three to four weeks to prepare us for two weeks at sea."

Gonzalez nodded his head. "That's a real challenge, I have to admit. Please tell me a little more about your ship and crew."

"Sure, Chief. Selfridge is a heavy destroyer at 1,850 tons compared to the norm of 1,500. She has eight five-inch guns, several triple-A guns, torpedoes, and depth charges. Our crew is currently about ten officers and 140 enlisted. When we head out to sea from dry dock, we should have fourteen officers and 170 enlisted."

"Commander, it doesn't look like I'll be going back to USS California anytime soon. Any sea duty beats counting bandages all day. How can I get orders?" Gonzalez questioned.

"Go to the main base Admin Building and find a First Class Yeoman named Smith, nicknamed Smitty. He can cut your orders out of here and back into the real Navy. Or better yet, I'll give you a lift," David replied.

Gonzalez offered his hand. "Chief Don Gonzalez, Captain. I'm pretty sure they will release me in a couple of days if I have orders."

"Chief, welcome onboard. I still need a gunnery chief and a Machinist Mate E-7. Have you seen any of those on light duty?"

"Chief Billy Wilson from California is here; he's one of the best snipes in the fleet. They have him locked up in the looney ward."

"What's his problem?" inquired David.
"He can't talk, not a word. He opens his mouth, and nothing comes out. He can write and use his hands to indicate what he wants. There's a complete disconnect between his brain and his mouth."
"I've heard of things like that as a hysteria that can go away with no notice or last for years. Where's Wilson located?"
"The psych ward is above us on the fourth deck. You can walk in, but they have to buzz you out. If you walk directly to room 413, no one will stop you if you look like a visitor."
"Thanks Chief, I'll give it a try. Pack up your gear and meet PO Brown in front of the hospital in the ship's car. Wait for me, and you'll have orders today."

David went up a flight of stairs, down a passageway, and located a set of double doors that said: NEUROLOGY AND MENTAL HEALTH. He went in and found Chief Wilson reading a book in his room.
"Chief Wilson, I just met Chief Gonzalez downstairs. He's joining the crew of my destroyer. We need a third section Engineering Chief in the worst way. He recommended you," David said.

Wilson motioned for David to take a chair, pointed to his mouth, shook his head, and threw his hands up in a helpless gesture.
"Yes Chief, I heard you were having trouble speaking. Do you have a brain injury?"

Wilson shook his head no.
"I have heard about this kind of reaction. You have a mental block for speech because you're feeling very guilty or have had a tremendous shock to your nervous system. Is it the first one? Do you feel extremely guilty because you survived in one piece while so many others were badly wounded or killed?"

Wilson shook his head no.
David continued, "Then it could be the second choice. Do you think that it was the bombing threat to your life? The fear that you could die at any moment, which caused you extreme mental anguish?"

Wilson shook his head no.
"That leaves us a third possibility. What you saw, was so horrifying that it has struck you dumb."

Wilson sat still, tears welling up in his eyes.
David spoke softly. "You saw men injured and killed, right?"

Wilson nodded his head.
"You saw men who were dying. Men badly mangled, burned, or missing limbs?"

Wilson, sobbing, nodded yes.
"You saw men bleeding to death, men partially dismembered, men trying to put their guts back in their abdomens. You saw men on fire, charred corpses, and frightened boys crying for your help and their mama. All of this overloaded your senses, and you probably passed out."

Wilson got up from his chair and lunged at a nearby trash can as he got violently ill.
Pelham was relentless. "Now you're punishing yourself, Wilson. Because of the horror you've seen, you've shut off your voice as your penance, your punishment!"
While kneeling, Wilson spit into the trash can, looked at David, and made a sound, "Uhhh."

David spoke in a commanding voice. "Get back in your chair, sailor. I've come here to redeem your soul and save your career. I don't care if you can talk or not, that's your problem. I want you in my Engine Room managing the ship's power plant. Your voice will recover now or in the future, because I will make you two promises. The first promise is that you won't have to worry about what you've seen, done, or didn't do. That's because our ship is on a suicide mission that will probably kill us both. You will have nothing to worry about but my engines.

"The second promise is if we're in a combat situation where those same horrifying scenes may happen to you, I will come down to Engineering and personally shoot you dead. Or I'll have someone else do it. See, your problem is solved. No need to punish yourself because, one way or the other, you'll be dead in a few weeks. Now stand up and lock your arm in mine; we're going outside for some fresh air!"

Pelham took Wilson outside the hospital. Brown drove them and Chief Gonzalez over to Smitty's office. David told Smith that Gonzalez and Wilson were on light duty at the hospital and wanted to volunteer for duty aboard the Fighting 357. He said Gonzalez would be in "S" Division, and Wilson would be in Engineering. PO Smith was surprised to see Chief Wilson in his pajamas, wearing a hospital bathrobe and cotton slippers. He typed orders to place the chiefs on six months TAD onboard Selfridge, and David signed them.

David had Yeoman Brown drive the chiefs and himself over to the ship. Lt. (jg) Carlson was onboard from "S" Division to welcome Chief Gonzalez and show him the supply spaces. Pelham accompanied Chief Wilson down into the engineering spaces. Chief Wilson inspected the boilers, the fireroom, the turbines, and the piping while communicating with David using a pencil and pad. Chief Romano saw them and came over.

"Chief Romano, this is Chief Wilson from the California engineering team. I found him over at the hospital, just itching to get back at sea and give the Japs some of their own medicine. He's had a wound that requires him to rest his voice, so he uses hand signals and that pad for the time being."

David asked Chief Wilson if he had uniforms and received a thumbs down. Brown drove them to the Uniform Shop, and the Chief found a seabag of CPO uniforms that would fit him. David bought the proper insignia for Wilson's rating, and they drove back to the hospital. On the way, they packed a large paper sack with a summer working khaki uniform: shoes, socks, trousers, belt, short-sleeved shirt, underwear, and a CPO hat. Pelham escorted him back to his room and instructed him to put on the uniform, grab any personal items that he needed, and leave a copy of his orders on the bed.

David went to use the head. When he returned several minutes later, he stood at the door and spoke loudly, "OK, Chief, that's all the time we have for visiting Wilson today. We need to get back to the ship most quick!" They walked out together, and the orderly buzzed them through the double doors into the hospital passageway. David and Brown dropped Wilson off at the ship with a note to the Officer of the Deck to see to his berthing and messing needs.

After Pelham was piped onboard, the OOD told him that Lt. Wolfe had returned and was up on the bridge. David ran up the three ladders to the bridge, eager to meet his Ops Officer. The Bridge itself was a pleasant surprise. It was dry and clean. The floor had been polished. Several patches of primer paint covered up combat damage or where marine growth was removed. Two officers were on the Bridge going over charts of Pearl Harbor and the channel out to the ocean. Lt. Donald Wolfe introduced himself and the Lt. (jg) with him. David asked them if they were qualified as Officers of the Deck Underway. Lt. Wolfe said he was qualified, and Lt. (jg) Carpenter was in training.

"Mr. Carpenter, please excuse us. I need to talk to the Lt. about our future operations."

After they were alone on the Bridge, "Mr. Wolfe, Admin tells me you are selected for Lieutenant Commander. Is that correct?"

"Yes sir, I should be putting it on in about six months."

"Excellent. I hear Lt. Lehman is also selected and is six months your senior."

"Yes sir, that's correct."

"Good, my dad taught me to seize the day, to pick lofty goals and make them happen. My immediate goal is to win a battlefield promotion to Commander and continue as CO of this ship for several months. If I can swing the promotions, I'll request that you and Lt. Lehman get frocked for O-4. Do you have any problem with me as CO, Lehman as XO, and you as Ops?"

"No sir, I've been assistant Ops for six months. I'm ready to take over."

"Good man. I have a risky operational plan to discuss with you. It has many moving parts; I need your help to shape it into a credible set of actions that'll stand up to Flag scrutiny. There are several phases, critical components, and weapons to identify. We need to define and refine the schedule of events. For the next three months, the only players knowing the plan details will be the two of us and the CinCPac Intelligence Officer, Lt. Comdr. Layton. A lot of people will know there is a special operation in motion. But, they won't know what we're really going to do. We need a cover story to throw the busybodies off on the wrong track. Mr. Wolfe, if too many people know our real mission and it leaks to the Japs, we can kiss this ship and our asses goodbye. Will you swear to me that you'll discuss this plan only with myself and Lt. Comdr. Layton?"

"Yes sir, I so swear. Could the Captain give me an idea of the phases as a starting point?"

David handed him paper and a pencil. "Here, you're about to create the first record of what I call 'Operation Slingshot.' We are in Phase 1; it's the Defining and Planning Phase. It includes defining the mission and planning the steps to carry it out. Phase 2 is the Cover and Deception Phase. It entails convincing the enemy we're about to do something that we aren't going to do. Phase 3 is Defining the Operational Units Required, Training Requirements, and Building the Operational Schedule. We'll be using some new destroyer capabilities and need assistance from other Navy units. Phase 4 is Final Preparations and Deployment. We need a detailed step-by-step plan to brief CinCPac by the middle of April."

Wolfe said, "Do we have an operational schedule, sir?"

"We deploy to Johnston Island sometime in April as part of an amphibious landing exercise. Then Selfridge returns to Pearl for final tweaks before sailing to Wake. Phase 5 is Confirmation of Enemy Intentions and Neutralization of Wake Island. Here we confirm the enemy has taken the bait and is heading for Wake to stop us from doing what we're not really planning to do. We find a way to put the Wake Island Airfield out of operation for twelve hours. Phase 6 is the Search and Approach Phase. We confirm the enemy's position and course. Then we infiltrate his battle formation in the dead of night. Phase 7 is the Concentrated Attack Phase. We put ourselves in a firing position to bring all of our weapons to bear on an enemy capital ship and sink her in five minutes. Phases 8, 9, and 10 are germinating in my

gut. Put them down for Exploiting the Tactical Situation, Withdrawal, Reporting, and Cleanup."
"Yes sir. How big is the capital ship?"
"It will be a carrier. A full-size Jap carrier and nothing less, or we'll not risk our ship and this great crew."
Wolfe looked confused. "So, we'll sink a Jap carrier in five minutes. How are we gonna do that?"
"Details Ops, just a few details for us to figure out. Are you back on board full-time?"
"Yes sir, I'm back on the watch schedule and bunking in the base BOQ. I inspected the Ops office today and found it pretty much as we left it. No damage at all."
"Good. Tomorrow morning, take my car and find some Army artillery types. Tell them who you are and that you're on a special project for Admiral Nimitz. You need to know if they have a gun that can fire a more damaging shell than our five-inchers. We need one that can penetrate an armored ship's side and blow up amidships. Have Brown type up your clearance data to take with you. See if you can look at a manual or the actual gun. Your mission is to find a gun we could mount on our ship for a short period, say a month. Determine if we could operate it or need an Army gun crew. Also, how forceful is the recoil? How can we mount it on the port side of the aft torpedo deck? If any of them are helpful or friendly, take them to lunch and pay the bill. I'll reimburse you from the Captain's slush fund."
"Aye, aye, sir. If I find a gun, how would we get it?" Wolfe questioned.
David smiled. "Details Ops, just more details, first things first."

Pelham glanced down at the Quarterdeck from the Bridge and saw Mr. Danny Price marching down the gangplank with a very pissed look. He excused himself and went down to meet the Yard rep. Price took him to task for telling his men they would reinstate ship services over the weekend and that some would be moving back onboard during the coming week.
David called the Bosun over. "Boats, Mr. Price and I have been discussing bringing the ship back to life. Have you any thoughts on that?"
"Yes sir, I have a plan of sorts on the back of an envelope. First, we hook up hoses with screens into the bay to pump seawater to the heads. Next, we connect sewer hoses from the ship to the harbor to get rid of the sewage. Next, we pipe fresh water, test all the fittings, and fill the freshwater tanks. For these steps, we're still using shore power. Engineering has good feed water for the boilers and will finish cleaning the fuel tanks by next week. We've been testing the ventilation and pump motors all over the ship. Nine out of ten are good to go once they're flushed and treated. Others have been sent to the electrical shop. We should have all of those back this coming weekend. We plug them in and run safety tests. If they pass, we start providing our own ventilation, still using shore power.
"Next, we give Engineering the green light to fire up the boilers and test the power plant. We cut over to the ship's power a week from now and retest every circuit, communication system, switch, and fuse box. Once we have ventilation, heads, lights, freshwater, and communications, each division brings five men onboard and can add five more each day. They stand the watches and man the damage control stations if necessary. Supply says they've located all of the mattresses, bedding, and furniture that we need."
"What watch routine are you recommending?" David asked.

"Skipper, with your approval, the XO wants to drop the twelve and twelve when we have half the crew onboard and go to six hours on and six off using two sections. The day before we sail, we go to a three-section double-dog watch."
Mr. Price spoke with sarcasm, "It's just hunky-dory that you two have this all worked out, but that's not how the Yard operates. We will tell you when you turn on the services that you just mentioned. It usually takes a week to test each service. We tell you when men can safely move back on the ship. We tell you when the ship is certified safe to generate steam and power. And we tell you when you'll float and be judged ready for sea."
David was intense. "Mr. Price, I'm sure your procedures made a lot of sense when you were running a peacetime Navy Yard. Look around; this is a new ball game. You and your men are doing a fantastic job repairing our damage and keeping us from getting gassed or injured. I've told my men that we'll sail before the end of the month. They have a plan that has us doing that. So be it if it involves some additional risk to the ship and the men. By the way, we'll use shore or ship power to man our guns next week with live ammo as required by the Commandant's standing orders."
"You can't do that, not while you're in dry dock, no sirree. It's not allowed," Price argued.
"Again, a peacetime rule. I bet the Shaw would've had all her guns manned if she had a second chance," David snapped.
"I'm not going to argue with you. Tell me why you have so many men working on the bottom and what they're doing?" demanded Price.
"The men are cleaning the bottom and knocking off all marine growth. Next comes grinding and sanding off all paint. Then they rough the bare steel with low-grit sandpaper before applying the bottom paint. After twelve hours, the paint is lightly sanded, cleaned with warm water, and the second coat of bottom paint is applied."
"Why all the effort? Why not just clean it and paint it Navy gray? Price asked.
"This is a special paint from Dupont. It has a polymer that reduces surface friction and has copper in the formula that discourages marine life."
Danny Price was angry. "This is news to me. Did you get it from the Navy supply system?"
"No, we bought it directly from Dupont in New Jersey. They gave us a good price because we're testing it for them. We have fifty to seventy-five men working around the clock. She's 381 feet long with a wetted surface of over 15,000 square feet. It's a tremendous job."
"So, you're making an unapproved modification using an unknown product purchased illegally, right?"
"There was no illegal purchase. I just used the Captain's slush fund."
"I never heard of any such money! How many other unauthorized changes are you planning to make?" asked Price.
"Only a few, and all of them cleared by Admiral Nimitz. Also, we'll be implementing approved and planned mods for the Porter Class. They include removing the rear mast and gun director, eliminating the torpedo storage units, and augmenting our .50 caliber popguns with dual Bofors 40-millimeter mounts."
Price exclaimed, "You can't make these engineering changes. They have to be made by Yard personnel!"
"Can you get them done in three weeks?" David asked with a challenging look.

"Of course not. That's impossible. We don't even have the blueprints and engineering drawings necessary."
"Then I rest my case. My ship and my crew need these changes. We can't wait for the Navy peacetime bureaucracy to make it official. We have places to go and enemies to kill. By the way, we'll make some classified changes behind canvas covers. We are posting sentries who will shoot anyone trying to take a peek."

Price muttered a muffled growl of rage, spun around, and marched off the ship. Later in the afternoon, David was called to the Quarterdeck to receive a message from Lt. Comdr. Layton. Layton wanted to meet in his office at ten hundred the next day. That afternoon, he and Lt. Wolfe added more detail to the first two Phases of Slingshot in preparation for the meeting.

The following day, when they arrived at Layton's office, he had a Navy Commander, Commander Joe Rochefort, sitting in on the meeting. David thought, *Great, now four people know the details of Slingshot. Hope we can keep it at that until we sail for Wake.* He covered the plan Phases in reverse, deciding that if the two staff officers couldn't buy off on a destroyer sinking a carrier, then there was no need to brief the initial phases.

Pelham stood at a blackboard with chalk. "Imagine yourself as an aircraft carrier in this Jap formation approaching your fly-off point in support of Wake Island. You know the Yanks will attack Wake on 16 May, so you arrive two days early. It's pitch black on the 14th, no moon, and 0130 in the morning. The lookouts posted at midnight are getting tired. Your carrier steams ahead slowly at three to five knots; most of the crew is asleep. Suddenly, two rockets fly over the ship from the aft starboard side. They arch over the top of the island and start falling in front of the carrier. All eyes are on the fireworks. Men on the deck, the forward lookouts, and the entire bridge crew focus on the two falling objects, leaving a trail of blazing sparks through the night sky. Then, bam and bam!"

He slapped his hands loudly together. "They explode within a second of each other. Two one-pound magnesium bombs so bright that they will completely blind anyone looking at them for five to ten minutes; full night vision doesn't return for fifteen to twenty minutes. USS Selfridge, a heavily armed Porter Class destroyer, is 500 yards from the carrier and closing on its starboard side. After the bright flashes, guns are free for two Marine sniper teams on our destroyer. They will be plinking as many aircraft and fuel bowsers as possible on the carrier's Flight Deck. Once they've ten to twenty hits leaking Avgas, the aft team will plink aircraft and aviation fuel containers in the Hangar Deck. The forward team will shoot anyone they see on the signal bridge, the bridge wings, and in the catwalks.

"After the rocket blasts, four Marine mortar crews drop several rounds per minute onto the Flight Deck. They get the Avgas burning, and two mortar crews switch to heavier shells that will open holes in the Flight Deck, so burning gasoline cascades down on the planes in the Upper Hangar Deck. The first mortar fire is the gun's free signal for our four dual five-inch gun batteries to open fire on the carrier's waterline. Gentlemen, eight, five-inch guns will fire into the carrier's side for four minutes. If they shoot every six seconds, they will blast over two hundred one to three-foot holes in four minutes. Thousands of gallons of seawater will flood the lower starboard decks each minute. The Bofors guns and our other AA guns will rake the bridge and the open areas to suppress any return fire. The first five-inch firing will be the signal to Big Bertha, our 6.5-inch Army artillery piece that can shoot every twenty seconds. Bertha will fire ten to thirteen armor-piercing shells fused to explode amidships in the carrier - right in her gut."

Rochefort interrupted. "Where are you getting the gun and ammunition?"

"The Army or Marines." David continued in a low, firm voice. "Picture the scene, ships in the formation have gone from peaceful tranquility to watching in horror as one carrier catches fire and starts exploding. If you're on the target carrier, all hell has broken loose. Your Bridge crew is blinded, fires are burning out of control on the Flight Deck and probably the Upper Hangar Deck. Planes and ammo are exploding; gut-wrenching explosions in the bowels of the ship are taking out vital services. Water is rushing in the starboard side; she's starting to list. I'm counting on the Japs to be surprised, slow on damage control, and confused about what's happening to them. Once the list starts, the only thing that can save the carrier is immediate massive counterflooding on the port side. Failing that, we keep punching more holes higher up the starboard side as the waterline submerges. More and more seawater rushes in until she rolls over."

Layton looked over at Comdr. Rochefort, who shrugged, seeming to say it could work.

"What about this bogeyman thing? You hope that'll make the other units run, right?" Layton questioned.

"Yes, I have ordered a large Chinese dragon head, the kind you see in the parades, but much larger. We are going to use a flamethrower for flames from the mouth and rig CO2 fire extinguishers for smoke out of his nostrils. The eyes will be two spooky, moving red lights. We will mount him on the bow on a ball joint so we can swivel it - raise and lower the head like he's looking right at the next target."

"Why do you think they'll run? If they can see you, they will engage." Rochefort challenged.

"Yes sir, that's possible, but we're so close to the carrier that the escorting cruiser can't fire on us. Also, we hope to have subs with us firing a couple of spreads at the other carrier. We give them a few short views of Fred the Destructive Dragon, and then we start firing on the cruiser."

Rochefort was surprised. "This dragon thing, where did that come from?"

Pelham said, "A very wise Japanese lady told me that Jap sailors would be afraid of anything supernatural. They will be far from home and very superstitious. So we will give them some strange lights, scary sounds, and short glimpses of fire-breathing Fred. Those sights, plus losing a carrier in five minutes, will put the fear of bad spirits and the gloomy underworld into them."

"Hold it. You're going up against two carriers and a land airbase. How will you handle their aircraft and avoid being sunk?" Rochefort demanded.

"Sir, with all due respect, we'll lure them in close and shoot them down. You don't want to know the details."

Rochefort said, "Pelham, I've had destroyer duty. After you attack Wake, how are you going to steam 150 to 200 miles and find the carrier group in the dark?"

"Sir, we'll have submarine and aircraft scouts. We'll see the formation before they see us by using Radar and an aerial observer," David replied.

"So, there will be other ships and aircraft involved?" Rochefort asked.

"Yes sir, I'd like to have two subs and two or three PBYs. We will need one tanker to top us off before and after the operation."

Layton broke into the discussion. “An interesting initiative, to be sure! Thank you, gentlemen. If you'll excuse us, I have some personal comments for Mr. Pelham.”

“Mr. Wolfe, I’ll see you outside in the car,” David directed.

Layton continued, “Pelham, now that we’re alone, I can tell you that Joe Rochefort is in charge of radio intelligence. He may be of great help to you. Also, your trip to San Francisco is on. You’ll leave in two days at 0900 from the Hickam Operations Building on our only Marine C-46. You’ll be traveling with Admiral Nimitz. Be ready to brief him on more details. Have you named it?”

“Operation Slingshot,” answered David.

“Good enough, I guess. You’ll need to convince Nimitz before you request the Jap signalman from our special friend.”

“In that case, I’d like to take Lt. Wolfe. He’s capturing the details of Slingshot and probably can brief it better than I can.”

“If you don’t hear back from me, that’s approved. Plan on four nights and three days,” Layton responded.

When he and Wolfe got back to the dry dock, they could see two Marine Sergeants waiting on the top deck of the dry dock.

David went over and greeted them. “Sergeant Hortsbock, Sergeant Giles, welcome to the Selfridge! To what do we owe this pleasure?”

“Sir, we got to thinking about your idea and wanted to see where you would place the sniper teams,” replied Gunny Giles.

David pointed upward, “I'm not sure. How about up on top of the Bridge and up on the aft superstructure?”

Giles looked suspiciously at Lt. Wolfe. “Eh, sir, to shoot the er, the objects you mentioned, it would be best for us to be prone with a spotter next to us - the higher, the better.”

David quickly responded, “Don’t worry about the Lieutenant; he knows what we’re talking about. Why higher? I don’t get it.”

“Captain, if we’re looking up at the target Flight Deck, we may not have much to shoot at,” said Giles.

“How high up do you think?” asked Pelham.

Gunny Hortsbock responded, “At least forty to fifty feet above the waterline, I’d think.”

“I’ll go talk to the Bosun about that. In the meantime, Lt. Wolfe will show you around. See if you can find a space where seven or eight Marines can survive in comfort starting in April for one month.”

David went onboard the ship, to the usual piping, and asked the OOD to make an announcement. “Bosun MacGregor report to the Bridge.”

Pelham went up to the Bridge Plot and found it freshly painted. He plopped down on his new bunk mattress in his At-sea Cabin and rested his eyes and body. In about ten minutes, he heard, “Captain, are you awake or just getting a beauty sleep?”

“Mostly awake, Boats. Just had some Marine snipers throw me a curveball.”

“How’s that, skipper?”

“I have a mission for them where I want them to shoot at another ship. Instead of climbing up in the rigging like in old times, they prefer a shooting platform where they can fire from the prone position with a spotter next to them. They want us to put them up forty to fifty feet or more above the waterline so they can shoot down or almost level with the target. I want one team fore and one aft. I'm stumped on how to do this. Any ideas?"

"Are they shooting close in or long distances?" Boats questioned.

"Both, especially close in at aircraft passing down either side."

"Skipper, we could build a platform on the Main Mast. I hope they know that even a small swell will have them gyrating like a hula dancer. For the aft team, we're already dismantling the aft mast. We could put them on the Gun Director platform, but it's not that high."

"We need a way to raise and lower them depending on the target," said David.

"Sort of like an elevator, up and down," Boats mused.

"Yes and no. The Marines have to be able to change position on the platform to shoot aft, starboard, or port, maybe fifty or more feet up."

"Holy cow, skipper, how would we get them that high off the deck?" MacGregor exclaimed.

"I don't know. That's why I have you to figure it out."

"I was thinking about some elevators that only go a few floors."

"Boats, we can't put them in an elevator." David admonished.

"I know, I know, but some elevators for three or four-floor buildings are pushed up and lowered using a hydraulic cylinder or a series of smaller cylinders. You know, sort of like the lift in a garage that lifts a car for service."

"Boats, that might just be crazy enough to work! We get a lift like that and put a shooting platform on it. We install it on the aft superstructure and get them fifty or sixty feet above the waterline!"

"What will these guys be shooting at?" MacGregor queried.

"Airplanes mostly. I have a scheme to entice the Jap planes to fly alongside for pictures. When they do, our snipers put a couple of bullets into the pilots, and our AA guns finish the job."

"Captain, you're a very sneaky person. I'm glad I'm on your side," sighed Boats.

"See if you can locate one of those elevator lifts. We can pay for it from my slush fund," directed David.

Boats touched two fingers to his brow, "Aye, sir. For your information, we have the saltwater pipes working in one head per deck fore and aft. We start with the freshwater tomorrow. I hear you're going on a trip. What do we do with Mr. Price and his kind while you're gone?"

David thought, *Amazing, I just came back on board! Wolfe must have canceled a watch or asked for orders. There is nothing like a ship's grapevine for lightning-fast good and bum gouge.*

He instructed MacGregor, "Boats, the gangplank is closed to Yard people unless they're making repairs. Rig some canvas on the aft torpedo deck like we have a secret weapon onboard. Tell them I'm on the West Coast with Nimitz and stick with your plan to bring the ship back to life."

A 3-section dogged watch

Name	Time	Day 1	Day 2	Day 3
First watch	**2000–0000**	Team 1	Team 2	Team 3
Middle watch	**0000–0400**	Team 2	Team 3	Team 1
Morning watch	**0400–0800**	Team 3	Team 1	Team 2
Forenoon watch	**0800–1200**	Team 1	Team 2	Team 3
Afternoon watch	**1200–1600**	Team 2	Team 3	Team 1
First dog watch	**1600–1800**	Team 3	Team 1	Team 2
Last dog watch	**1800–2000**	Team 1	Team 2	Team 3

Source: Wikipedia

M2 60mm Mortar
Source: Misc. for sale ads

Chapter Twenty-four

RETURN TO SAN FRANCISCO – C-46 FLIGHT

February 1942

The flight to San Francisco went by quickly for Pelham due to a brisk tailwind and repeated discussions with Admiral Nimitz. David introduced the Admiral and Lt. Wolfe to each C-46 Ginger Snap crew member before they launched. Lt. Wolfe briefed the Admiral on the Slingshot Phases during their first hour airborne. Nimitz was noncommittal about the operation and went back to work on a mass of messages and paperwork.

The flight crew had installed a desk for him to work at, and his aide had endless paperwork to sign, messages to read and send, and briefing folders marked SECRET and TOP SECRET. Before landing, David asked Nimitz if he should see their contact in San Francisco and continue with Slingshot plans. The Admiral said yes on both accounts and handed him a sealed envelope containing Layton's three questions for Kimicho. After that, the Admiral departed in an unmarked car.

Captain Harry "Rats" Mueller took David aside outside the aircraft. "Great to fly with you again, Commander. The crew is sleeping here tonight at NAS Alameda. You might want to join us for drinks and dinner. Also, we fly to San Diego tomorrow to get some work done on the plane. You're welcome to come along."

"Thanks Rats. Please take Lt. Wolfe under your wing and get us both rooms or a room if you can. I have to make some phone calls and will join you later. On your flight tomorrow, can you drop the Lieutenant and me at the Burbank airport and pick us up there on the way back?"

"Yes sir, no problem. We leave at 0900. How about we pick you up at 1400 from Burbank on the return flight in three days?"

"That would be perfect, Harry, much obliged," David answered.

Pelham went into the Ops Building, called the contact number for Kimicho, and asked for a meeting that night or early in the morning. A gruff voice with an Asian accent said he'd get a call back in thirty minutes, but he wanted a different number. When the phone rang in one of the other phone booths, David found himself talking to a young lady. She invited him to join Mr. Kimicho for dinner at nine o'clock. The lady instructed him to be carrying a briefcase at the NAS Alameda main gate in civilian clothes at eight o'clock sharp. She said he should look for a four-door black Cadillac and asked if Admiral Nimitz would also be free to join Mr. Kimicho for dinner. David replied in the negative.

His hand shook as he hung up the phone. *My God,* he thought, *How much do they know about us? We just landed, and they know I'm across the Bay. More importantly, they're aware that Nimitz was on the plane. How much more intel do they have? Is the Admiral just here to see his wife, or is there more? If the Yakuza have the information, then the Japs can know.*

Pelham thumbed a ride back to the BOQ and found he was rooming with Wolfe. He showered, shaved, shifted into civilian clothes, and decided to nap. David told Lt. Wolfe to find a briefcase, wake him in one hour, and arrange transport to the main gate. One minute later, he was out like a light.

The ride over to San Francisco proper was uneventful. Pelham tried to engage the chauffeur in conversation to no avail. They arrived at a different building from the one for the first meeting. He was patted down before ascending several floors in a service elevator. There were no markings on the elevator buttons. David was searched again by two menacing men outside two double doors with wooden dragons carved on each side. The doors opened, and Kimicho, in traditional Japanese dress, welcomed Pelham.

After exchanging three bows, he shook David's hand and welcomed him back to the "City by the Bay." They moved further into a penthouse with a commanding view of the blacked-out city. Kimicho introduced his wife and two granddaughters, who wore beautiful kimonos. After much bowing, giggling, and introductions with Japanese names that David couldn't remember, they brought him slippers and took his shoes. Then they wrapped him in a pink and white kimono with elaborate gold dragons embroidered on the front and back. Kimicho's wife ushered him into a dining room and served the men tea and scones filled with raspberry preserves.

"How was your flight David-san? I trust it was not as hectic as the one to Hilo?" Kimicho inquired.

David thought, *He seems to know more about me than I'll ever learn about him and the Yakuza.* "Yes, this flight was a piece of cake. We had a good tailwind and a perfect landing."

"And your ship, the Selfridge, were you disappointed when you finally found her?"

You knew she was sunk and didn't tell me, you bastard. Of course, neither did Nimitz, Lamar, or Tiny Wright.

"Yes, I was pretty down for one day. Then the task began to get her back afloat and in fighting shape."

"You have done an amazing job. I have seen pictures of the Selfridge and her weapons. Very impressive!" Kimicho raised his teacup. "To USS Selfridge, may she be victorious in battle and carry you safely through the war."

David raised his cup and clicked against his host's. "Hear, hear. Please tell me how you knew I was at Alameda and that Admiral Nimitz was on the same aircraft."

"One has ways of finding out these things if one asks in the right places," Kimicho slyly responded.

David showed his irritation, "My concern here is if you and your organization know his whereabouts, so can the enemy. What is there to stop the Japs from killing Nimitz when he visits San Francisco."

"Not much, my friend. He has only one bodyguard. I assure you that no one in my organization would take on such an assignment. He is here to meet Admiral King, who always stays at the Francis Hotel. His aides order specific suites, bourbons, cigars, and 24-hour room service. Half of the hotel service people in town know when King is coming. It is only a small leap to figure he'll meet with Nimitz and then monitor the handful of flights from Hawaii to this area."

"But military flights and who is on them is classified. No one in the general public should know what will happen," David asserted.

"Yes, David-san, that is true. But the Navy and Marines tend to be sloppy with their paperwork and more solicitous of their flag officers. They send advance information on car transport, billeting, and alert base commanders of their impending arrivals. Now the Army Air Corps is a different matter. We rarely know who is flying in and out of the Army's Hamilton Field in San Rafael."

"Thank you; we'll fix this security matter," David declared.

"Now that you have visited me twice in two locations, you have become a security matter yourself."
"How so, Kimicho-san?"
"The FBI is quite aware of who we are and what we do. They photograph everyone who enters the two buildings you have visited. It's only a matter of time before they identify and target you for questioning," the criminal boss replied.
"Thank you for the heads up. I guess I'd better start coming up with a cover story."
"Good luck. The FBI can be very persistent. What's in the envelope?"
"I think it can wait until we finish dinner; that and my special request."

Kimicho's wife came in, bowed, and served them wontons with a small bowl of soy sauce for dipping. They were delectable. David asked what they were eating, and Kimicho replied that it was Dungeness crab combined with Philadelphia cream cheese wrapped in a thin sourdough crust and fried in sunflower oil. One of the granddaughters then poured them a portion of white wine, and the other served them a salad plate of sliced duck, plum sauce, and thin pancakes. Next was a few ounces of freshly caught halibut in lemon sauce served on white rice. David asked how many more portions were to come, as he didn't want to insult the ladies by getting too full and unable to finish the dinner. Kimicho replied that there would be two more modest servings. His wife cleared the plates and wine glasses.

One granddaughter poured some deep red wine into crystal goblets, and the second served them small thick steaks that Kimicho identified as Kobe beef. It was the tastiest steak David had ever had. He could cut it with his fork. The finale was a scoop of vanilla ice cream with plump, juicy blackberries on top. Next, the ladies cleared the table and placed a brandy snifter before each man.

David held up his brandy. "That was truly a feast. Thank you and your family."
Kimicho responded by clinking David's glass. "That was no feast. But, you are truly welcome, now to business."
Pelham opened the briefcase. "This envelope contains questions from Lt. Comdr. Layton, Nimitz's Intelligence Officer."
"What do you think of the questions?" Asked Kimicho.
"I don't know what they asked."
"Why not?"
"I'm getting my ship ready for combat. I don't want to know more than what I need to know in case I'm captured."
"OK, and your special request, what is that?"
"It's vital to me and a test of the Yakuza's powers. I'm not just interested in information only. I want a specific type of person provided to me to serve on my ship."
Kimicho was intrigued. "What kind of person and when?"
"I want an Imperial Japanese Navy experienced signalman. He must be a five or six-year veteran who will serve with my crew and fight against Japan. I want him by the end of April."
Kimicho presented a poker face. "And if we can't provide such a person?"
"It will greatly hinder an operation I'm planning and increase the risk of losing my ship in combat."
"Of course, anything is possible with the right amount of pressure and money," Kimicho replied. "Are you prepared to compensate us for our expenses?"

"I have $5,000 authorized for the question answers and $5,000 for the Navy man."
"I'll see what we can do. Thank you for joining me for dinner. Please tell Admiral Nimitz that I would like to host him someday. My granddaughter will see you to the car."

He shook David's hand, bowed once, and was gone.

Ron Strong met Pelham and Lt. Wolfe at the Burbank airport and chauffeured them to USC the next day. Ron dropped Pelham off for an appointment with two scientists in the Chemistry Department arranged by Professor Nagy. David had previously briefed Lt. Wolfe on what they wanted from the General Electric lab in Pasadena. Strong drove Wolfe to the GE Lab and waited for him. Then he returned for David, who finished a late lunch with Michael and Rachael Nagy.

Strong and the two officers spent the night in the Beverly Hills Hotel company suite. Strong drove them down the coast to the Scripps Ocean Science Lab in La Jolla the following morning. After the two officers completed their meeting, he took them to the NAS San Diego Officers Club overlooking the Pacific Ocean. The three men had lunch and went to the BOQ, searching for Rats and his crew. Instead, they found Whitlock and Tex sunning themselves by the pool. Whit said the Chief was home with his wife and kids. Corporals Fitzpatrick and Jenkins, who shared the Plane Captain duties, were working on the plane. And Mueller was at San Diego Teacher's College, having lunch with his fiancée.

David thought, *"Whit, you may think she's his fiancée, but I know better." And I'm technically engaged to her sister.*

Whit and Tex invited Strong, David, and Wolfe to a beer bust at a waterfront house in Imperial Beach that night. Strong begged off and said he had to get back to L.A. David thanked him and asked Ron to arrange a shipment of equipment from the GE Lab to USS Selfridge a few days later.

That night, the four naval officers piled into a taxi and headed south for Imperial Beach with a couple of cases of cold beer in the trunk. They found the bachelor pad where big band music blared from giant speakers. Everyone in the small crowd had a beer bottle or a glass of beer filled from a full-sized, iced-down keg. Rats and Fitz arrived an hour later, escorting two attractive young women. David met Corporal Fitzpatrick and Rhonda Bessemer, lead engineer on the Marines' C-46 transport program. Fitz had told her that Commander Pelham probably saved the Ginger Snap and her crew during their third flight to Hawaii.

Rhonda shook David's hand and then hugged and kissed him on the cheek as she said, "Thank you for saving my friends."

David smiled at Rhonda and thought, *I know lots more about you, young lady, than you can guess. You were the one at the Curtiss-Wright factory who told Rats and his crew about the glaring deficiencies in the C-46 design. You helped them take an out-of-date testbed for the aircraft and turn it into an honest-to-goodness flying machine with critical safety modifications. Ginger Snap saved many lives during the fall of Wake Island. No doubt, she will be shuttling men and hardware to Midway during the coming months. And rumor has it you're married to Fitz, something that wouldn't go over well with the Marine Corps.*

Captain Mueller caught Pelham's eye and brought Ginger White forward; she hugged David and kissed him on both cheeks.
"My sister's dreamboat. I've heard so much about you!"

As the evening progressed, more young men and women arrived until about forty or fifty were dancing, drinking, and making out. Ginger found David in the crowd and asked him to dance. She wanted to know everything that he knew about her sister. When the song ended, she grabbed two beers, handed him one, and led him out on the veranda, where they could see the waves breaking on the beach. He described Betty's work after the Pearl Harbor attack and the Wake Island flights. He left out the affair with a married surgeon.

She said she moved to San Diego against her parents' wishes and told him about her classes at San Diego State. She was excited about her engagement to Captain Harry Mueller, but left out that they were married. He learned that she and Rhonda had an apartment in Old Town. Ginger took the bus to school. Rhonda had an office in the Navy aircraft maintenance Buildings on the Naval Air Station. She commuted on a motorcycle to Coronado via the ferry.

Pelham could see that Whitlock and Johnson had found a couple of young ladies to dance with and take on a stroll down the boardwalk. He asked Ginger if the Commando crew had any wheels. She said they had an old Packard, a late-model Ford coupe, and a Chevy sedan that the Chief drove.

Corporal Reggie Jenkins arrived with an older woman. Jinx immediately came over to Pelham, saluted, and they shook hands as both remembered the harrowing flight to Hilo. Jenkins was the sixth "Rat Pack" member who traded off with Fitz as Plane Captain. Jenkins crewed on some trips but usually worked with Miss Bessemer at the Assembly and Repair Depot (ARD). They were responsible for modifications to Ginger Snap and technical requirements for the Navy/Marine C-46 aircraft procurement. Jenkins said he was billeted in enlisted quarters at NAS San Diego. David asked him if the Ford had a base pass and could Jenkins drive him to the Marines' Camp Eliot the next day. The answer was yes on both accounts, so they set a time to meet at the BOQ after breakfast.

David spent some time with Harry Mueller and asked about the Chief. Rats said Chief Watkins was not much of a drinker and spent every hour he could with his family. He advised David that the flight to San Francisco in two days would take off at ten hundred, and they would RON at NAS Alameda before flying to Hawaii. David razzed him on whether they would have enough gas. Rats replied that they had added a 150-gallon self-sealing bladder tank to the lower forward cargo bay. And he added that the Chief, Fitz, Rhonda, and Jenkins were upgrading the water/methanol injection system. The new rig would give the Commando more combat power on demand for a few minutes longer than the first installation. At 2200, David decided not to bother Lt. Wolfe, who had a lovely young lady sitting on his lap and laughing at his jokes. David said goodbye to the Commando crew, found a ride back to the base, and gratefully fell into his rack. Momentarily, he wondered how long the youngsters would be partying before sinking into a deep sleep.

Jenkins drove Pelham and Lt. Wolfe north on Highway 395 to Camp Elliot the next morning. They showed their ID to the Main Gate sentries and talked their way onto the base. After several queries to base personnel, they drove to the artillery range. They found a couple of Gunnery Sergeants willing to brief them on the Marines' inventory of mortars. First, they received a classroom blackboard chalk talk about the different types of tubes and ammo. Then the Sergeants took them to a hangar where various mortars and shells were displayed. They said it was hard enough to hit a target on land; they couldn't imagine firing from a heaving and rolling ship. David and Lt. Wolfe thanked them and left the base with a good idea of the mortars and types of ammunition they could use but no solution for accurate firing from a pitching and rolling destroyer deck.

When the C-46 crew and passengers landed at Alameda the next day, The BOQ Desk Clerk handed David an envelope. An invitation read, "Admiral Nimitz and Lt. West request the pleasure of your company at a cocktail party in the President's Suite, Francis Hotel, at 1700 hours."

Pelham handed it to Wolfe, who read it over and said, "I hope you're going to have a good time with the brass. Do you want me to arrange your transportation downtown?"

"Make that transport for both of us. Keep this on the QT, but I suspect that it's not just our Admiral at the hotel," David replied.

They got cleaned up from thc flight, pressed their service dress blue uniforms, and arranged for the BOQ van to take them into San Francisco and return. On the way into town, David briefed Wolfe on how to present to Admiral King in case that was the real purpose of the invitation. He told the Lieutenant to keep his sentences short and to the point and expect to be interrupted, even rudely interrupted. He said to wait for King to finish, then continue with the brief as if nothing had happened, and not get upset or angry.

Nimitz's Aide, Lt. West, greeted them at the hotel suite. He made them a drink from the well-stocked bar, and they made small talk when the inside double doors opened. Admiral Nimitz and King in civilian clothes, with King's Flag Lieutenant following behind, joined the other officers. Admiral King made a beeline for David, offered his hand, and said, "Commander Pelham, good to see you again. I hear you're cutting a wide swath through Pearl Harbor just as you did in D.C."

"Thank you sir. Congratulations on your promotions. May I present Lt. Donald Wolfe, my Operations Officer on the Selfridge."

King shook Wolfe's hand. "Nice to meet you, Lieutenant. How are you getting along with your new CO?"

"He sets a fast pace, Admiral," Wolfe replied. "My job is to turn his best ideas into feasible action.".

Nimitz introduced Wolfe to the two Aides.

King announced, "We're going to make ourselves some stiff drinks; then I'm expecting you Selfridge men to brief us on this special operation you are proposing to CinCPac." He poured himself a hefty whiskey on the rocks, "Who says you can't mix business with pleasure?"

David frowned. "Admiral, let's make this a flag briefing only if you please. The more people that know about this plan, the less chance we have of succeeding. We already have five officers briefed, counting Admiral Nimitz. So you will make six, and that's how I want to keep it until we're one week from sailing."

King took a swig of his drink, shrugged, looked at Nimitz, and winked. "Same old Pelham. He acts like a four-striper, and we're green midshipmen who just reported onboard. If he weren't so effective, most of his seniors would like to court-martial him for insubordination."

David laughed. "Admiral, I'll make you a deal. After Lt. Wolfe briefs you, we will do so if you believe we should brief more people."

"You are going to pass up the chance to brief two four-star Admirals and have your Lieutenant speak for you?" King said with surprise.

"Absolutely Admiral. I describe the mission, technology, and players. Lt. Wolfe works on the specific tactics and documents what we plan to do."
"Pelham, you never cease to surprise me. Lt. Wolfe, what did your CO tell you about briefing me?"
"Sir, he told me you would interrupt the brief with your comments, to hear you out, and then proceed where I left off."
King's face turned red, he looked from Wolfe to Pelham, who nodded in agreement, and the four-star burst out laughing. "Let's get to it. I think I'm going to enjoy this!"
They entered the master bedroom anteroom, where a blackboard and several chairs were in place. Wolfe started the Slingshot brief while David drew a diagram of the ships in an IJN dual-carrier battlegroup on the board. Wolfe had no more than stated the operation's intent to sink one Jap carrier and its name before King asked, "Why Operation Slingshot? What does the name signify?"
Wolfe was surprised, so David answered while rapping his chalk on the diagram. "This dual-carrier Jap battlegroup is Goliath; I'm David in my feisty Leader Destroyer. I mean to slay Goliath with Operation Slingshot."
King laughed, sipped his drink, and said, "Carry on."
Wolfe covered each planned phase, and Admiral King took it all in without comment.
At the end of the brief, David commented, "If we pull this off and put real doubt in Jap sailors' minds about the supernatural aspects, we need to capitalize on that by not crowing about the victory. There should be no press releases about a U.S. Navy victory. Instead, we should leak a rumor to several large papers that Japan lost a carrier to forces led by King Neptune and Davy Jones."
King lit a cigarette and looked at David with disdain, "That's the biggest crock of shit I've heard since the war started!"
"Yes sir."
"Why would the Japs believe in such nonsense? Davy Jones, my ass!" King declared.
"Has the Admiral ever heard of the Columbus Effect?" David asked.
King dragged on his cig and blew smoke into the room. "Can't say that I have."
"Columbus kept two navigation logs. The first showed how far his three ships had sailed. He didn't share that with his crew as they were highly superstitious and worried they could sail off the edge of a flat earth. The second log he shared with his crew showed much less progress and helped allay their fears."
King looked bored. "So what? How does this relate to the Pacific?"
"Think about it, Admiral, the attack on Pearl Harbor probably is the farthest the IJN has ever sailed in full strength to give battle - nearly four thousand miles from home. I figure that even 2,000 miles to Wake makes these superstitious sailors nervous and jumpy. They aren't afraid to die, but they're petrified that they may end up in their version of hell. An underworld of purgatory, gloom, and continual unhappiness. Showing them some strange lights and scary sounds, plus the terrifying loss of a carrier in five minutes, kicks them off balance and feeds their irrational fears."
"You really think you can sink a carrier in five minutes?" King challenged.
"Actually, Admiral, that may be the easiest part of the mission. Once we're in position, close aboard on her starboard side, we can blind and strafe her bridge crew. We can set her flight

deck aircraft on fire and punch so many holes in her side that only immediate counterflooding can save her from capsizing. Getting to that ideal firing position and fighting our way back out of the formation is tricky and dangerous."

Nimitz added his voice. "Admiral King, you wanted aggressive action. Here it is, pitting a destroyer against a battle force in the dead of night. We might lose the destroyer but also bag or damage a carrier and make the enemy think twice about attacking Midway and Hawaii again."

King responded, "Pelham, if it were anyone else but you, I'd think this is a fool's errand. You have a knack for bending the rules. No doubt you're going to commit a lot of court-martial offenses to get this done. Don't answer that. Tell me about these technology initiatives you alluded to."

"Sir, we're making changes approved for the Porter Class, like removing topside weight by shedding the rear gun director and mast, removing torpedo storage lockers and most of the torpedoes, adding improved anti-aircraft guns, removing our damaged aft torpedo tubes and substituting a larger gun and a gyro glider…"

King held up his hands, "Whoa! Whoa there, where are you going to get this gun?"

"We've found a six-inch artillery piece with the Army that fires a shell that can penetrate the side armor of a carrier and explode amidships. In five minutes, we plan to put ten or more shells into the CV's gut. The gyro glider is being tested as we speak. It will fly to 300 feet high or more, greatly extending our visual horizon. Our ship is in dry dock getting a new treatment on the bottom, a slippery compound from Dupont that should up our speed by nearly twenty percent."

King interrupted, "From what to what?"

"From thirty-seven knots to forty-two knots. We're adding shooting platforms for Marine snipers. During the attack, they will start the carrier planes burning and kill anyone in the open, trying to shoot back at us. We just negotiated for the latest sonar gear from General Electric. Under certain water conditions, we can detect a carrier group as much as fifty miles away. We're replacing most of our slower ships' boats with speedboats. They can harass the enemy at up to thirty to forty miles an hour, play amplified strange sounds, and display weird lights to heighten the supernatural effect. USC scientists have devised a combination of liquids that we can inject at the bow and stern to dampen the brightness of our bow wave and propeller wake below fifteen knots."

King finished off his drink and lit another cigarette. "How in the hell are you going to do that? Get rid of the wake?"

"Sir, we must fine-tune the solution, but it involves pumping crude oil and liquid detergent overboard together in an 80/20 proportion. It dampens the plankton phosphorescence and produces tons of dark bubbles on the surface that block most of the glowing light," David answered.

"You have more, don't you, Pelham? Things that will get your Admiral and this one in trouble if you tell us," asserted King.

"Sir, certainly we have some dirty tricks up our sleeve that you don't need to know about."

King looked at Nimitz, "They're going to have all the fun while we are stuck in ivory towers pushing chess pieces around the oceans."

Nimitz smiled. "Yes, it's their turn to take risks and reap the glory. Lt. Wolfe, that was an excellent brief. The proof of that is that your boss and I've never heard a complete brief that Admiral King did not interrupt. Now that most of the Slingshot cards are on the table, I suggest we adjourn to the bar and enjoy the company of our fellow officers."

They enjoyed a couple more drinks and the light snacks by room service. Then, as David and Wolfe left, Admiral King walked them to the door and shook their hands.

He said, "If Operation Slingshot is a go, you have my blessing to tell the Selfridge crew at some appropriate point that you briefed me, and I said, 'Give them hell, Selfridge, remember Pearl Harbor!'"

Ginger Snap landed at Hickam AAF Base at twenty-three hundred hours the next night. David was carrying his bags from the plane and noticed the Army coupe beside the Ops building. He wondered if it could be Betty and waved. The lights blinked, and he went over to the car.

"Hi sailor, need a lift?" Betty asked.

"God, you're gorgeous, but I'm beat and not up for a Honolulu hotel tonight."

"Good, you're in luck. I'm now sharing an apartment near Tripler with another surgical nurse. She's gone for the weekend, and I just finished a forty-eight-hour shift. You're not the only one who's tired. My bedroom has two beds, and tomorrow is Sunday, so I suggest we sleep in, and you can make breakfast."

David tossed his bags in the back seat, slid into the passenger's seat, and sleepily mumbled, "Do we have to sleep in separate beds?"

Betty looked over at her slumbering sailor and quietly said, "No, my love, unless you start to snore."

The following day, he got up before 0700, made coffee, and prepared scrambled eggs, bacon, and toast. She joined him in the kitchen, filled a cup of coffee, and hugged and kissed him on the neck as he flipped the bacon. Then she squealed with delight like a little kid when she saw the Sunday editions of the Los Angeles Times and San Francisco Herald on the table. Martial law in Hawaii had throttled news flow from the Mainland to a dribble.

David decided to give himself the morning off before boarding the ship at 1300. They shared items in the papers and read the comics to each other. David looked at Betty and thought, *This is nice, almost like a husband and wife relaxing on a Sunday morning. She's a striking woman; looks damn fine in the morning, even without makeup. I hope we make it back from Slingshot. I want more Sunday mornings with this gal. I'm sure even Lorraine would approve.*

Betty picked up on his pensive look. "Hi there, Captain; what are you thinking?"

"That this is very nice, relaxing with a beautiful dame on a quiet Sunday morning," he responded mockingly.

"That's all well and good, David Pelham. You take a shower and shave while I pick up the kitchen. You'll then find me in bed nakedly reading the paper."

He looked her up and down in her bathrobe, "That sounds very nice, but those beds are rather narrow."

She swatted him with a rolled-up newspaper and said, "I trust that an intelligent and resourceful naval officer can figure out how to push them together. Now git!"

After Betty dropped him off at YFD-2, David started climbing the ladders in the dry dock. As he reached the bottom of the last one, he saw a sailor lookout on the top deck scurry away. He thought, *Nimitz was correct; there's nothing like being CO of a Navy ship. Whatever they've in store for me will be unusual, and I'll accept it as what I expect from an outstanding destroyer crew.* He walked out on the top deck and found the ship's Quarterdeck was now aboard Selfridge with several crew members waiting for him in service dress whites. He was in short-sleeved working khakis and felt underdressed as he saluted the American flag on the fantail.

Captain Pelham then turned and saluted the OOD, stating, "Reporting my return aboard ship, sir!" The bosun piped him aboard, and a 1MC broadcast could be heard in all parts of the ship, "Selfridge arriving," as the ship's bell rang four times.

David shook hands with all the men on the Quarterdeck. There were two that he didn't know or could not remember. He asked their name, rate, division, and home city, something he'd do with every man on the ship until he knew them all by sight. David looked around and admired the shiny brass bell, braided white decorative ropes, and polished brass shell casings roped together to form a walkway with braided manila rope.

He spoke to MacGregor. "Boats, the Quarter Deck looks outstanding. It's good to be back onboard. How are we doing?"

"Captain, as far as creature comforts, we have all the heads operating. We're ready to make our own freshwater and have verified the generators and boilers are operational. All the ventilator motors are running, and the snipes say they're ready to bring up ship's power. We have over half the crew moved onboard. The other enlisted men are now in the floating barracks. We had a rope yarn Sunday this morning, but the watch crews will be working the rest of the day and night."

"That's an excellent report, Boats. And the bottom, how goes the bottom?"

"Sir, we have a single coat of the new paint on two-thirds of the bottom and two coats on one-third. We should finish by Friday night."

"Outstanding Bosun MacGregor. If we give the last coat forty-eight hours to cure, we should be able to float her Sunday night and sortie midday on Monday."

"I think that will work, Skipper. The XO and Division Heads have more details for you on the Bridge. They've put together a welcome back buffet up there."

"Well, I'm starving. I better not keep my officers waiting," David said over his shoulder as he walked aft on the ship's starboard side. He noticed that the green slime was gone, and there were many new paint or primer areas as he bounded up the ladders to the Bridge.

As Pelham entered the Bridge, the XO sounded off, "Attention on deck!"

David responded, "As you were gentlemen," and shook hands with each of his Division Heads and Chief Gonzalez from Supply. "How are you settling in, Chief? Any regrets?"

"None sir, this crew is the best, and we should have the ship fully stocked with food and sundries this week. He pointed to the narrow table filled with sandwich makings, condiments, and salads, "Now if you gentlemen will excuse me, I need to get down to the Chief's mess and get some real chow."

David and his six Division Officers started building sandwiches, helped themselves to salads, and grabbed an iced tea, lemonade, or water before they all settled around a narrow

table. They all were interested in any news from the states and asked David and Lt. Wolfe what they observed that had changed with the war.
After finishing his sandwich and washing it down with lemonade, David said, "It's a real pleasure to be back onboard. We'll make this our first Department Head meeting. Lt Wolfe, please take notes for the record. I commend all of you on the condition of the ship and our progress in reinstating ship services. I think we should start with manning. We'll have each of the Divisions report their current manning and how many you'll have onboard when we sail on Monday. XO, could you start us off with an overview?"
"Yes sir. We've been gaining new men nearly every day. We have a target of sailing with 177 enlisted and fourteen officers. We've been allocating nonrated men and strikers to the appropriate Divisions so they can advance in their desired rates. While you were gone, we evaluated combat manning for all Divisions. We need more men, probably thirty more spread amongst the Departments."
"Thank you. We can go into more detail later. First Division report," David said.
"Skipper, 1st Lieutenant, and Damage Control Officer Jeffrey Jones speaking. First Division is planning on twenty-four men next week; we have them onboard."
"Ltjg Bob Campbell, Captain, Second Division is expecting thirty men and will probably be three short."
Pelham said, "Just for my review, I have the ship's organization manual here. Let's see, First and Second Divisions are responsible for manning the deck crews, maintaining general equipment throughout the ship, operating the small boats, and construction. You have the Boatswain's Mates, Quartermasters, the Coxswains, ordinary Seamen, Carpenter's Mates, and Shipfitters, right?"
"Yes sir, and some other miscellaneous strikers looking to advance to Petty Officer," Jones answered.
"Great. Gunnery, which apparently will be called Ordinance soon, report."
"Captain Pelham, I'm Lt. Randy Kindall, your Gunnery Officer. We have an allocation of thirty men who should all be here by Monday. To meet our combat watch standing and gun requirements, we really need thirty-nine to forty-five men. During combat, we augment the gun crews with other rates on the ship."
"Noted, and the billet list tells me you have the Gunner's Mates, Torpedomen, Fire Controlmen, Radarmen, and nonrated that want to go into those rates. If that's all correct, we move on to Engineering."
"Engineering will sail with sixty men, twenty in each section. We need nine to twelve more for sustained operations," reported Lt. Comdr. Haldane.
"Thank you. Your crew is Machinist's Mates, Water Tenders, Electrician's Mates, Firemen, Boilermakers, and Metalsmiths, correct?"
"Yes sir."
David continued, "Navigation or the new term, Communications?"
"Sir, I'm Lt. Skip Gehres, Assistant Navigation Officer to the XO here. We expect to have twenty-seven men on sailing. When we add more Radars, we'll need more men."
"Roger that," David replied. "You have the Radiomen, Radio Technicians, Sound or Sonarmen, Yeomen, Signalmen, Pharmacist's Mates, and the new Radarmen Strikers. Our last Division, but not least, Supply."

"Lt. James Baxter here, Captain. Thank you for recruiting Chief Gonzalez; he's a real gem. We will sail with fifteen men. Eighteen would be better for our next at-sea period and would be best for longer cruises."
Pelham responded, "I understand that more sailors will make it easier to fill all the billets and wartime watch standing requirements. We'll plan on 197 enlisted and 15 officers in May. You'll all have to lobby the XO for your fair share of that increase. Back to "S" Division, you have the Stewards, the Cooks, the Bakers, Storekeepers, Mess Attendants, and Supply strikers, right?"
"Yes sir, that's right, and we have the Temporary Additional Duty men from other Divisions. The TAD men man the mess deck, Chief's mess, Wardroom, and run the ship's laundry. They also take care of the officer and chief's quarters." Baxter replied.
"Think about it," David said, "We've made tremendous progress in three weeks. We'll be back at peacetime manning by Monday. In the meantime, we're getting the ship into fighting trim. MacGregor tells me that we should have full ship services this week and the bottom paint done by Friday night. Is there any Division that won't be ready to sail at ten hundred hours on the coming Monday?"

He looked around the table; none of his Division leaders raised their hands.
"Good, excellent, now I want a status report from each Department. Make it a how gozit to share with your counterparts, so we all know what is going on. I plan to have this meeting every week. Let's start with Engineering."
Lt. Comdr. Haldane spoke. "Skipper, we're ready to bring the boilers and evaporators back online; that'll give us ship's power and freshwater. We tested the pressure integrity of the boilers and the steam fittings while you were gone. They tested satisfactorily at low pressure. We won't know about the high-pressure integrity until we have her at sea and go All Ahead Full."
"Excuse me, Engineer, how did you test the boilers without my knowledge?" David challenged.
"Captain, we used an old trick not to use the oil burners, which would've violated your standing order. We cleared it with the XO and built fires under the boilers with wood. It was a little smoky, but it worked like a charm. With your permission, we will test the entire power plant tomorrow."
"I'm all for it except for the men working on the bottom. Will we be putting them at risk?" David questioned.
"No sir, only if we have to have an emergency release of steam and hot water."
"Thank you. We'll wait until Saturday morning to light off and shift to the ship's power. Anything else, Engineer?"
"No Captain, we'll be ready first thing on Saturday. We've tested all the connections to the Bridge and Secondary Conn. We have all of our instruments back from the Yard, and my men could use some liberty."
"Good point. XO, what's our watch schedule this week?"
"Captain, I recommend we shift to eight on and eight off with two section liberty on Friday and Saturday. We give the last section Cinderella liberty Sunday and move to our double-dog watch and four-hour watches at 2400 Sunday night."
"Sounds good, make it so. First Lieutenant, your report, please," David responded.

"Skipper, we should be ready for sea by Sunday night. There will be gaps on the bottom where we're resting on the blocks. To take care of them, we'll need to return to YFD-2 after our at-sea period for five days."

"Thank you. Any flak from the Yard and our friend Mr. Price?" David asked.

"No sir. They fixed the battle damage and got us safely started on our cleaning and routine repairs. I think they're ignoring us, so whatever we do on our own is not their fault," Lt. Jones answered.

"Is there anything we need from them?"

"Just the use of the dry dock in three weeks. I asked Lt. Kordowski to save us five days. While we're gone, they will bring USS Shaw back in, put a false bow on her, and send her to the States. Oh, one more thing, Lt. K. would like duty onboard Selfridge."

"Engineer, what say you?" David asked.

"Captain, I'd love to have Kordowski in Engineering. He's smart as a whip; no one works harder."

"XO, check with Smitty and see if we can transfer him in after our next stint in dry dock. Gunnery Officer, report."

"Yes sir, we've disassembled the forward guns, completely cleaned them, and put them back together. There was a small amount of seawater that got into the forward handling rooms. We have them cleaned up, and we're starting on the aft guns. We have an unusual situation where we would like all the ordnance manpower available during the day to work under the direction of Gunner Phillips. If you and the XO agree, we would like to stick with twelve on and twelve off until we sail."

Lt. Lehman answered, "OK by me. But, remember, your men will still have other watches to stand at night."

"Yes sir. One more thing. I think we're gonna need the Yard's help to install the Bofors guns. None of my crew has experience with them, and they won't arrive with an installation manual."

"Noted. I'll talk with Mr. Price and see if they will help us out during our next inport period. Navigator, what kind of shape are we in for sea duty?" Pelham asked.

"We'll be ready, Skipper. A couple of questions, did you pick ten hundred on Monday after looking at the tides? And do we plan to use a Pilot through the Pearl Channel?"

"No on the first one, and yes on the second. We draw less than thirteen feet; do we need to worry about the tides here?"

"Normally no, Captain, but there is a standing order that all ships will enter or exit port near high tide unless there is an emergency or operational necessity."

"What else do you have?" David asked.

"On the Admin side, Yeoman Brown and his striker are keeping up with the new men and their service record requirements with help from Petty Officer Smith, who is still ashore. Will he be sailing on this at-sea period?"

"I don't know. XO, where do we stand on the gyro glider testing?"

"Skipper, we had some stability problems, and Smitty got pretty banged up at first. We were pulling him with a car and had a couple of handlers running beside him with lines on the aircraft to keep him steady until lift off. The copter crashed nose down, right-wing down, and left-wing down on the first three tries. Smitty now has a couple of banged-up ribs and a cast on his left arm to heal a hairline fracture. We called the factory and found we were supposed to put stops on the controls to limit their movement.

After getting patched up, Smitty insisted that we try again, so we set the rig up in the back of a stake truck. We had four strong handlers lift him up on five-foot poles. At about fifteen miles per hour, he could start to climb a bit. The good news is that the bird was stable, and we could reel him back in before hitting the end of the Hickam runway."

"Good work, XO. That would be about twelve knots for the ship; any kind of headwind would make it easier still."

"Yes sir. We added some padding and a seat belt. Boats found him a football helmet, and we got him up about thirty feet on one run," Lehman said.

"Let's let him heal and install it on the ship when we return. Navigator, I think Smith can make the biggest contribution by staying ashore and lining up the rest of the crew. What else?"

"Sir, we got the first radar set installed, and we're training two men to work the display in the Nav Plot aft of the Bridge," stated Gehres.

"What ranges can we expect?"

"Hard to say sir, until we get more experience with it. Our new Third Class Radarman says probably ten to twenty miles on a medium-sized aircraft and ten miles or more on ships. We put it up on the forward mast, which is our best chance for longer range. Our Division is thin on medical personnel. We rate a Chief or First-Class Pharmacist; so far, we only have a Second-Class and a Seaman Striker."

"How about Signalmen and Soundmen?" David queried.

"All the billets are filled with the authorized rates," Gehres answered.

"Great. Plan to install some newer sonar gear this week if it's air shipped, as I expect. Supply Department, what's your status?"

Baxter replied, "Captain, we have most of our food onboard; we have a good crew to man the galleys. We were lucky that many of the ship's supplies, other than perishables, are undamaged and reusable. One question is about ammunition. We have seventy percent of our five-inch shell load. Are we going to be firing the guns a lot? Enough that we should bring that up to 100 percent?"

"No, we should be OK with a single gun practice per day, but see if you can get us up to ninety percent for the next cruise period and 110 percent for our May cruise."

"One hundred and ten percent, sir? How are we going to do that?" Baxter complained.

David was annoyed. "I don't know. You're the Supply Officer. Find a way to get those five-inch gun supplies, and don't forget to order the Bofors rounds. Gentlemen, you've made fantastic progress during the past two weeks. One more week should see us fine-tuning the ship and the men into a hard-hitting fighting unit at sea. In April, we'll be training with a carrier battle group.

Lt. Wolfe and I met with CinCPac Fleet staff about a possible combat operation in May. It's in the planning phase; that's about all we can say about it."

Early Monday morning, David was sitting in his Captain's chair, sipping his second cup of coffee, when the OOD approached him and said, "Good morning, Captain. XO reports no serious problems, Engineering has steam up and reports ready to sail. We have light breezes, good visibility, and it's seventy-two degrees with seventy percent humidity. Ops says we will have two tugs by 0730."

Pelham knew it had been a hectic night as the crew took dozens of actions to ready the ship for floating before midnight. The 2400 to 0400 Duty Section completed the separation from all dry dock equipment connections and services. The Special Sea and Anchor Detail were set at 0600 as the first light crept into the eastern Oahu skies. When the detail was called away, David was on the bridge observing the OOD and JOOD working through the Pre-sail Checklist. Selfridge had been floating for six hours without any serious problems. He knew that the XO and 1st Lt. were moving about the ship, ensuring no leaks, electrical short circuits, or fires. No doubt, Haldane and his snipes were extra vigilant in Engineering and monitoring adjacent compartments for leaks and equipment failures. Additional watertight integrity and fire watches inspected all the accessible spaces below the ship's waterline.

The Morning Watch OOD approached David. "Sir, I have to admit I don't know how getting out of dry dock works. Any recommendations?"

"Lieutenant, that makes two of us. Ops has researched it and will be on the Bridge to advise you. You will have the conn unless we formally take it away. Also, we may have a Harbor Pilot up here before 0800 that could have experience. Light winds are a big help. As I understand it, we set our rudder amidships and engines ready to reverse. One tug has a line to our stern, and we winch in on our stern lines to start moving backward. The towing tug slowly eases us out of dry dock, and the Bosun's crew retrieves and stows our lines. We don't use our props until we drop the tow line and have room to maneuver. If necessary, the second tug will be on our leeward side to nudge us into the wind. It's a piece of cake. Don't you worry! I'm going to my head and shave. Be right back."

The OOD pointed forward toward the bow. "Sir, it looks like Boats is trying to get your eye."

David looked down from the Bridge and saw MacGregor up forward, standing near a bowline. Boats pointed to the two lines on each side of the bow and then aft with both hands. He held one hand up with four fingers upraised. Pelham understood that four lines kept the ship tied to the dry dock and gave Boats a thumbs-up. David had the JOOD report on whether the starboard gangplank was still in place while he checked the unobstructed calm water aft of the ship from the Port Bridge Wing. It was an idyllic tropical morning; the sparkling turquoise waters of Pearl Harbor beckoned.

USS Selfridge was slowly pulled out of RFD-2 on a sunny February morning by a fleet tug. Another was standing by to push against the ship's side in case the crosswind was too strong. The destroyer had flags and pennants strung from bow to stern. Red, white, and blue bunting was coiled around the eight five-inch gun barrels as she slowly sailed north along Ford Island. Once at the end of the East Loch, the ship stopped at the red buoy marking her former resting place on the bottom, where fifteen crewmen died. Selfridge sailors placed a large wreath on the floating marker. Then, after the First Division secured the small boat back onboard, Selfridge swiveled 180 degrees and pointed east. The Bosun's whistle shrilled, and the 1MC speakers announced, "All available hands in Sections One and Two, man the starboard rail, working uniforms allowed, covers required."

Bosun MacGregor boomed as the ship approached the buoy, "All hands salute our lost shipmates in turn." Each man bid his goodbye with a salute as Selfridge passed by the buoy. When the last honor was rendered, the men on deck heard Pelham's voice from the 1MC speakers. "Good morning, men of Selfridge, on deck, parade rest! We will be sailing by

Destroyer Row. As we reach each ship, we will fire our ceremonial gun. We will come to attention when they fire theirs and salute them for their selfless support to our ship."

The celebration ceremony arranged by XO Lehman and Boats MacGregor went off without a hitch. Each of the six destroyers in the nest had a large group of men on the bow or stern who saluted back, waved their hats, and gave a great cheer to their fellow destroyer men. David had a lump in his throat as they approached Phelps, and he saw the CO, XO, and Commodore Early return his salute from their Bridge Wing. As DD-357 sailed south along the east side of Ford Island, the bunting and colorful flags were removed and stowed. A Pearl Harbor Pilot had boarded the ship at the floating dock; he now had the conn. The Pilot guided the OOD and Helmsman out through the open submarine net at the harbor's entrance and into the channel along Hickam Field leading to the Pacific Ocean. Once they crossed the bar and started picking up the rolling ocean waves, he handed back control to the OOD, who kept the ship on the starboard side of the ocean channel marked with green buoys. As they reached the second-to-last green buoy, the Pilot Boat came alongside and retrieved the Pilot.

USS Selfridge (DD-357) and her crew were back at sea!

Chapter Twenty-five

HAZE GRAY AND UNDERWAY – DD-357

February 1942

Lt. Comdr. David Pelham, the acting Captain of USS Selfridge, was perched in the Captain's Chair on the starboard side of the 357's Bridge. He had a clear view of the forward guns and the ship's bow knifing through two to three-foot ocean swells. David could smell the spray of frothy white saltwater flung from the bow when it dipped into the azure Pacific water.

It was a piece of cake! She sailed out the channel like a dream. We dropped the Pilot, cleared the channel buoys, and Ops announced we were in international waters. The OOD secured the Special Sea and Anchor Detail and ordered All Ahead One Third. That's all we'll ask from the engines for the first 24 hours. My standing orders take us around Molokai and the north side of Maui at one-third throttle or less. Place the ship south of Kahoolawe at the entrance to Kamohio Bay at 0700 on Tuesday. Remain at all times ten miles from all land until the approach to Kahoolawe after sunrise. The crew needs a few days to get comfortable with each other, the ship, and the daily routine. Nice to have leaders like Boats, the XO, Ops, 1st Lt., and Haldane. By now, the dock lines and boats are secure, and the ship's weather decks are prepared for cruising. No doubt, engineering has already pumped the bilges, blown the tubes, and is evaporating seawater at full speed to produce the maximum ship's freshwater.

His phone let out a shrill sound. "Captain on the Bridge," David responded.

"Captain, this is the XO in Secondary Conn. Just wanted you to know that everything is going smoothly, almost too smoothly."

David smiled; he only saw the XO once or twice a day as they both agreed that being in the same place at the same time for extended periods was to be avoided. "Thanks XO. We have a great crew. Think about it; you and I benefit from hundreds of years of training and experience, Captain, out."

He picked up the Plan of the Day and thought, *The ship's POD puts everyone on the same page, but it's only a plan. Dogging the watches gives every man a different daily routine and helps Supply feed the entire crew dinner and supper during the same hour. One-third of our men operate the ship around the clock, and the rest are on call. We're cruising at Wartime Condition 2. The Bridge is fully manned, and lookouts are posted. Engineering operates on half boiler power, and separate engines for our two screws. Guns, depth charges, and torpedoes are ready for immediate firing. Training occurs everywhere in Selfridge as the Chiefs and Petty Officers show the men what is essential and what to avoid. The XO laid out the underway plan and daily routine for the entire crew in the POD:*

"USS SELFRIDGE - DD-357 - Plan of the Day, 16 February 1942

2400 – Ship is floating

0000 – 0400 Midwatch – Section One, secure from dry dock services

0400 - 0800 Morning Watch – Section Two

0545 – Reveille

0600 - Set Special Sea and Anchor Detail, man guns as directed by Standing Orders

0630 – Set Condition YOKE throughout the Ship

0630 – 0730 Breakfast, Sections One, Three and Four
0730 – Morning Quarters
0745 – Officers Call
0800 – 1200 Forenoon Watch – Section Three
0800 – 0830 Breakfast Section Two
0800 - All Hands Turn to and commence ships work
0815 - Ship is underway
0845 – Wreath-laying ceremony in East Loch
0900 – Man the starboard rail, render honors to Destroyer Row
0930 – Set Condition ZED
1000 – Ship is at Sea
1115 – Knockoff ships work
1130 – 1230 Dinner, all hands
1200 – Twelve o'clock Reports
1200 – 1600 Afternoon Watch – Section One (be ready for GQ drill)
1300 – All hands continue ships work
1600 – 1800 First Dog Watch – Section Two
1600 – Knockoff ships work
1730 – 1830 Supper, all hands
1800 – 2000 Last Dog Watch – Section Three
2000 – Eight o'clock Reports
2000 – 2400 Evening Watch – Section One
2015 – Movies on Mess Deck and Wardroom
2200 – Taps, lights out, evening prayer."

After receiving the Eight o'clock Report summary from the XO and an update from the Engineer on the power plant, David was drained and tired. He made sure the OOD and JOOD understood his Standing Orders. They were to wake him at the slightest hint of any ship problems or closing targets. He took a shower in his Inport Cabin, returned to the cramped cabin behind the Bridge, stripped to his skivvies, and was immediately asleep. He heard eight bells as the First Watch ended. He got up, took a piss, put on his robe, and stepped into the tactical plot from his cabin.

"Navigator report."

"Captain, we're thirty miles east of Maui, heading 180 at twelve knots," replied the Assistant Navigator.

"Very well. Radar report."

"Quiet night, Captain, no contacts.

"Can you see the land?"

"Yes sir, I have a good return on the volcano and a couple of hills."

"Show me," David directed.

The Radarman pointed to the glowing screen. "This glob of radar hits is Haleakala on Maui."

"Why do your track plots make it seem to be heading north?"

"Sir, that's our relative motion to land. As we sail south, a stationary target appears to move opposite our heading and at the same speed."

"So if a target, say a ship dead ahead, is coming directly at us at twelve knots, it would seem to be tracking the reciprocal of our heading at twenty-four knots?"
"Yes sir. We may be closing him at twenty-four knots, but after a few minutes, we can estimate his twelve-knot speed."
"And if a target was overtaking us directly astern at twelve knots?"
"We can compute that his speed is twenty-four knots by applying our relative motion."
"Train me on a target ninety right at twelve miles, speed unknown," David challenged.
The Radarman marked a small x on the scope. "When we first see him, we mark the echo or hit and note the time. As we track him, we don't want to see constant bearing and decreasing range."
"Because that would mean we're on a possible collision course. In that case, what do you do?"
"Skipper, as soon as we see a high-speed target, an aircraft within thirty miles, or a ship within ten miles with constant bearing decreasing range, we notify the OOD and provide the bearing and range to the target. We can estimate his course and speed by taking out our relative motion if we've tracked him for five or six minutes."
"Thank you, sailor. It's the first time I've seen Radar. Damn impressive."
David stepped onto the Bridge and heard the Messenger announce, "Captain's on the Bridge!"
"As you were, men. OOD."
"Sir!"
"I have three things for you. One, don't let me sleep past three bells on the Morning Watch. Two, make sure the lookouts are changed every hour. Have the JOOD go chat them up and check on their technique. Three, spend some time in Nav Plot on the Radar and learn how it can prevent a collision."
"Aye, aye, sir."
"Good man, good night."

David Pelham rarely had trouble falling asleep. If he did, he only had to work surface target relative motion problems in his head, figuring which way he'd turn the ship. He never made it past three examples.

The Selfridge Captain heard a knock on his door, and three bells rang on the Bridge.
The compartment door opened, and the ship's Senior Steward, Petty Officer First Class Jackson, entered with a tray. "Good morning, sir. I gots some fresh coffee here and a couple of rolls. Would you like breakfast now or later?"
David stretched and yawned. "Later Jackson, in fact, I'll have it in the Wardroom at 0800."
"Yes sir. I've got three uniforms outside on the bulkhead, working khaki, service khaki, and service dress whites. What would be the Captain's pleasure?"
David smiled and thought, *A damn fine perk to have you're a personal steward.* "Make it the working uniform. I plan to move around the ship a lot today. No doubt I'll get it dirty. And thank you for the coffee."
"No need to thanks me, Captain. I'm just doing my job like you doin' yours. Doubt that folks is coming up to you and thanking you all the time."
David smiled inwardly. No mistake he was being trained by an old hand. "This is new to me, Jackson. What's the best way we can work together?"

“Captain tells me what he wants, and I gets it done. I don’t get it done you make sure I knows it. I’m getting older. I don’t always remember everything.”
David looked at the steward’s dark brown skin and wrinkled face. “Just how old are you, Jackson?”
“I be close to forty, but not there yet.”
“That makes you the oldest man on the ship. Your voice sounds familiar. Are you the preacher at Taps?”
“Yes sir, I take a prayer from my daily book for the men each night, and I preach on Sunday in the Chain Locker.”
“What are your other duties?”
“Yes sir, I wants to talk with you about that. For our lost Commanders, God rest their souls. I helped both of them with their staterooms, uniforms, and meals if they couldn’t get to the Wardroom. My other duties are to look after the colored cooks, bakers, messmen, and laundrymen. I makes sure they is clean, do a good job, and they are respectful to the crew. The Supply Chief puts my comments in their evaluations and they knows it. I is like a daddy to those men.”
“Are you sure it’s not too much to ask for you to work on the mess deck, climb up here to tend to me and go forward to take care of the XO and his stateroom?”
Jackson looked perturbed. “There you go, Captain, off in the wrong direction. Your job is to make sure I’m doin’ the work of two men. I will ask for shore duty when I can’t climb these ladders. Until then, I wanna end each day dog tired, knowing I gave my best.”
“Where are you from, Jackson, and what’s your full name?” David asked.
“Birmingham, Alabama sir. My name is William Clayton Jackson, the third.”
“OK, request approved, you take care of the XO and me. Do you think he’ll go for it?’
Jackson smiled, “Yes sir, when I tells him it’s an order from you. You see his job is tougher than yours. You has the thinking and worrying job. He has the pounding muscle job all over the ship, plus he got to impress you to make Captain. If I can make his day a little better, I wants to do that.”
“To recap for my memory, you’re William Jackson, a First-Class Steward in the “S” Division, from Birmingham, Alabama. Do you have family there?” David inquired.
“Yes sir, some. My main family, wife, three kids, and two grandkids are in Chicago.”
Pelham shook Jackson’s hand. “We’re both a long way from home. Pleased to learn more about you. Carry on.”

As David dressed and combed his hair, he thought about the ship’s officer complement and their roles dictated by the Navy and tailored by the XO:
Captain – Command and Morale
Executive Officer – Navigator, training, personnel, ship’s organization
Operations Officer – Long-range schedule, operations planning, readiness, threat assessment, sensor management
Communications Officer – Radio/Radar maintenance and training, transmission policy, message handling
First Lieutenant – Damage Control Officer, First Division Officer, deck and hull maintenance, personnel records, medical

Engineering Officer – Engineering Department machinery operation and training, Supervision of Electrical Division Officer
Gunnery Officer – Supervision of guns, torpedoes, and depth charges, weapons training, and target aiming
Supply Officer – Supervision of food services, ship's stores, supplies, laundry, and crew pay

Pelham checked his appearance in the head mirror, put on his cover, and stepped out into the navigation and radar plot room. David strode to the Radar display, "Good morning, Radarman. What do we have?"

"Good morning, sir. We have a small fishing boat 070 at six miles. He's hardly moving."

"When you give me a bearing, is that magnetic or relative to the ship?"

"Relative to the ship sir. 360 or 000 is straight off the bow, 090 is ninety degrees right, 180 is off the stern, 270 is ninety degrees left. In A school, they taught us that it gets the guns aimed faster to do it that way."

"Very well." David looked at the grease pencil plot, "Hmmm, in that case, from my relative motion lesson last night, and looking at this surface target plot, we must be heading west at about twelve knots."

"Yes sir, we're passing between Maui and Hawaii. We are painting the volcanoes on both islands here and here."

"Thank you, carry on."

After being announced on the Bridge, David asked the Officer Of the Deck when they would arrive at Kamohio Bay. Given the on-time response of 0700, he instructed the OOD to bring the ship to a dead stop at the bay entrance, set condition YOKE, and deploy Sonar. He told the OOD to pass to his relief that they would launch small boats at 0815, and all OODs and JOODs need to report to the Bridge at 0830. David knew that the word would quickly spread to those involved like Boats MacGregor, the 1st Lieutenant, Ops, XO, and officers in the Wardroom.

After Reveille sounded, David punched the XO's cabin button on his phone.

"Good morning, XO here."

"Good morning, Mr. Lehman. We'll be arriving at Kahoolawe on time. It's a grand day up here on the Bridge. Steward Jackson has asked to take care of us both, and I've agreed."

"Yes sir, he was just here with my morning coffee, rolls, and a clean uniform. You're sure that this is OK? He's your Steward."

David laughed. "I don't think we could stop him. Relax and enjoy it. You deserve some perks."

"Thank you Skipper. Just want you to know how proud I am of the ship and how you got her back on top of the waves."

"Thanks XO. Plan on coming to the Bridge at 0830. We're going to have some Sonar training and take turns maneuvering the ship. Warn Boats Mac that we're launching the small boats at 0815, and two of them require rowers. Plan on a Man Overboard Drill after the midday meal. We will launch and recover a whaleboat for that. Once we have it back and secured, we'll go to GQ for an hour or two. Check with Ops. We should have a target area for gun practice late in the afternoon."

"Aye, aye, Captain."

David punched the E1 button on his phone controller and heard, "Forward engineering, Chief Romano speaking."
"Chief Romano, this is the Captain. I'm checking my comms. How are you doing down there?"
"Everything is burning and turning, Captain. We had a couple of minor leaks that got fixed. Based on your standing orders, we lit the superheaters, tested them, and had them on low heat, about 300 degrees. We're running the boilers about half speed and could use a few minutes' warning if you want two-thirds or more."
"Good to know, Chief. Good to know. Is Commander Haldane available?"
"Yes sir, standby."
After a brief pause, "Engineer."
"Captain on the Bridge. I hear things are going well in Engineering. Congratulations. Come up here and give me a first-hand report before you turn in tonight. Plan on low-speed maneuvers this morning after chow, including backing down. Have the Chiefs tell everyone on engineering watch the Captain sends his regards and well done!"
"Thank you, sir. I'll be topside after Evening Reports."
"Works for me, Engineer. Please bring the XO with you. Captain out."
He punched the E2 button on his phone controller.
"Aft Engine Room, Chief Wilson speaking."
"Chief, this is the Captain. Good to hear your voice. When did you get it back?"
"Captain, last night, I was playing checkers with Chief Romano. I didn't like one of his moves and found I could cuss him out. Go figure!"
David chuckled. "I have a question for you. At All Ahead One Third, it looks like we're making about fourteen or fifteen knots. I think the slick bottom gives us more speed than in the past. If I want to slow down to twelve knots, what's the best way, start talking in quarters and eighths?"
"Sir, try asking for twenty-five revs less. It should be about two knots for each change," Wilson responded.
"So Chief, this morning, if we're at All Ahead Slow and I want slower, what should I say?"
"Captain, you could say cut the revs in half. To start off real slow, you could say All Ahead twenty revs or make turns for two knots."
"Thanks Chief, we'll try that. Pass the word that we'll be doing slow-speed maneuvering starting at 0830, Captain out."

David sat back in his chair and appraised his communications setup. He had a 1MC unit that let him broadcast to speakers placed around the entire ship or select a specific Department and critical working spaces. The duty Bridge Talker could relay orders and information on the 2MC sound-powered circuits to every part of the ship. The ship's phone system connected him to the XO, Engineering spaces, Gunnery Officer and Gun Director, Repair 1 Damage Control, Navigation and Tactical Plot, Communications, Secondary Ship Control, and Admin. He could call one station, select a set of spaces, or punch the All button. Backup Bridge communications included Voice Tubes, electrical telegraphs to Engineering, and a messenger on duty. The Captain, XO in After Conn, Communication Officer, OOD, and JOOD could monitor or transmit on the short-range radios except when the ship was in radio silence. One deck down, there were two radiomen poised at typewriters listening for Morse code or voice transmissions to the Selfridge on long-range radios. The Signal Bridge was manned with Signalmen who could send and receive messages by signal light, Morse code, flags, or pennants to other ships or aircraft.

His thoughts were interrupted when Lt. Wolfe arrived on the Bridge. "Good morning, Captain, I heard a rumor about maneuvering practice this morning."

"Good morning, Ops. That's no rumor. We will come to a dead stop in front of the Bay and do a sonar sweep for subs. Then we launch the small boats for several reasons. One, we're going to send them out to sea to test our new Sonar gear. Two, we'll benefit from small boat handling practice. And three, as we practice maneuvering the ship, we'll use them as stand-ins for docks and narrow passages. Please have Boats and the 1st Lt. report to the Bridge so I can brief them on what I want."

"Aye, aye, sir. Mind if I join the brief?"

"Sure, join in and feel free to add some ideas and inputs."

With the small boat and maneuver exercise briefed to Boats MacGregor and the two officers, David left the Bridge and went below to the Wardroom for breakfast. He enjoyed the company of the other officers and especially the XO. There was no discussion of politics, sex, or religion – a tradition in all Wardrooms. Friendly conversation was encouraged, and some good-natured kidding. Praise in public was always recommended. Pelham and Lehman agreed it was the wrong place to take an officer to task, to complain about their performance or problems with their division. The JOs asked their Captain about the maneuvering training. He replied that they all needed to see how she handled with the new slippery bottom.

At 0830, Pelham, Lehman, and five teams of OODs and JOODs crowded into the bridge. Two of the ship's boats headed out to sea under power. The other two rowed a short distance into the bay, separated, and strung a 100-foot line from one to the other. David announced that the first event would be practice docking using the boats and the line as a target dock. Ops went first with Lt. (jg) Carpenter as his JOOD. They approached the makeshift dock at about a twenty-degree angle and slow speed. With 100 feet to go, Wolfe ordered a port turn, and then All Astern Slow on the engines to stop the destroyer's forward momentum. Selfridge's stern kept swinging starboard so that the closest boat had to row away not to get banged by the ship.

Lt. Wolfe maneuvered Selfridge away from the pretend dock by backing down on the port engine and ordering Ahead Slow on the starboard. Each team tried their luck until it was time for the CO and XO to show their skills. David positioned himself on the starboard flying Bridge and conned Selfridge a quarter-mile seaward. He swiveled her 360 degrees port until the bow pointed at the two whaleboats. Pelham relayed orders to the Helmsman via the XO as they slowly approached the boats at a ten-degree angle. They backed down with the port engine at fifty feet and ordered the starboard engine at Full Stop. As the ship's stern approached the aft boat, David ordered a brief Slow Astern on the starboard engine, and then Full Stop on both. Selfridge came to rest five feet from the whaleboats and the line. They merged as the southern wind pushed the ship to the make-believe dock. David gave Boats a signal for the next event and swiveled the ship broadside to the open sea. He had the signal bridge shoot a flare into the air that caused one of the boats five miles at sea to start their engine and head back to the ship.

Two Sonarmen were on the destroyer's seaward side, listening on the recently installed GE hydrophones. They each reported the relative bearing of the new contact via a 2MC talker to the radar plot adjacent to the Bridge. David secured the ship's maneuver drill. He had each of the OOD/JOOD teams exit the Bridge after reviewing the radar plot. They then proceeded down to the weather deck to listen to the sonar contact.

Pelham had a second flare launched from the Signal Bridge. He knew the whaleboat about three miles to sea would secure their engine and signal the other boat to motor across the bay opening. Meanwhile, the boats closer to shore placed three small red buoys roughly 200 feet from the far shore and beached the oar-powered craft. After the Sonarmen tracked the second whaleboat across the bay's mouth, David had the third flare fired that signaled one hour of free time for the four whaleboats.

The OOD set Condition Zed throughout the ship, and the JOOD announced, "Standby for an announcement from the Captain."

David thought for a minute, pushed the 1MC ALL button, and said, "This is the Captain speaking. We've completed our OOD and JOOD training exercises. We'll stow the whaleboats in about an hour. In the meantime, we have set modified ZED because we will take the ship into shallow water. Boats MacGregor has marked the sixteen-foot depth with three buoys close to shore. Selfridge weighs about 2,100 tons with our current combat load. That means we need fifteen feet of draft beneath the screws. It's possible, but not probable, that our bow may briefly touch bottom. We'll be testing our fathometer and our ability to maneuver in shallow water. All hands are welcome on deck to get some fresh air and sun. Remember to reset ZED if you come up on deck. Duty deck crew call away a leadsman to the bow and an anchor detail fore and aft. That is all."

David took the conn from the OOD and ordered All Ahead Slow. He had the Helmsman aim for the middle buoy, and the ship slowly sailed up the broad bay. Pelham ordered the engines to stop about half a mile from the floats, and the destroyer continued slowly toward shore. He briefly backed the screws at a quarter-mile, and Selfridge was barely moving forward. The bow leadsman reported ten fathoms or sixty feet via his 2MC talker. At two hundred feet from the buoys, he backed down again with twenty revs until forward movement stopped. The Leadsman held up five fingers, and the Bridge Talker relayed five fathoms or thirty feet. David had the JOOD call away the aft anchor. He moved the ship forward until the anchor was reported set. Then the bow anchor was dropped, and the destroyer slowly backed down until it took hold. Pelham knew the tide was rising, the wind was light, and the bay was not known for having severe currents. Once both anchor lines were taut, the OOD announced that the ship was at anchor and set Material Condition YOKE.

David was surprised to see the Engineer enter the Bridge. He was one of the officers who did not have to stand OOD duty on the Bridge since he and the Assistant Engineer stood their watches in the Engineering spaces.

Haldane motioned for David to follow him out on the Port Bridge Wing. "I just came topside to see for myself where we ended up. How do we get out of here?"

"The XO will raise the anchors and back us out from Secondary Conn," David replied.

"OK, skipper. I'm wondering why we are doing this. Remember, we need to be careful we don't suck in any sand or debris into the pumps," Haldane admonished.

David pointed to the four beached whaleboats and the exuberant sailors exploring the beach, "Just exercising the ship's Engineer, and giving the men some time ashore. If the water were warmer, we could have a swim call."
Haldane spoke with emotion, "You're missing the point. Is it important we do this? Is it worth the risks?"
"What risks?" asked David in a terse voice.
"The risk of grounding and you losing your command. The risk of ingesting sand, rocks, and seaweed and damaging my pumps. The risk of setting a bad example for your Bridge crew that seamanship is a casual profession and even fun!"
David responded, "There it is, there's the nubbin of this discussion. Conning a ship is fun; no doubt it's a serious business, but still exhilarating and satisfying. If you think this evolution today is risky, you may be serving on the wrong ship. This drill is child's play compared to what we'll be doing in two months. Let me ask you something. On 7 December, this ship took two bomb hits and tried to run ashore. Why did it not make it?"
"I don't know, I was down in Engineering," replied Haldane
"Excellent. Did the Bridge crew ring up for full power to make land?"
Haldane hesitated, "No, there was no call for more power."
"Thank you. You're making my case. As the tactical situation changes, new decisions are required. On December 7th, an aggressive OOD and JOOD could have saved the ship from going to the bottom. That's what I'm doing up here, first evaluating the OODs to see which ones I want on the Bridge in combat. Second, demonstrating some unusual ship handling shows that we control the ship, not vice versa. Third, we're working on team training with small boats and trust."
"Trust, how are we working on trust?" Haldane asked.
"Every man you see onshore and on deck now knows that I implicitly trusted Boats to place those buoys correctly. If he screwed up, the ship could be aground. They're absorbing the lesson that their actions count even when they don't imagine that they might."
"I'm sorry, I may have spoken out of turn."
"Don't be sorry, Engineer. We've both worked with several Captains, but you're the one with far more experience. I need your input and counsel, but there will be no harm, no foul if you want off this ship. Give me your decision tonight. Dismissed."

That afternoon, GQ was set, and Selfridge conducted gunnery firing against targets on the southwest shore of Kahoolawe. Upon completion of the gun drill, she went back to Condition 2 and set a course for the west coast of Kauai. David had the OOD warn Engineering to expect one hour of two-thirds throttle starting at 1630. They completed the higher speed run without incident. With her new bottom, DD-357 charged through the waves at twenty-nine knots on All Ahead Two Thirds. When Pelham turned in, his standing orders advised the Bridge crew to position the ship in an exercise warning area near the Army's Mana Airport on the west coast of Kauai at 0600 using no more than one-third throttle.

For the next two days, the ship conducted antiaircraft gunnery in the morning against towed target sleeves, and antisubmarine drills in the afternoon with the men at General Quarters. Each department was also working on maintenance and operational training for their assigned equipment and spaces on the ship. They then returned to Kahoolawe for two more days of daylight and night gunnery practice. Pelham was pleased with the crew's performance and

announced on the ship's 1MC that they would be going to Hilo on the Big Island of Hawaii for the weekend and two days of liberty. He requested the XO come to the Bridge and warned the ship's company that in one hour, there would be a full-power run for two hours. When the XO arrived on the Bridge, David asked him the previous All Ahead Full speed.
"She's an honest ship, Skipper. You can anticipate thirty-four knots at full power and thirty-seven at flank."
"Good," David replied, "We'll give Boats and the Engineer an hour to make the ship ready. Then we'll run her south of Hawaii full ahead for two hours if everything holds together. You'll have the conn. I'll be down in Engineering observing the first hour and then roaming around the ship. Standing orders for tonight have us sailing up the east coast of Hawaii at one-third throttle. We'll anchor in Hilo Bay north of Mokaoku Point and inside the long breakwater."

David was on the Port Bridge Wing as the XO called for more power, first two-thirds for a few minutes, and then All Ahead Full. Selfridge surged forward like a thoroughbred in the home stretch. She was running with the waves, and David was sure he had at least thirty knots of wind in his face. As he went below, he saw a massive wake wave building up behind the ship and could hear the forced draft air blowers wailing. The full speed test was positive; the pit log topped out at thirty-nine knots, and there were no problems with the engineering plant. The crew mightily appreciated liberty in Hilo. On Sunday, David hosted Big Mama, Little Mama, and Jordy Manelo, the Airport Manager, onboard the ship for dinner in the Wardroom. They each told different versions of the harrowing flight of Marine 78744 to Hilo that had the junior officers alternately entertained and dismayed.

Selfridge slipped anchor late Sunday night and commenced ASW drills against a target sub off the east coast of Maui for two days. They returned to Kahoolawe to exercise the five-inch guns and then to Mana Airport to shoot at more towed aerial sleeves. On their second-to-last day at sea, the Engineer said his equipment was ready for All Ahead Flank testing. After evening Eight o'clock Reports, David conferred with the XO, the Engineer, Ops, and the Assistant Navigator. They decided to make the run from ten miles offshore of northeast Kauai to ten miles offshore of the northeastern shore of Oahu.
The next morning, after the 0800 Bridge crew reported, David addressed the watchstanders. "OOD set course for the high-speed run in my standing orders, bring us up to Two Thirds, and warn Engineering."
"Aye, aye, sir, Helmsman, starboard to course one-five-zero, ring up All Ahead Two Thirds. JOOD inform Engineering we will be requiring All Ahead Full and then Flank speed in a few minutes."
The OOD came over to David in the Captain's chair and nervously asked, "Skipper, I've never conned a ship at this speed. Any recommendations?"
"OOD, I've been thinking about that. We have good weather, light wind on the bow, and four to five-foot swells. We should warn the crew that high winds may be over the deck. We'll run into the waves, which should give us a good ride. No one is going to catch us from our rear quarters. We need the lookouts to call out any ships or small craft ahead and in the forward quarters. If we see any small boats, it would be smart to turn away and reduce speed to keep from swamping them. Make sure Radar is on their toes, looking ahead for ships. New thought, have the lookouts also looking for whale spouts. Running into a thirty-ton Humpback at forty

knots could ruin our whole day. Pay attention to how she rides down the waves. You may need to change course to go down them square if she starts to dig in and veer away. Other than that, relax and enjoy the ride."

"Yes sir, thank you, sir!"

The OOD stepped over to the Bridge 1MC station, selected the All position, and announced, "Attention all hands, this is the Officer of the Deck. We will be commencing a high-speed test run for one hour in a few minutes. There will be high winds over the deck; all hands use caution. Any unusual vibrations, equipment failures, or problems should be reported to the Bridge and Damage Control immediately. That is all."

David noted that the OOD gave the lookout instructions to the JOOD for relaying on the 2MC and then stepped out on the Starboard Fly Bridge to survey the sea ahead with his binoculars. Soon they were surfing down the waves at over forty knots. David nodded with satisfaction when the JOOD called out the pit log at forty-two knots. After the hour run, they slowed the ship to two-thirds and turned into the wind. Once again, the OOD warned the crew that there would be a thirty-minute high-speed run, high winds over the deck, and possible heavy pounding into the waves. At All Ahead Full, they were making thirty-five knots into the wind, and the ship was taking the waves well. She'd fly over two or three swells at flank speed and then bury her bow into the next one, causing the whole ship to shudder. David had the OOD bring the pace back to two-thirds and realized that high-speed runs into Sea States 3 and above could be a real problem.

Selfridge finished off the daylight hours with several drills, and Captain Pelham held a Department Head meeting in the Wardroom after supper. Each officer covered the pluses and minuses experienced by their Division during the two-week cruise period. The XO provided his comments; then, it was David's turn.

"First, I want you to know how pleased I am with the crew, with this magnificent ship, and with your Divisions' performance. I roam around the ship to meet the men and build a mental image of every space and passageway. I still have much to learn. This has been a pretty relaxed at-sea period on purpose. We all needed to see if the ship would respond properly to our repairs and maintenance. She's a beauty! We needed the crew to get used to the new men and back into a combat ship's routine. We needed to exercise the weapons, the power plant, damage control functions, sensors, communications equipment, the ship's boats, and a hundred other systems and capabilities. Good for us. We've done all of that and more. Now we're ready to get down to the serious business of warfare and kill some Japs. We have several ship mods to make before our next two-week at-sea period, the last two weeks in March. We're going to install the gyro glider, complete painting the bottom, and integrate the new Sonar so we can control it from Bridge Plot. There were ambitious plans to add a larger caliber deck gun. That has been canceled and dropped into the Too Hard Bin."

Lt. Baxter asked if more of the ship's schedule was firm so he could order supplies.

Pelham said, "We plan to be back in Pearl the first two weeks in April and back at sea for the last two weeks for a major Third Fleet exercise. Before that exercise, we will be adding two sensor platforms that extend forty or fifty feet upward. Schedule-wise, we have two more workups and three more inport periods before we sortie for combat about 7 or 8 May. The last period in Pearl will only be a week-long and dedicated to topping off supplies, ammo, and men. Most of you know that Ops and I've met with the Pacfleet staff about an unusual

operation in May or June. We'll need to accommodate several more men onboard Selfridge, probably about twenty-five, if it's a go. We will also need to change out most of our whaleboats for swifter small craft. First and Second Divisions are responsible for determining how we can substitute four smaller boats for three of our twenty-six-foot whaleboats."
1st Lieutenant Jones said, "Sir, can you tell us what the boats are for and what capabilities are needed?"
"Ops, can you answer that?" David asked.
Lt. Wolfe responded, "Yes, on our first cut, we think they need to carry four men plus the Coxswain at a speed of twenty to thirty knots in sea states one and two. They must have basic navigation equipment and have a range of sixty miles. My thought is they would be twenty to twenty-four feet long and have multiple outboards, say like two or three twenty horsepower outboards."
David added, "For our next at-sea period, we plan to sail to Midway Island and operate there for roughly a week. We'll be testing the new equipment, doing underway refueling, and completing more drills, including longer periods of GQ. I believe we can improve our gun accuracy and increase the firing rate. We will have competitions for the gun crews to encourage both. We'll have at least five days back at the Hawaii firing ranges to finish the gun crew training. Any questions?"
Haldane raised his hand. "Captain, can you tell us what these twenty or thirty new men will be doing?"
"Until we get flag approval, not in any detail, Engineer. I can tell you that some of them will be Army and Marine specialists that we may put ashore using the new boats."
David thought, *A heck of a deal when you can't tell your own people the straight skinny. Ops will let it slip that the operation might be in the Marshall Islands. That should keep the scuttlebutt on the ship and the bars ashore from focusing on Wake until we're ready.*

Selfridge returned to Pearl Harbor and tied up alongside a destroyer on the east end of the destroyer row. David, the XO, and Ops called on the DesRonOne Commodore onboard Phelps and laid out their plans for the following three months. The next day, they placed the ship back in Dry dock YFD-2. During the five days in dry dock, Navy Yard personnel came onboard and helped them replace their older AA guns with the more accurate Bofors mounts. Two tech reps from GE wired the updated Sonar gear to a display next to the Radar plot, adjacent to the Bridge.

David was able to break away from the ship a couple of times during the two weeks in Pearl to see Betty White. On her one day off the first week, he drove her out to Hanauma Bay to snorkel and relax on the beach. They were stopped three times by sentries along the Kalanianaole Highway. David talked his way through each of the checkpoints. They drove down a single-lane dirt road to a parking area above the Bay and changed from their uniforms into swimsuits using the two abandoned outhouses in the parking area. David showed Betty how to use the fins, snorkel, and mask when they heard a shot. They saw a man in uniform a quarter-mile away who had shot in the air to get their attention. As he reached them, they could see that he was a young man in National Guard garb with Corporal stripes.
"Hold it there. No one is allowed on the beach. You have to go." the soldier ordered.

"Good morning, Corporal, I'm Lieutenant Commander David Pelham, United States Navy, and my lovely companion is Captain Betty White, U.S. Army Nurse Corps. Who might you be?"

"Corporal Kino Haina, Hawaiian National Guard. You have to leave the beach. We're authorized to shoot anyone on the beaches."

"Corporal Haina, this is our one day off, and we heard that the snorkeling here is fantastic. What would it hurt if we paddle out to the reef while you stand watch and make sure we aren't spies or the like?" David asked.

"Sir, I have my orders. You have to leave, or I have to arrest you, even shoot you if necessary," Kino replied.

"Fine," David said. "Tell me about your orders."

"No one is to be on the beaches. Remain vigilant to resist anyone coming ashore and sound the alarm," Haina replied.

"OK, we'll get off the beach by going out to snorkel. Your new orders are to stand guard until we return. You can see we're unarmed, have no cameras or radios. Like you, we're serving our country and deserve a brief swim after driving out here from Pearl. We're going into the water now. If you decide to shoot us, you better come up with a damn good story why you shot the Captain of a U.S. Navy Destroyer and an Army nurse from Tripler in the back - both unarmed and presenting no threat."

David and Betty waded into the bay and put on their snorkel gear when they were chest high in the ocean. The bright colors and variations in the sizes and shapes of the tropical fish were amazing. They both delighted in each discovery. For an hour of snorkeling in the warm, shallow Pacific, they forgot the war and enjoyed the wonders of nature. Corporal Haina escorted them to their car when they returned to the shore.

"Corporal, I mean no disrespect when I ask that you might be of Japanese descent?" said Pelham.

"Yes sir, my grandparents were born in Japan," Kino answered.

"Let's see, I learned from a Japanese friend in the States that makes your parents Nisei and you Sansei. Can you speak, read, and write Japanese? I'm looking for people who can do that."

"Yes sir, I would be classified as Sansei. I can speak Japanese, but not read or write in it."

"How many National Guard here are of Japanese descent?"

"At least 25 percent, probably closer to one-third," Kino said.

"Do you have young men like yourself who want to serve in the Army and Guard that can't get in?"

"I'm sure there are a ton of them. My uncle is in charge of recruiting for the Hawaiian Guard. He could tell you," replied the soldier.

David showed a keen interest. "Please give me his name and where I can reach him."

Pelham wrote down the uncle's name and contact information. The two officers changed back into their uniforms, and David dropped Betty at her apartment on his way back to the dry dock. He noticed a hospital ship, USS Solace, tied up at one of the piers. A new idea was forming for Slingshot, but he was not yet sure what it was. Pelham made a mental note to contact Kino's uncle and visit Solace the next day.

Back onboard his ship, David told Wolfe to plan meetings with the Sub Base and PBY aircraft operational commanders. Lt. Wolfe and the XO were working on a draft Operation Order for Slingshot that included the forces involved, the sequence of events, annexes for SECRET, and TOP SECRET activities like Cover and Deception, Communications, Neutralization of Wake Air Units, and Enemy Strengths and Weaknesses (Supernatural). The three of them knew that Lt. Comdr. Layton and the CinCPac staff were working up plans for an air and sea assault exercise against Johnston Island in the last part of April. It was tailored to make the Japs think it was preparation for attacking Wake. Yeoman Smith sent a memo to the ship assuring David that manning for the April at-sea period would be 190 enlisted and fifteen officers.

As David completed receiving the XO's summary of Eight o'clock Reports that night, he was handed an envelope from Layton marked: "CO Selfridge, Eyes Only." Opening the folded sheet, he read, "Nagy family has arrived. Michael Nagy has met with Admiral Nimitz, Rochefort, and me. Very impressive. I believe we should brief him on Slingshot. Plan to meet with me at 0900 tomorrow to discuss his role and our mutual friends on the Mainland."

Hanauma Bay circa 1990s

Source: Wikipedia

USS Solace Hospital ship

Source: Wikimedia Commons

Chapter Twenty-six

OPERATION SLINGSHOT – CinCPacFlt HQ

March 1942

Lt. Comdr. David Pelham and Lt. Wolfe were waiting in Lt. Comdr. Layton's office at CinCPac Headquarters while he plowed through a stack of messages and memos. Layton tossed David some stapled pages and said, "Here's some light reading while you're waiting."

Pelham started reading:

"**DRAFT BIO FOR CinCPac:**

Admiral Chester William Nimitz was born in Fredericksburg, Texas, on February 24, 1885. He was raised in the Texas Hill Country and Kerrville, Texas by his mother Anna, his grandfather, Charles Nimitz, and his stepfather, William Nimitz. At a young age, he was called on to do chores in the morning before school and in the evening to help his family manage a small Kerrville hotel. In his mid-teens, Nimitz was accepted into the Naval Academy and graduated in 1905. Ensign Nimitz served on several Navy ships in the early 1900s and was the Commanding Officer of USS Panay in 1907 and USS Decatur in 1908. He was assigned to a series of submarine commands before World War I, including Commanding Officer USS Plunger, USS Snapper, USS Narwhal, and USS Skipjack.

In 1914, Lieutenant Nimitz married Catherine Freeman, and they traveled to Europe to expand his knowledge of diesel engine technology. In subsequent years, they became the proud parents of three girls and a son. Lt. Comdr. Nimitz led the design and integration of the first diesel engine in a U. S. Navy ship, the USS Maumee oil tanker. He was a proponent of converting submarines to diesel power and profoundly influenced the design and conversion process.

Comdr. Nimitz served in several increasingly important Navy staff assignments and ship tours during the 1920s and 1930s. During his career, he also developed new methods for underway refueling and replenishment, established the first NROTC Unit at the University of California, Berkeley, and supervised the construction of the Pearl Harbor Submarine Base. In 1929, he was selected for Captain and excelled in several ship and shore billets that included command of the USS Rigel destroyer tender and the cruisers Chicago and Augusta. Captain Nimitz served as assistant chief of BuNav for three years before sea duty as a Cruiser Division Commander.

Upon promotion to Rear Admiral in June 1938, Chester Nimitz commanded a Battleship Division. His flagship was the USS Arizona. In 1939, he returned to shore duty as Chief of BuNav. Rear Admiral Nimitz was selected by President Roosevelt to assume the responsibilities of CinCPac in mid-December 1941, and was promoted to the rank of Admiral on 31 December 1941."

Layton looked at David, "So what do you think?'

"Pretty impressive," David replied. "He's had duty on just about every type of ship except carriers and amphibs."

"Don't worry," Lamar answered, "He'll guide the grand strategy and have seasoned subordinates manage the weapons. Of course, he didn't let us say he graduated 7th of 114 in his Academy Class. And only a few people know that Roosevelt offered him the CinCPac

Fleet billet a year ago. Nimitz turned him down because he considered it unseemly to jump ahead of so many senior admirals. When FDR was Assistant Secretary of the Navy from 1913 to 1920, he developed complete confidence in Nimitz as a leader and innovator. Last month, FDR did not take no for an answer. He ordered our boss to get out here and lead the war against Japan."

Layton looked over at Lt. Wolfe and back at David. "Pelham, can we talk about Professor Nagy and his planned responsibilities freely?"

David responded, "Mr. Wolfe, Professor Michael Nagy reads, writes, and speaks Japanese. He's written a book about Japanese history and their modern social values that should benefit our operations. Please excuse Comdr. Layton and I while we discuss his specific roles on the staff."

After Wolfe left his office, Layton said, "Admiral Nimitz wants Nagy on the Planning Staff. We both think he should know everything about Slingshot and be our go-between with our friend in San Francisco."

"I concur. My time is tied up with the ship and planning for Slingshot. I don't see myself flying back and forth to the States," David agreed.

"Good, that's settled. The boss wants a detailed brief on Slingshot after your next at-sea period. He'll give his opinion to Admiral King; they will give you a go or no-go decision."

David said, "We are lining up the other players needed. You may be surprised by some of the other units. If they come across your desk complaining, please bear with us until all is explained in the detailed briefing."

"OK, I can do that," said Layton. "And you can find Professor Nagy three doors down to brief him on your operation. Best of luck!"

David introduced Lt. Wolfe to Michael Nagy, and they briefed the Professor on their current plans for Slingshot. David asked Wolfe to step out of the office for a few minutes and told Michael about the meetings with his great uncle, the need for more intelligence, and his plan to recruit an IJN Signalman. David warned Nagy that he'd most likely be sent to San Francisco to meet with his mob boss uncle in April.

The March two-week at-sea period was a blur of activity for Selfridge's crew and their acting Captain. Nineteen new crew members reported onboard. They had to learn the ship's routine and their watchstanding responsibilities. Selfridge conducted mock gun attacks on Midway Island, anti-submarine exercises, practice torpedo launchings, and simulated depth charge attacks. The gun crews competed for the most hits and shells fired against towed surface and air targets.

As they headed back to Kahoolawe for shore bombardment practice, the officers inspected all compartments and spaces on the destroyer. The ship was clean, equipment discrepancies were low, and the crew was at the top of their game. David and the XO privately discussed executing the Strip Ship Bill while they were back pier-side in Pearl Harbor. They both knew it would be disruptive and damage morale. There was no choice. They were going into combat.

After the ship pulled back into Pearl, David finalized his plans with the Captain of the AH-5 Hospital Ship, the Marine sharpshooting team, the Operations Officer for Pacific submarines, and the Squadron Commander of long-range search aircraft. Wolfe confirmed the ship's schedule for the oiler refueling Slingshot units. Pelham had a second meeting with

Corporal Kino Haina's uncle, who introduced David to a Hawaiian Army recruiter. They provided David a roster of several young Nisei and Sansei who wanted to serve in the Army but had been turned down due to their ancestry. Captain Pelham turned the list over to Lt. Forest, Petty Officer Smith, and Yeoman Brown, along with his interview criteria. He wanted fifteen Sansei who could speak and read Japanese. Michael Nagy would be the final interviewer to assess their language skills and make an informed judgment on whether they would be seasick.

On the final day for Selfridge pier-side, Pelham and Yeoman Brown were about to board a liberty boat out to the Phelps. David saw a civilian onshore who waved at him. He had Brown hold the boat while he went to see what the man wanted. The short, stout man was about forty years old and Japanese.

"Lt. Comdr. Pelham, I've been sent by a mutual friend."

"And who might that be?"

"The man is my boss who lives in San Francisco. I believe you recently dined with him in his penthouse. He wants you to know that the package you asked for will be arriving in one week. It will arrive at noon on April 6th, ten miles due south of Navy Town on Attu Island in the Bering Sea. Oh yes, there will also be about 500 pounds of cargo for my boss, along with your package. Do you completely understand what I've said?"

"Yes, I understand, but I have questions," David tried to reply as the man walked away, was picked up by a Navy truck, and left the pier area.

David and Petty Officer Brown met with the DesRonOne Commodore onboard Phelps. They reviewed Selfridge's personnel manning, training status, and additional personnel requirements. The following morning, the Bridge crew expertly sailed the ship over to the west end of destroyer row and tied up alongside a destroyer on one end of the seven-ship nest. The day after, David took a boat ashore and drove to CinCPac HQ, where he found Layton at his desk. Layton got up and shook David's hand, "And to what great honor do I owe this visit, Captain Pelham?"

"We've heard from our friend in San Francisco. Quote, the package that I requested will arrive south of Attu Island in the Bering Sea at noon on 6 April, unquote."

"Damn!" responded Layton, "Do you think it's our signalman?"

"I don't think it can be anything else. There's more. The ship will also have 500 pounds of cargo for delivery to Kimicho in San Francisco; the meeting point is ten miles south of Navy Town on Attu. What do you think?"

Layton pulled out a map of the Northern Pacific. "Holy cow, he's not doing us any favors. Attu is way out on the Aleutian Chain. It's closer to Russia than it is to the Alaska mainland."

David used a pencil length to measure the scale at the bottom of the map. "It looks to be over 1,000 miles. We won't be flying him out of there."

"There are plans to develop airfields on several of the islands. We will have to use a ship or a sub," Layton replied.

"What do we have up there in the way of ships?" David queried.

Layton thought for a moment. "We have some defensive older cruisers and several tin cans. We could have a destroyer out there by the 6th.

"Better make it two in case one breaks down. We have to have that so-called package to make Slingshot viable."

"What about the cargo?" asked Layton. "What are you going to do about that?"
"What cargo? I don't know anything about any cargo, and neither do you. I'm meeting with Captain Mueller and his C-46 aircrew about another matter later today. If we wanted them to pick up our package in Alaska, where would they do it?"
Layton stabbed the map with his finger. "Here at Anchorage, Elmendorf Army Airfield if construction is complete, or the civilian airport."
"OK," David said, "They could fly to San Francisco from Oahu and up to Seattle, or to Anchorage from San Francisco, which looks to be about 2,000 miles."
Layton said, "It would be safer to fly to Seattle and then to Anchorage. The weather up there can be pretty hairy."
"Who is going to handle the package and the cargo? How are we going to get this done?"
"Leave it to me. I know the Navy intel officer up there. No doubt he'll want to be on one of the ships to see firsthand what is going on with those islands. What do you think is included in the 500 pounds?"
David said, "Your guess is as good as mine. I'm willing to bet my next paycheck that it's a currency or even gold. The Army and the secret police are making things difficult for the Yakuza. They're probably moving some assets out of the country while they can. If the pickup is on the 6th, figure about three days sailing back to Anchorage. We should have the aircraft there on the 8th so they can fly to Seattle and San Francisco on the 9th. They could take a day of rest there on the 10th and fly here on the 11th."
"Sounds like a plan, Pelham. I'll run it by the boss this afternoon. If you don't hear from me by tomorrow night, it's approved, and I'll contact Lt. Comdr. Roger Blankenship and get the destroyer ball rolling. Plan on briefing the Admiral in three or four days. By the way, who are these people you have Professor Nagy interviewing?"
"Believe me; you don't want to know. Gotta go to my next meeting," David responded.
"Hold it, Pelham. I have more for you. From your comments about shooting down Jap search planes, I have the suspicion that you're going to take a page from the Royal Navy and make Selfridge look like a Jap destroyer. Am I right?"
"It's possible that we might do that."
"Have you picked a ship to imitate?"

Shit, thought David, *How could I have let this slip?*

He replied, "No sir, I've not got to that yet."
"You may be in luck. Selfridge does not look like any of the mainline Jap destroyers, but one under construction at the Maizuro Naval Arsenal has a striking resemblance. She's the first in class with a bow number 104. If you make some minor changes, such as the same gun configuration, the same stepped deck, and even the same stack appearance. She's bigger, has an upraised bow and curved stern. With the right markings, Selfridge would be the spitting image of the Akizuki. She's 2,700 tons, 440 feet in length, and we believe the first one will come down the ways in June."
Layton opened a locked desk drawer and pulled out a manila envelope marked TOP SECRET. "Here are some photos of her construction, and we even have the Jap's visual identification card that will be distributed to all their ships and squadrons. Here is our U.S. recognition card for Selfridge. See a resemblance?"

"Yes sir! This is fantastic. If we build up our bow, round the stern, and fair the forward stack into the Bridge superstructure, we're an excellent match. Regarding the markings and flags to fly, our IJN Signalman should coach us on that. This is great! How much of this can I take away?"
Lamar pulled out a large sheet of drawings. "None of these first ones, I had an artist show how Selfridge would look with cosmetic changes. Note the paint scheme on your Bridge area to make it look smaller to a plane flashing by at 150 miles an hour. You can take this with you."
"What a day I'm having," David enthused. "First, the odds of getting the Signalman have shifted in our favor, and now we have an enemy ship to mimic."
"Good, so these young men of Japanese descent you seem to be recruiting, they're going to be your deck crew looking like Japanese sailors, right?"
"That's possible. We aren't sure how to use them yet."
"Pelham, don't try to bullshit a bullshitter. You are sticking your neck way out here. If one of these civilians gets killed or wounded, you'll be facing multiple felony charges, from kidnapping to manslaughter."
"There will be no kidnapping. Each one of the young men will know the risks and sign on as ship's company," David shot back.
"How in the hell are you going to do that?"
"Like you say, taking a page from the Royal Navy, I'm going to impress them into the U.S. Navy. Each man will sign that they're serving under their own free will, and Smitty will issue them orders to Selfridge as Seaman Recruits."
"Jesus Christ, what a can of worms that is!" Layton yelled. "You've no authority to recruit people of Japanese descent. You can't issue them service numbers. They don't rate medical treatment or uniforms. There is no way to legally pay them. etc., etc."
"All I know is that I need them for this mission, and they want to serve. As Captain of a U.S. Navy ship, I believe I have the right to impress them into the service and sort out the details later. It's not like we'll grab them when they're intoxicated or knock them out and drag them onboard like days of old. As for uniforms, I'm having two sets of dungarees made for them and a variety of Japanese Navy outfits. I'll make sure they get paid from my own funds."
Layton spoke softly, "Captain Pelham, we never had this conversation. I know nothing about your illegal plans. Do we understand each other?"
"Perfectly." David held up the sheet with Selfridge drawings on it. "Thanks so much for this. See you at the Admiral's brief."

Petty Officer Brown took David over to AAF Base Hickam Operations, where he left a message for the C-46 Ginger Snap crew. They then went to the Tripler Army Hospital, where David had lunch with Betty. Later as Pelham and Yeoman Brown were in the liberty launch on their way back to the anchored Selfridge, David was lost in thought. *Things and people are coming together. We have to convince the Japs that we're going to retake Wake precisely on the 16th of May using only one carrier. This next at-sea period is our time to test some new equipment and the new men. We need a plan for evaluating the loyalties of our turncoat signalman. One false move, and he kills us all. Michael needs to go see Kimicho and see if we can get the IJN battle plans. It would be ideal to know the ships involved and their cruise* *plan. If we strike Wake at dusk, will the carriers sortie planes to search for us at night or wait for pre-dawn patrols?*

Hmmm, Michael goes to San Francisco, then to Alaska with Mueller's C-46 crew. Then he takes the Signalman in tow, speaking his language. Then flies back to San Francisco. He sees his great uncle with the signalman and delivers the cargo. Kimicho puts the fear of the Yakuza into the signalman. Rats and his crew return to San Francisco after a couple of nights in San Diego and fly them here. What's left? We have to brief the detailed plan to Nimitz and get his and King's approval. We have to pay Kimicho and get him to confirm the Jap battlegroup is on its way. We have to fine-tune the timing at Wake and play on their fears.

David's thoughts were interrupted as the small boat pulled alongside his ship. He climbed up the liberty ladder and touched base with the XO and Lt. Wolfe, telling them about his meeting with Layton, and asked them to meet him after supper on the Bridge. Then he finished a pile of paperwork and took a nap. After supper, he reviewed Slingshot plans and requirements with Lt. Wolfe and the XO. They worked well past taps before knocking off for the night. David told the OOD to wake him at 0700 if he was not up and about. He knew that the watch standers would find a way to relay his instructions to Steward Jackson. The following day, he was in the Wardroom finishing breakfast and his second cup of coffee when he heard the bosun's whistle, and the OOD announced, "Captain, United States Marines arriving."

Pelham picked up the Wardroom phone and called the Quarterdeck. "This is the Captain. Do you have a Captain Mueller up there?"

"Yes sir, and a Chief Watkins, USN."

"Great, have them escorted down to the Wardroom, find Ops and send him down here, make an announcement if you have to, Captain out."

Harry Mueller, Chief Watkins, and Lt. Wolfe all arrived about the same time in the Wardroom. After introductions, coffee was served, and David asked Rats what got him up so early in the morning.

"Commander, I got your message that you were looking for me. We have a flight tonight, so I thought we better get together this morning."

David asked some of his officers lingering after breakfast, to clear the Wardroom, and told the galley crew to take a smoke break on deck. He winked at Mueller and Chief Watkins. "It's always great to see you transport aviators. I think we have some work for you that involves flying from here to San Francisco, then up to Anchorage, Alaska, on the 8th of April. You would be flying from there to San Francisco on the 9th with a passenger and some cargo. You can layover at Alameda or in San Diego on the 10th and 11th and return here on the 12th."

Rats Mueller figured on paper, "To be in Anchorage on the 8th, we'll need to leave Oahu on the 6th, RON in Alameda, and fly to Seattle and then Alaska on the 8th. Depending on the passenger's arrival on the 9th, we fly to Seattle or on to San Francisco. Next day, we take the plane to NAS San Diego for two nights and return to Hawaii via Alameda. How am I doing so far?"

"Real good Rats, except I have some additional requirements. The passenger you're picking up is coming in on a ship. He's Japanese and will be in a prisoner status. I'm sending an interpreter with you from here to Alaska and back to San Fran. The interpreter, Professor Michael Nagy from CinCPac Fleet staff, the prisoner, and the cargo will be delivered to the

San Francisco airport. You can go on from there to Alameda or San Diego. A couple of days later, you need to fly Nagy and the prisoner here."

"Picking them up from where, Commander?"

"I suppose the San Francisco airport. Nagy is a civilian of Japanese descent; I don't think he could talk his way onto NAS Alameda," David answered.

"Commander, is there a deadline for us to get back?" Chief Watkins asked.

"Not really. It looks like you will be back on the 14th or 15th. We sail on the 16th for the exercise. The prisoner may sail with us, or you bring him to Johnston Island with the other men during the fleet exercise in April."

"Yes sir," the Chief responded. "How many men on that flight will we be moving?"

"Eight Marines and fifteen new Navy recruits and their gear," David responded.

"No problem, Commander. There's a rumor in the wind that you're heading up a combat operation. Can we be part of that?" asked Mueller.

David replied in a low voice, "You must treat what I'm about to tell you as highly classified. If too many people know what we're going to do, it will sink my ship and destroy your aircraft. Lt. Wolfe and I have been meeting with CinCPac Fleet to refine our proposal to pull a surprise attack on Wake Island."

Captain Mueller was all ears. "Yes sir, now you're talking. How can we get some revenge on Wake? What is your objective?"

"We plan to put them out of commission for twenty-four hours. We want to make the runway and taxiways unusable, destroy every aircraft, and explode their fuel supply."

"We want in, but I'm not sure how we can help," Rats said.

"We want to know if there is some way you could fly from Midway to Wake, drop bombs on the runway, and return safely to Midway."

"Drop bombs?" Mueller asked with a curious look.

"Yes, we'll be offshore and bombard the runway, taxiways, and airplanes. The initial indication to the Wake Island Japs that something is up will be your first bomb hitting the runway. I need you to find a way to hole the runway every 50 to 100 feet so no planes can take off and attack us."

"Why us? You know the Commando is not a bomber. So why not a B-17 or a B-24? They both are here in Hawaii," Harry interjected.

"It's simple," David responded. "Trying to coordinate with the AAF is just too hard. This plan only works if it's closely held, and the whole attack is a complete surprise. The AAF will want to issue formal orders, stage to Midway, send more planes than we need, etc. Worst of all, they will want to bomb from a high altitude. We need an airplane that can make the round trip, attack at a specific time after sunset, and bomb the runway from low altitude. We need an aircrew that will keep their mouths shut before and after the attack. One that we can trust to arrive on time and make the runway unusable in one minute."

"One minute? What's with one minute?" Rats demanded.

"Let's assume the runway is 5,000 feet long, and you're at your slowest speed of sixty miles per hour. You're flying at one mile per minute, so you get less than a minute over the target. I want the runway holed every 100 feet. That equates to one bomb out the door nearly every second!" David exclaimed.

Chief Watkins responded, "That's impossible! I could see how we could shove something out of the airplane every five or ten seconds, maybe use some kind of rails as they do for depth charges."

David winked at Lt. Wolfe. "See, I told you so. They're already trying to figure out how to turn their plane into a bomber. Chief, we've done some of the research for you. At first, we thought of dropping depth charges, but they're too bulky, heavy, and hard to arm. We think the best-fitted explosive is the MK1-M30 – a 100-pound bomb. It provides forty-eight pounds of TNT that will make some nice big holes in the runway. Two men can handle them easily, and you should be able to carry more than fifty of them. You need to kick one out of the main door or the starboard rear opening every two seconds. If ninety percent of them hit the runway, that will do the job."

Mueller complained, "How do you expect us to move them out of the aircraft? The starboard opening is probably not big enough for this. How many people will it take?"

"All things for the Ginger Snap crew to figure out, my talented Marine friend. You have to shove the bombs along some kind of delivery system that has them vertical, horizontal, or nose down, so you can quickly shove them out the opening while pulling the arming lanyard. Put them on roller skates, roll them horizontally, and shove them down a slide, or put wheels on each one that also go out the door. It would probably take a minimum of four to six men to move two lines of bombs to the openings, push them out, and pull the arming wire. Get Miss Bessemer, Corporal Jenkins, and the Maintenance Facility working on the problem. I bet they can figure it out," David asserted.

Harry smiled, "OK, we'll carry sixty of these beauties and get fifty out of the aircraft. How do we see the landing strip at night?"

"Rats, we need to fine-tune the timing. After sunset, we want it as dark as possible to mask your plane, yet we want you to line up on the runway from a few hundred feet in altitude. You've been to Wake and Midway. Are the landing fields visible at night?"

"Commander, you can see the outline of the cleared area, but not the exact runway itself. Wake has a triangle of taxiways and the main runway that runs mostly east and west."

"Harry, we could put some aerial flares up, but they would make you too good of a target."

"Commander, while it's twilight, we can hit the runway, but they might still be able to use the runway edges or taxiways for flight ops."

"Selfridge will put a few hundred shells into those to shut them down. After your first bombs explode, we will shell the island right behind you. During the ten minutes after you finish, we'll lay down 300 five-inch shells and destroy any gun emplacements that attempt to fire back."

"How do we get the bombs?" asked Chief Watkins.

"They're available at NAS San Diego as we speak," David responded.

"In that case, we're onboard. So when do we do this mighty deed?" Rats enthused.

"Plan on setting aside three or four days before mid-May. Then, as we get closer, I'll get the exact dates to you."

They shook hands all around. Lt. Wolfe escorted the two Commando crewmen to a waiting liberty boat. David sat by himself in the Wardroom, sipping his coffee thinking, *Another piece of the puzzle falls into place. We have to finish the ship mods, convince Nimitz to back us, and be sure about our IJN signalman. He can make the difference between defeat and victory.*

We'll brief Nimitz, Layton, and probably Rochefort tomorrow. Do we tell them about Ginger Snap's role? Probably not; getting approval for the units we have already identified will be hard enough. Three months into the war, and all of the victories are on Japan's side. They've taken Guam, Hong Kong, Bali, Timor, and East Java. They occupied Rabaul in New Britain and turned it into an air and sea stronghold. First, Kuala Lumpur fell, and then supposedly invincible Singapore. MacArthur has his back against the wall on the Bataan Peninsula, and the Nips are consolidating their control over the oil-rich Dutch East Indies. They've severed the Burma-Chinese logistic lifeline and shocked the Aussies with a bombing attack on Darwin.

That night, after Eight o'clock Reports, the Engineer came up to the Bridge to meet with Captain Pelham – it was a nightly ritual. Lt. Comdr. Haldane enjoyed seeing the Bridge crew in action, the fresh air, and moderate temperatures compared to engineering. David eagerly wanted a daily report on the engineering plant and the men operating it. And he often asked for Haldane's advice and counsel. At David's request, they went out on the Port Fly Bridge.

"Engineer, tomorrow we'll brief Admiral Nimitz on the secret operation in May. Is everything in your bailiwick going to be ready? Can we count on Engineering to deliver high speeds for several hours?"

"Yes Skipper, we're ready. We've made several changes to the burners and boilers to support sustained flank speed. We're cutting some corners here. All Ahead Full will be much safer than Flank. It will be your call. Please remember we'll be using fuel like it's going out of style. We'll have to think about ballasting the fuel tanks when we use more than sixty percent of our stores."

"Understood. We should get a top-off from an oiler before we start two days of intensive ops. I think we're ready to brief it tomorrow. In a way, I feel I've lost control between you and me," David confessed.

"How so? Is someone not accepting your authority?"

Pelham said, "Oh no, it's not that. It's the size of this plan and the number of critical pieces. Normally, I would've met with everyone involved and have first-hand knowledge of each player and their capabilities. To keep up with all my responsibilities on the ship, I had Wolfe meet with several key people in Slingshot or units supporting it. I'm not used to relying on other people."

"Captain, what is happening down in Engineering right now?" Haldane demanded.

"Well, we're at anchor, so I expect some equipment is offline, and some are at low heat. Your men are providing the ship with steam, electricity, and freshwater."

"That's it," Haldane said. "You have people down there you must trust. There's an officer of the watch on duty; he has an experienced Chief in each Engine Room working with the men. Throughout the ship, you have men ready to do their job. You tell us what you want done, and we'll get it done. Imagine you're the Commodore someday. You have six or seven ships like this one, a ton of requirements, and you don't have time to visit each ship daily. As our Captain, your direction must be clear, and then you have to rely on other people. There's no other way."

"Thank you, Engineer. That's sage advice, to be sure. I'm turning in now. Have a good sleep."

David sat on his at-sea toilet and thought about Haldane's words. *Of course, he's right. Even now, I'm dependent on the watch standers and senior leaders to prepare the men and ship for combat. Luckily, I have a great XO who prowls the ship day and night to ensure our material condition and encourages the men to keep their spaces shipshape. We have an outstanding team, excellent enlisted leaders, and one damn fine ship. If I have my way, we'll make the Nips pay big time for Pearl Harbor, Guam, Wake, the Philippines, Hong Kong, Singapore, China - all of it. It's my destiny to kill a shit load of Japs, maybe thousands. It's my revenge for the hand that fate has dealt me. I just need to do that without getting my crew killed and losing the ship.*

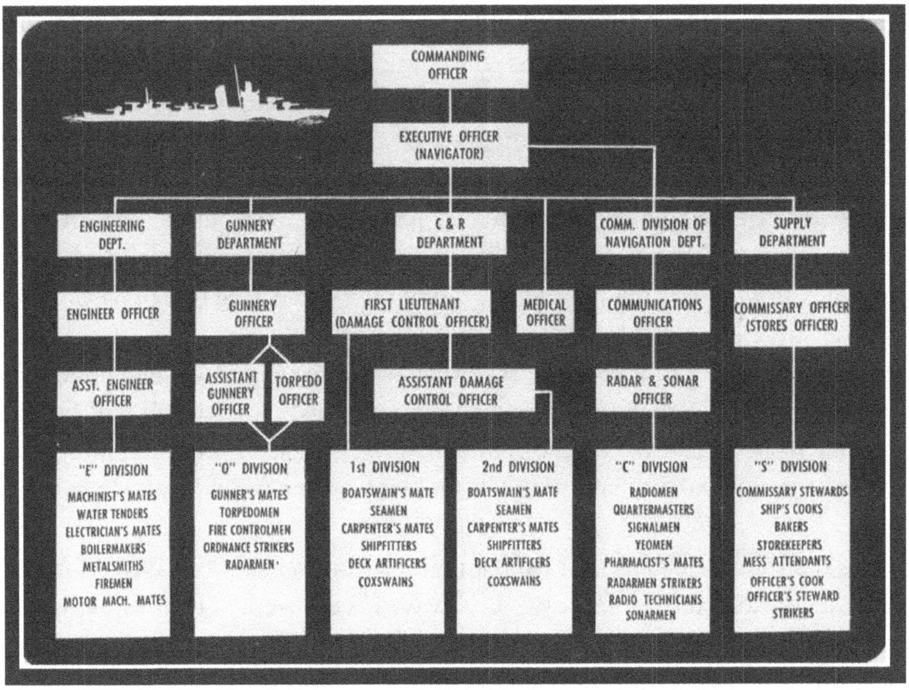

U.S. Navy Destroyer organization circa 1940s

C & R stands for Construction and Repair

Source: Navy training publications

Chapter Twenty-seven

DECISION BRIEFING – CinCPac Fleet War Room

March 1942

Pelham, Wolfe, and Lehman worked late into the evening on the presentation and the final details of the Slingshot plan. David instructed Wolfe not to cover the C-46 bombing flight, the Japanese decoy recruits, or some of their ideas for Phases after sinking the carrier. The following day, the three Selfridge men were waiting in Layton's office before the briefing.

Layton greeted them. "Morning, gents. I expect we'll be called to the War Room shortly. I have a blackboard and some maps set up for you there. Also, there will be some aerial shots of Wake that you can use.

"Do you know who will be at the briefing?" questioned David.

"Nope. I know that Comdr. Rochefort and I will be there, and the Pac Fleet Chief of Staff at a minimum."

There was a knock on Layton's door. Rochefort stuck his head in and said they were ready for the brief in the War Room. As the five of them filed in, David was surprised that three other Admirals flanked Nimitz.

Nimitz shook David's hand. "Captain Pelham and Lt. Wolfe, good to see you again. Captain, I believe you know Rear Admiral William Furlong, who is taking over the Navy Yard command. You will also be briefing Rear Admiral Claude C. Block, Commandant of the 14 Naval District, and my good friend, Rear Admiral Raymond Spruance, Commander of Cruiser Division 5."

David and Wolfe shook hands with each Admiral as Nimitz introduced them. Pelham introduced his XO, Lt. Lehman, to Nimitz, and Lehman shook hands with the other Admirals. The senior officers took their seats, facing the three blackboards that Layton had arranged. One had maps, a second diagrams, and aerial photos of Wake Island; the third was for chalk drawings.

Nimitz spoke first, "Gentlemen, I want you to know that Commander Pelham and Lt. Wolfe briefed Admiral King and me on Operation Slingshot two months ago. I believe they have much more detail to share with us today. Ernie King was impressed with this bold proposal. If we give Slingshot a thumbs-up today, he'll back us 100 percent - no matter the outcome."

Nimitz took his seat as Pelham stood to give his opening remarks. Layton displayed the first briefing poster board with large lettering. David noticed that Spruance was actively taking notes.

"Admiral Nimitz, gentlemen, thank you for sharing your precious time and naval experience with us today." He looked at Admiral Furlong. "As some of you know, I'm the acting CO of USS Selfridge, DD-357, a Porter Class leader destroyer. Lt. Lehman is the XO of Selfridge and has made many inputs to this OP Plan. Lt. Donald Wolfe is my trusty Ops Officer. He's become the Slingshot expert and will brief the details. Don, don't be overwhelmed by all the stars in the room. After all, you survived briefing this to Admiral King!" The comment elicited some smiles and polite laughter.

David picked up a three-foot wooden ruler and pointed to the first poster board. "The mission of Selfridge and her supporting units for Operation Slingshot is four-fold. First, we will bombard the Wake Island airfield and put it out of operation for twelve hours. Second, we will place the Fighting 357 next to a Japanese carrier on a moonless night and sink her with a five-minute concentrated bombardment. Third, combined with the shock of rapidly losing a carrier, we subject the battlegroup to strange sounds, mysterious lights, and other supernatural events to rattle their decision-making and cause the combatants to flee. Fourth, we will maneuver through the enemy's Service Group ships and take them under fire. Lt. Wolfe will now brief the Slingshot Phases. Please hold your questions until the brief is complete. Thank you."

David sat down and watched the Admirals' demeanor as Wolfe took them through the details of each phase. The three flags seemed to agree with the mission as they heard the progression of tactical steps and the ship modifications supporting Slingshot. Spruance was hard to read as he appeared to write down every word in his notebook. He raised his hand when Wolfe asked for questions.

"Your thesis is that you're risking a destroyer and her crew to destroy a Jap carrier." Spruance looked at his notes, "What about these other units, the submarines, patrol aircraft, and refueling ship? Aren't they also at risk?"

David stood to field questions. "Yes sir, they are, but to a much lesser degree. The aircraft's mission is to shadow the advancing enemy battlegroup and remain hidden as much as possible. One sub will fire torpedoes at the second carrier to cause more fear and confusion. The oiler will top off Selfridge and the subs before the surface action and remain 300 miles east of Wake."

While waiting for the next question, David thought, *There are a few things we aren't telling you, Admiral. Bet you would like to know about the single bomber we have on tap, how we would use the hospital ship, and where we got an IJN Signalman. Nagy is back, and Kimicho has delivered. We now can estimate the Jap battlegroup composition and their Planned Intended Movement, their PIM. What we don't know is the accuracy of his data.*

"I think your plan has a lot of merits. However, one sticking point is the whole operation can fail if the Japs conduct disciplined aircraft searches, especially at night," Spruance asserted.

"Yes sir, our intel is that Wake is short on Avgas. We believe they're searching near the island, emphasizing eastern and southeastern approaches. We plan to approach them during a high-speed dash from the south."

"High speed means a significant stern wave that is visible for miles. What speed is your approach to shelling Wake, and for how long?

"Sir, we will sprint from 100 miles south at forty-two knots for two hours before sunset," responded Wolfe.

Admiral Bloch protested, "No destroyer that I've ever seen can make that speed."

David said, "Admiral, Selfridge can. We've made some changes to the power plant by removing some fail-safes. The slick bottom paint that Lt. Wolfe described gives us a theoretical top speed of 44 knots in sea states one to three."

"How fast have you sailed her?" Nimitz asked.

"Admiral, we've done All Ahead Full for two hours at thirty-nine knots, and flank speed for an hour at forty-two knots," David replied. He thought, *We aren't going to cover impersonating a Jap ship to lure aircraft close aboard so we can kill the pilots.*

"You shell Wake at dusk; they will fire back on you. How do you survive that?" Admiral Furlong questioned.
"We'll be firing from the edge of the visual horizon using a spotter in our airborne gyro glider. So they will see some flashes and smoke, but not our exact location."
Also, he thought, *we'll have one of our fast boats in front, a quarter mile away, and one in the rear setting off fireworks and making smoke. From Wake, it will look like an armada is shelling them.*
"You shell Wake at nightfall and then expect to sneak up on a carrier group that believes there is a U.S. carrier in the area? It doesn't make sense. They will be at a high state of readiness," Spruance challenged.
David looked at Comdr. Rochefort, "Yes sir, I agree. It would be very supportive if we could have a credible radio transmission from Lexington or a ship mimicking Lexington 300 miles southeast of Wake that night."
"What is your strategy with that?" Nimitz asked.
"If the Japs on Wake and in the carrier group believe our carrier is that far away, 500 miles from where we expect their carriers to launch, they will concentrate on getting ready for dawn searches. They will feel certain that no surface group could reach them that night. So they might get complacent," David said.
"And you'll be closing them at nearly fifty miles an hour!" Spruance declared.
"Yes sir."
"In four hours, you'll be near their expected dawn launch point."
"Yes sir, if we luck out on the sea state."
"In one or two more hours, you plan to intercept them and complete your approach to your firing position," Spruance summarized, looking at his notes.
"Yes sir."
"How do you know where they are and what course to set?"
"We hope that the PBYs are keeping rough track of our targets, and we've requested that a returning sub from long-range patrol shadow the Jap battle group." David thought, *Better than that, Admiral, we know who is coming and their PIM course and speed.*
Spruance continued, "Let's say you get in the same general area as the battle group. It's pitch black at night. How will you identify them and get close to a firing position?"
David went to the blackboard and picked up some chalk. "We expect the formation to look like this: two carriers abreast of each other, with about half a mile separation, with a cruiser half a mile out from each carrier. Three destroyers out in front, roughly a mile, here, here, and here. There should be two or three destroyers on the rear flanks at just under a mile, here and here. We expect the service group will bring up the rear about four to six miles with three destroyer escorts. Our job is to slip between these two destroyers on the starboard side of the formation or to swing around the aft one and approach the starboard carrier from astern," David said while drawing the two approaches on the board.
Spruance spoke abruptly. "Commander, you didn't answer my question. How do you find the battle group, identify them, and determine their course and speed?"
"Sorry, sir, I guess I got a little carried away. Every time we run through this scenario, we learn a bit more and get some refinement. When we're in the battle group's immediate area, based on projecting their reported position and PIM, we'll slow to fifteen knots and launch

our gyro glider. We believe our airborne observer will see the ship wakes at ten to fifteen miles. We will see them before they see us."

"And if you're detected, what then? You are hopelessly outgunned!" exclaimed Admiral Furlong.

"Yes sir, no doubt about that. On the other hand, we're the only ship that can retire from those guns at forty knots. Please remember our original premise. We're willing to risk a destroyer and its crew to damage or sink a Jap carrier."

"You really believe a destroyer can sink a carrier?" asked Block.

"Yes sir, Admiral Bloch, I do. How would you like to be on that carrier in the dead of night with Selfridge ready to fire from less than a quarter-mile away? How would you like the Marines plinking your planes and fuel bowsers on the flight deck and hangar decks and then setting them on fire with mortar rounds? How would you like to be blinded on the Bridge and feel our five-inch guns putting over 200 shells into your starboard side as our automatic Bofors guns rake your bridge and open areas?"

Admiral Bloch responded in a low voice. "Captain Pelham, I wouldn't like that one bit. One thing is bothering me. How do you expect the mortars to be accurate firing from a moving ship?"

"I'm glad you asked that, sir. Bosun MacGregor rigged up a crude pitch and roll indicator. It's a two-foot-high open pyramid with a penlight hanging from the top. The light shines a small dot down on a grid of circles and lines. As the ship pitches and rolls, the penlight dot shows the amount of movement on the grid. We'll be testing it in the upcoming exercise. The mortar crew makes corrections and fires when the penlight dot is roughly in the same place on the grid as it was when they fired their last round."

"Ingenious. I'd like to see that device in action. Raymond, what do you think?" Block queried.

"I think the whole darn operation is just crazy enough to work, although it's a long shot." Spruance pointed at Pelham. "And I think he's just devious enough to make it work!"

"Pelham, what happened to the larger Army gun?" asked Lt. Comdr. Layton.

"It fell into the too-hard bin. Too big, too much recoil, and Army red tape sank the idea. It turns out that our five-inch guns at close range are ideal for the job. We will have two of eight guns firing maximum fuse delay, armor-piercing shells that'll explode after penetrating two or three spaces."

Spruance questioned, "Please explain this gyro glider, and who makes it?"

Pelham answered, "Yes sir, it is a prototype from Pitcairn Aircraft in Willow Grove, PA. In 1938, they bought the U.S. rights to produce it from Focke-Achgelis, a German company. Harold Pitcairn took their plans and used lighter wood and aluminum to drop the weight by thirty percent. It has three rotor blades that are tough to mount on the German design, but dead easy on Pitcairn's. Put in two locking pins on each one, and you are good to launch."

Nimitz stood and closed the meeting. "Selfridge, another excellent briefing. We move ahead with the April exercise, whether Slingshot gets approved or not. We will take this under advisement and recommend approval or not to Admiral King today. Dismissed."

Layton, Rochefort, and the Selfridge men filed out of the War Room and said goodbye to each other in the downstairs lobby. Petty Officer Brown drove his three officers back to the boat landing, where the four of them boarded a launch out to the destroyer line.

On the boat ride out to Selfridge, David had time to think. *The brief went well. We'll have a green light if King agrees. The ship is raring to go. Most of the mods are complete. Daniel Price has a yard crew working on improving the AA guns. The gyro glider is installed and ready for testing; Smitty will be airborne a lot during the exercise to check it out. We have all of the pieces and parts to scare the Japs, including three fast boats secured in the whaleboat davits and covered in canvas. MacGregor fixed the clogging oil line problem on the anti-plankton system. We installed the hydraulic elevator lifts during the dry dock period. All of the Marine and American Japanese personnel are identified, signed up, and arriving at Johnston Island for the second week of the exercise. Harry Mueller and the Ginger Snap crew are taking care of that. In early May, the supply pork chops will have us at 110 percent of our full magazine shell load. We will load the mortars and shells, and the sniper gear next week. After that, it's time to hold the weekly Department Head meeting and take the rest of the day off.*

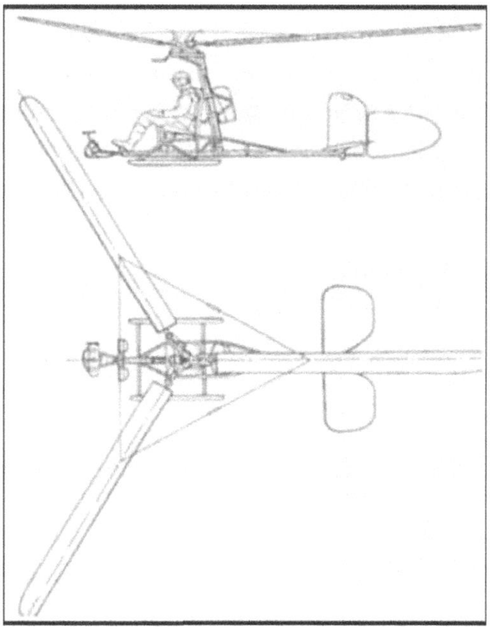

Focke-Achgelis Gyro Glider

Source: Wikipedia

Chapter Twenty-eight

SIGNALMAN GORO AKEMI – Pearl Harbor, Oahu

April 1942

Captain Pelham had two pages of personal action items after completing the Department Head meeting. He came ashore, drove the ship's car to Tripler Army Hospital, and parked next to Betty White's coupe. He fell asleep while waiting for her to finish her shift. After discovering the Navy car, she leaned in the driver's window and woke him with a kiss on the cheek. She put her hand on his shoulder and said, "Hi there, sailor. Could you give a girl a lift?"

David woke. "For a looker like you, anything. What would you like to do?"

"For once, I'd like to forget about the war. How about we go to the Princess Theater downtown? They're showing Hepburn, Grant, and Stewart in Philadelphia Story at 1700."

"Fine with me," David said. "A little air conditioning, popcorn, and Coke will hit the spot. Of course, they're bound to have a war newsreel. How do you want to handle that?"

"We'll wait in the lobby until the movie starts," she instructed.

They both enjoyed the show, the popcorn, and the ice-cold Cokes. They held hands and for two hours enjoyed being on a date. They relaxed, forgot about the war, and didn't think about the future. Betty's roommate did not go on duty until 2000, so they went to Fort D. for a couple of drinks and dinner. As they shared an ice cream dessert, Betty gave David a meaningful look and said, "Something big is coming, isn't it?"

"What do you mean?"

"You're planning something, maybe leaving, maybe picking a fight somewhere with the Japs."

"Whatever gave you that idea?" he protested.

"See, now I know that I'm right. You're covering up. You're such a poor liar! Don't leave me for a long time without warning - promise me."

"OK, I promise. I'll be at sea for the last two weeks in April, and I'll see you in May. Can we go now?"

"Yes, my Captain, it will be eight soon. I need a shower. I'm betting you need a shower. We could save a lot of water by showering at the same time, albino boy. What do you think?"

David got up and left cash for the bill. "I think that's an excellent idea, the water-saving that is. And for your information, I even have some tan lines now."

The following day, Pelham drove to CinCPac HQ and called Layton from the lobby. Layton told him that the item they discussed with the Admirals was a go. He returned to his Inport Cabin and found that Yeoman Brown had left him a ton of paperwork to plow through. David called Chief Gonzalez and asked him to set up a working lunch for himself, the XO, and Ops on the Bridge at noon. After about two hours of paperwork, he got up, stretched, and used the head.

The session with the XO and Ops lasted nearly three hours. They pored over the details of every Slingshot Phase, the logistical requirements, and each unit's job in the plan. As they were finishing up, the OOD arrived on the Bridge with a new officer who had just reported onboard. Lt. (jg) Grant Simpson shook hands with his senior officers and answered their questions. He was from Little Rock, Arkansas, graduated from the Academy in 1938, and had served on the USS Dahlgren during the past two years. Simpson said Dahlgren was an old four-stacker primarily assigned training and equipment testing duties. He had served as 1st Lieutenant and then Ops Officer on Dahlgren and was in gunnery school when the war started. Upon graduation, Simpson found himself assigned to Selfridge in mid-March.

David excused Lehman and Wolfe while inviting Simpson to stay and chat. During a few minutes of additional conversation, David decided that Simpson was very sharp and told him that he might be assigned as Assistant Ops Officer if Lt. Wolfe agreed. David informed him that they needed an officer to maintain the timeline of events and actions for a combat operation during the next two months. Simpson then reported to the XO for cabin and watch standing assignments.

David touched base with Wolfe, who agreed to take Simpson on as his assistant. Then Pelham told the OOD to wake him at 1700 and took a nap. He took his supper on the Bridge and queried the Engineer on the ship's power plant. Finally, after a good night's sleep, he enjoyed breakfast in the Wardroom with the XO and officers who weren't on watch. The following week was busy with new equipment installations and regular maintenance. Soon, Selfridge would be moving to the ammo pier to finish loading fuel, food, and ammunition for the upcoming exercise.

Pelham was working on another pile of paperwork when a messenger came to his Inport Cabin and announced that there was an Admiral's gig waiting for him. He took a piss, put on a fresh service dress white uniform, and grabbed his wallet, sunglasses, and cover before emerging on deck in the morning's humid air and bright sunshine. Lt. Comdr Layton sent the gig to take David ashore to meet with Professor Nagy. It took him to the Pearl City Ferry landing, where a car was waiting. After a short drive up Lehua Avenue, he got out of the car at a house on Ashley and Laniwai. Michael Nagy greeted his friend and led him to the patio behind the house.

"David-san, it's so good to see your friendly face. Please sit and share some of this lemonade with me."

"Good to have you back, Michael. What's with the house? Do you live here?"

"We aren't sure. This house is where Commander Layton had us bring Akemi. That is Goro Akemi, your signalman, who is in the house with Rachael and the boys."

"What do you think of him? How was your trip? Did you meet with Kimicho?" David gushed.

"Slow down, my friend, one thing at a time. The home was owned by a Japanese doctor, Dr. Yokichi Uyehara, born in Japan. He seems to have left suddenly because the furniture and many family personal items are still here. We don't know if they're coming back or if we get to live here. Regarding Petty Officer Akemi, he's an IJN professional Signalman and talks a good game, but only in Japanese. He grew up in the port city of Otaru on Hokkaido. His father was a history professor at Hokkaido University in Sapporo until the Army secret police took him two years ago for questioning. Goro believes his dad is locked up in a Yamagata prison on the main island. The Akemi family is Christian. That makes them suspect to the Japanese police forces and the Army. He's pretty sure it has hindered his advancement in the Navy."

"What's his history with the Navy?

"He joined the Japanese Navy in 1933. After boot camp, he was a seaman on an ocean-going tug based in Yokohama. After that tour, he graduated from the Navy's trade school in radio technology and Morse code. Then, he advanced as a Signalman Petty Officer for two years aboard a small destroyer. For the past year, he's been our equivalent of a Petty Officer Second Class Signalman on a Japanese cruiser."

"Please tell me he was not in the force that attacked us at Pearl," David said.

"No, his ship was in the group that took up defensive positions near the home islands."

"He sounds like exactly what we need!" David enthused. Do you think he can be loyal to the U.S.? What did the Yakuza promise him? Why did he defect?"

Michael held up his hands. "Please, slow down, let me think. You asked if I met our friend in San Francisco. I did, and I can tell you more about that later. I don't know if Akemi can be loyal to America. He has not been willing to talk about why he's here. I don't know what the Yakuza promised or all the ways he was threatened. I do know that the man you call Kimicho told him that death is the penalty for breaking their deal. I don't know why Akemi is willing to work for us. I do know he didn't plan to defect. Goro's ship pulled into Tomakomai on Hokkaido a month ago for repairs. He went home on two days' leave; the Yakuza put some kind of pressure on him and somehow faked his death. They delivered him to Attu in a fishing boat and thoughtfully sent us a seabag full of Japanese signal flags, national flags, and battle flags."

"Was that part of the 500 pounds?" David asked.

"No, it was not. That so-called cargo was a pain in the backside. It arrived in four thick, hardwood containers. Each one weighed over a hundred pounds and was difficult to move."

David said, "I think it was gold. Did you deliver the crates to Kimicho?"

"No, he met our plane at the San Francisco airport along with four men. They took the cargo to a hangar; that's the last I saw of it."

"I think I'm ready to meet Petty Officer Akemi now. Please tell him he is about to meet his commanding officer, and he must be completely truthful. Otherwise, I can put him in jail or send him back to Japan." David threatened.

Michael went into the house and returned with a young man wearing civilian clothes. David was waiting, wearing his cover. Akemi came up to him and bowed deeply. David acknowledged the bow with a nod of his head, the second one, and the third. Finally, Akemi came to attention and saluted. David returned the salute and motioned for the sailor to sit at the table. David removed his white cover and stared intently at Akemi when they were all seated. After a minute or two, Pelham spoke to Nagy. "Tell him that the answers he gives his commanding officer will determine his future. Goro must be completely truthful and hold nothing back."

Nagy translated David's direction into Japanese, and Akemi nodded in agreement.

"What did the Yakuza promise you if you came to the United States?"

Nagy translated the response. "They promised me that they would do their best to keep my father alive in the political prison and protect my family in Otaru."

"How many members of your family will they protect?" asked David.

Nagy translated, "My wife and two children and my mother live together. I am not afraid to die. I am willing to die if my family can live."

David thought, *If Otaru is a port city, you can count on it being bombed sooner or later, probably in three or four years. We won't worry about that now.*

"Are you still loyal to the Emperor and the Japanese Navy?"

Nagy translated. "No sir, I renounce both of them. They and the Army are ruining Japan. Now they are going to ruin many other Asian countries. They lie to the people while they take everything for themselves. The thought police beat my mother to get her to testify against my father. They failed in their mission and ruined her right arm and hand forever."

"Who do you think will win this war?"

Nagy translated. "America will win the war against Japan. My father has taught me that. He saw the war coming as soon as we signed papers with Germany and Italy. He warned against the war. He lectured against it and wrote a long paper saying it would be a great mistake. He thought he had the freedom to speak his mind as a professor at a national university. He was wrong. They threw him in jail with no charges and no trial."

"Why did your father think we would win?"

"My father is a student of history. He said that history is against Japan. He's traveled to the United States and seen the factories. He said that once men like Henry Ford started building tanks instead of cars, nothing could hold the U.S. back. He said the Army was wrong calling American men weaklings and afraid to fight," Nagy related.

"Did the Yakuza tell you that you would be fighting against your Navy? Did they say that you would be helping to kill your former comrades?" David demanded.

Nagy translated. "No, not in those words. They told me I would have an American commanding officer like you, that I would take an oath of loyalty and do my duty. I am a military man. It is all that I know. I spit on my Navy. They treated me like dirt because I am a Christian. I love my Japanese country. My country is no longer Japan. My son and daughter are learning nothing but worthless rubbish in their school. The Army spews out endless lies and controls the press. Many people will have to die for us to get our real Japan back. I'm sorry to do it. I see no other way to regain my family's honor."

"Petty Officer Akemi, I agree with you. The last thing I want to do is to kill the Japanese people. My honorary uncle was born in Japan. The Japanese military and diplomats have given us no choice. War means fighting and killing. My ship will take revenge on the Japanese Navy very soon. We will kill more Navy men than all of their planes slaughtered in the Pearl Harbor attacks in one night.

David waited for Nagy to translate and for Akemi to nod his understanding. "Your role in this mission is vital. You will stay here with the Nagy family for about a week and then be flown to Johnston Island to join my ship. You must search your soul during that week and make your peace with God and Jesus. I must be sure that you will be trustworthy and loyal to me and your new ship. We are going into combat. I can't have any hesitation on your part or misguided efforts to help the enemy. Do you understand me?"

Akemi listened intently to Michael's translation and answered in the affirmative.

Pelham had one more question he put to Nagy. "Please ask him if the Japanese Navy uses the same Morse code as the American Navy."

There was a long back-and-forth exchange between the two men in Japanese. Finally, Nagy said, "Yes, and no. Goro can send International Morse code, but that is not what the IJN uses afloat. They have upwards of seventy code groups of dots and dashes instead of our thirty-six. The groups represent kana codes or pictographs of their language. It is completely different than regular Morse code."

"Damn, Michael, I must be psychic. We have no one but Akemi that can communicate with Jap units. He's worth his weight in gold. Take good care of him."

David stood and put on his cover. Akemi got to his feet and stood at attention. He bowed to David, who returned the bow briefly as the senior person. Akemi saluted, and David returned the salute. He smiled and firmly shook hands with his new signalman.

As they went around the house and in the front door to see Rachael and the boys, David noticed a sign over the entrance arch, "Uyehara Sanitarium Rakuzanso." Nagy told him it meant Happy Mountain Retreat. After David thankfully hugged Rachael and the Nagy boys, they told him about their voyage to Hawaii. They all talked rapidly in English, which was probably confusing to Goro Akemi. David said goodbye to Rachael and the boys. He had Michael follow him outside.

"Michael-san, during this coming week, I'd like you to drill Akemi in typical English words and phrases. Get a list of basic Navy terms from Layton. When Akemi comes aboard the ship, he must understand terms like general quarters, the smoking lamp is out, and man overboard. Also, have him study the USS Porter class pages in Jane's Fighting Ships and explain the English terms. Try and teach him the American names for different parts of a destroyer. Make it clear to him that we will disguise Selfridge to look like the Akizuki class of destroyers. I'll provide you with sample drawings as soon as possible. It will be his job to paint the correct markings on our ship and fly the appropriate flags from the proper masts at all times. He'll be working with our Japanese-American crew of fifteen to make our deck crew look like that of an IJN destroyer. Tell him bluntly that we're going to lure Jap planes alongside and then shoot the pilots. Watch for his reaction and see if he has a problem with that.

"I'm giving you the name of a tailor shop near the base. Take him there for uniforms and have him make sure the ones they're making for our crewmen look like Japanese Navy uniforms."

Nagy replied, "I'll do all of that. I took notes on my visit with my great-uncle. He had more information for you on the Hawaii attack. He said the IJN has a plan to protect Wake Island with two carriers in May. Layton has my notes locked in his safe. Before I left for San Francisco, he gave me five grand in hundred-dollar bills to pay my great-uncle. Kimicho, as you call him, reminded me that if you're satisfied with Akemi, you owe him another five thousand."

"I'll know that after the exercise," David replied.

Pelham thanked Michael for shepherding Akemi from Alaska to Hawaii and meeting with Kimicho. Then, he got in the car and had the driver take him to CinCPac Headquarters. In Layton's office, he reviewed Nagy's notes and asked if he could take them to Selfridge.

Layton said, "You can have them after I copy them. There is some stuff in there that I want to brief to the boss."

David was thoughtful. "Yes, Admiral Nagumo was up against more constraints than we realized after his two-wave attack on Oahu. His small boys needed refueling. He lost twenty-six aircraft and had more than 100 that were no longer mission-ready. Two of his carriers had green flyboys that couldn't operate at night."

"Yeah, the same two carriers, Shokaku and Zuikaku, may be coming after you and your mythical carrier group. You can expect their air wings will be flying at night after five more months of training," Layton asserted.

"Maybe, maybe not," David said. "Kimicho says that many of the fliers on both ships were pulled from shore assignments and non-flying jobs to augment the Pearl Harbor attack. If they returned to their shore assignments, the two ships would have a new group of nuggets to train. But, of course, I guess they could put on a different Airwing."
Layton corrected him. "No, they don't rotate Airwings as we do. The planes and pilots on a Jap carrier belong to the Captain, and they stay on that ship - another example of Jap inflexibility. So you have two front-line, brand new carriers with the least experienced aircrews."
"I love it; the more weaknesses I hear about, the better our chances," David said.
"Guess what Admiral King sent my boss in a personal message about Slingshot?"
"I'm all ears. What?"
"He said, quote. 'Go ahead. It's your neck,' unquote."

David took the sealed envelope that Layton gave him and placed a call to Tripler from the CinCPac lobby phone. He left a message that he'd be available on Tuesday night and assumed Betty would figure out that he'd be sailing the following day. Selfridge would be at the Ammunition Dock all day and night on 14 April to top off her fuel and take on ammo. He returned to the ship and briefed the XO, Ops, and Lt. (jg) Simpson on the Kimicho intelligence without naming the source. The next day, he met with Captain Harry "Rats" Mueller and Chief Watkins in the Hickam Field cafeteria.
"Harry, I want to thank you and your crew for making the trip to Alaska and back with my cargo. When you were in San Diego, did you two get any ideas on that other matter in May?"
Chief Watkins replied, "Yes sir, we all spent a day on it with Rhonda, Jenkins, and my uncle from the Assembly Depot. We got some dummy MK1-M30 bombs filled with sand and tried loading them on the plane and testing what it would take to shove them out. After many ideas and discussions, my uncle drew a picture of something we all thought could work. Each bomb will rest vertically on a 12 by 12-inch steel platform with swivel rollers on each corner. There will be a row of twenty-five bombs for each opening. We built an upper and lower railing on both sides, so we can shove them vertically to the open doors, pull the safety wire, and slide them nose-first down steel plate chutes."
"How many and how often out the door?" David questioned.
Rats fielded the question. "We have to drop them from 500 feet to make sure the fuses get armed and protect the ass-end of the Commando. We can carry forty to sixty bombs on a 2,000-mile round trip. Picture that it's almost dark when we start our approach. We slow the plane down to sixty knots. This gives us fifty seconds over the runway and the aprons at each end. We think we can shove a bomb out of each side every two seconds. We have a good chance of blasting somewhere between thirty to forty nice big holes in their coral runway. We'll still have twenty more bombs at that point, so I recommend we do a second pass down the fuel farm, the parked planes, and the taxiway. Remember, their fighters can take off on that taxiway."
David was alarmed. "Negative Rats. You get one pass and get the hell out of there. We will be wiping out the taxiway, airplanes, and the fuel farm."
"Yes sir, aye aye,, sir."

They reviewed the planned passenger flight to Johnston Island during the upcoming exercise. David informed them they would be delivering the Japanese passenger they picked up in Anchorage, who would no longer be a prisoner. Then, David went back to his ship to prepare for the at-sea period. When the refueling and ammo storage were complete, he pried himself loose from his duties and met Betty at her apartment.

Selfridge sailed at 0700 on Wednesday morning. She had one tug nearby in case of trouble pulling away from the pier. There were no problems, and Captain Pelham didn't request a pilot because several large ships were sortieing. Admiral Fletcher's staff had briefed the task force units and formations at the pre-sail conference. There were six destroyers, two cruisers, the aircraft carrier USS Lexington, an oiler, two amphibious ships, and a tender for repairs. The destroyers sortied first and conducted ASW searches near the coast for sub contacts. Then the cruisers put to sea. Three destroyers joined on each of the larger ships, and they moved in a lazy box pattern about a mile apart. When "Lady Lex" finally sailed out of the channel and dropped her pilot, three destroyers fanned out and led her out to sea. Two cruisers took up station on each side of the carrier about a half-mile out, and the other three destroyers covered the rear quarters of the formation. Selfridge was directly behind the carrier maintaining station at one mile when she received a blinker message from the massive ship.

The Bridge's 2MC Talker reported, "OOD, message received by Signal Bridge, sir."

"Very well, let's hear it," replied the Officer Of the Deck.

"From Lex, close up and assume plane guard."

"Talker, have signals acknowledge, Wilco. Skipper, any advice?"

Close the distance by half. Talker, can signals see Phelps?" Pelham asked.

The 2MC Talker spoke into his microphone and replied, "Signals says affirmative, off our starboard beam."

"Very well, send to Phelps, quote 'we are assigned plane guard, any advice,' unquote."

After a couple of minutes, the 2MC Talker said, "Captain, we have a reply."

"Very well, read it out."

"Follow about a quarter-mile in her wake with no air ops. Then, move closer starboard of the wake for air ops. Stay on the outside of all turns. Good luck."

"OOD, now that we have some smart guidance, make it so. Watch her wake and her stack for acceleration or slowing down. Turn when she turns. You'll be fine."

Lexington started zig-zagging along a base course to Johnston Island at nineteen knots. There was no warning of course changes. She'd turn either way, and the formation had to adjust immediately. David decided that the OOD and his crew were doing fine, and visibility was excellent. He thought it was best to get some sleep and plan to spend the night in his Captain's chair. There were no flight ops for the first two days. They expected to be attacked by the "Red Nation" forces as they got closer to Johnston Island. Lexington put up Combat Aircraft Patrols at first light to counter planes from Johnston and longer-range bombers from Oahu.

Selfridge's Radar detected some of these aircraft groups at thirty miles. The "bogies" were reported to the carrier, and Selfridge simulated firing on the aircraft as they attacked the Lexington. Mock air-to-air battles broke out above DD-357, providing lookouts and gun crews with practice in aircraft recognition and the challenge of sorting out the "bandits" from the "friendlies." After two days of Plane Guard duty, David was anxious to have Selfridge complete several other drills and training evolutions. He appealed to the Commodore for relief, and Phelps traded positions with them at dusk on the third day out.

The next day, the battle force aircraft and ships simulated a coordinated attack on the Johnston Island Atoll. The carrier operated over a hundred miles southwest of the island in radio silence, guarded by two destroyers. At first light, two cruisers and four destroyers formed a battle line and simulated bombarding the main island from five miles at sea. They were protected by Lexington's Combat Air Patrol, or CAP fighters, circling overhead. Before the surface ship attack, two waves of Lady Lex bombers conducted practice bomb runs on the airfield and ships in the harbor.

After two more days of exercise events, David knew that Lexington, the cruisers, and five destroyers would be heading west to strike locations in the Marshall Islands or the South Pacific. Selfridge would remain behind at Johnston, do a high-speed run to the Midway Island area, and then finish her second week at sea in the Hawaiian Islands.

Before pulling into Johnston, David addressed the Selfridge crew using the 1MC. "Men of Selfridge, this is the Captain speaking. We're nearing the halfway mark of our two-week exercise. You men and this superb ship are doing an outstanding job. Our Bridge crews were magnificent in maintaining Plane Guard position without overrunning Lady Lex or losing her in the dark of night. You know that we have a special operation coming in May. Tonight, we embark several men who will be critical to our success. Their efforts may even ensure our survival during that combat. There will be a Marine sniper team that will use the two extendable platforms we just installed. And we will have Marine mortar crews.

"Now we come to an exceptional group of young men. They're Americans of Japanese descent. They all were born in the United States; they're American citizens. Each of them has sworn allegiance to our flag and taken the Navy enlistment oath. I call them Amjaps. These young men want to serve our country, just like you. They're putting their lives on the line just like you. As I said, their contribution to our mission could well save your life. So you will treat them with respect and welcome them as shipmates. That's an order. You have one week to teach them the basics of being a U.S. Navy sailor.

"Next, I want to introduce a unique person. We have recruited a signalman from the Japanese Navy with eight years of experience. He was not part of the Pearl Harbor attacking force and has never taken any action against our country. He's a Christian, making him a target for harassment in the Nip Navy. The secret police have tortured his family, and they hurt his mother. His name is Goro Akemi. He understands only a few words of English. He's enlisted in the U.S. Navy as a Petty Officer Second Class and sworn an oath to support and defend our Constitution. Now hear this. Any crew member who doesn't treat these new shipmates with respect and kindness will be court-martialed and left ashore when we sortie for combat in May. That is all."

Selfridge anchored in the Johnston lagoon the day David expected the C-46 Commando to arrive. A big twin-engine aircraft landed at the airstrip before dark. Boats MacGregor was already ashore with the whaleboat and tasked with moving the men to the ship with as little fanfare as possible. MacGregor brought the new men onboard in groups. Each man had a sponsor who took them below for a late supper and helped them locate their berthing assignment.

David was surprised to look down on the Main Deck and see Professor Michael Nagy escorting Petty Officer Akemi. He sent the Messenger to fetch them up to the Bridge. Nagy explained Akemi had never flown in an aircraft before their flight from Alaska and wouldn't board the plane to Johnston without him. Michael said they had made some progress with the English lessons and requested to stay onboard to interpret for Akemi. David said he'd think about it while they stowed the whaleboat, pulled the anchor, and tied up to the oiler to take on fuel.

Captain Pelham had his Administrative Officer and Smitty report to the Bridge to discuss Michael's legal status. They decided Nagy could stay as an observer for CinCPac Fleet, and Yeoman Smith cut some TAD orders to the ship. Pelham assigned Nagy and Akemi to his Inport Cabin. He had Jackson bring them supper and set up an Army cot for the Signalman. While they were both still on the Bridge, David spoke to Petty Officer Smith. "Smitty, are you going to be ready to fly first thing in the morning?"

"Yes sir, I can't wait!"

"We plan to leave port at first light; the wind should be westerly, so we will head west to launch you. Then you can guide us around the atoll reef to the north, keeping us about seven miles away. Get some shut-eye. We will call away the Sea and Anchor Detail and the Dumbo Detail at 0530," David directed.

"Yes sir, thank you, sir. See you in the morning."

The following morning, Selfridge hoisted her anchor at the first hint of light in the east and slowly cruised out of Johnston's half-moon bay. The OOD had the ship heading into a steady five-knot breeze and making turns for fifteen knots when Pelham ordered Dumbo's launch. David went out on the Starboard Fly Bridge to watch the first attempt, directed by the XO. PO Smith was sitting in the gyro glider, wearing a football helmet and a life jacket. Four husky sailors used stout poles to lift Smitty and Dumbo eight feet above the weather deck. They then walked him back about twenty-five or thirty feet so he was clear of the rear shooting platform. As Smitty eased back the control stick and lifted off, the XO slowly gave him more wire from a large drum until the flying kite was airborne abaft the stern of the ship.

Smitty tested the controls gingerly and moved right and left and up and down. After a thumbs-up to the XO, Dumbo was given fifty more feet of braided steel line and then fifty more. With 200 feet of line out, Petty Officer Smith was up over 100 feet. The conn was passed to Secondary Control that followed hand signals from Dumbo for a series of starboard turns that had the ship skirt clockwise around the Johnston atoll reef without ever seeing it. After an hour of circling the atoll, David had the OOD turn downwind and reduce speed to twelve knots. Dumbo started to settle, which was the XO's signal to steadily reel in the line until the deck handlers could insert their poles into metal cups welded under Dumbo's skids. All the Dumbo testing and mishaps at Hickam Field paid off. The first flight was a complete success.

Pelham settled back in his captain's chair, put his feet up on the forward bulkhead, and prepared to take a nap.

"OOD."

"Yes Captain."

"Set sail for Kure Island, K-U-R-E, fifty-eight miles northwest of Midway, speed All Ahead Two Thirds. Wake me if you need me."

"Aye, aye, sir."

David got about an hour's sleep before he felt a presence next to his chair. He knew it would be Yeoman Brown with the perpetual set of messages to read, personnel paperwork to sign, and outgoing reports to approve.

"Is that you, Yeoman Brown?"

"Yes sir, is the Captain sleeping? I can come back."

"No, no need to make another trip. What have you got for me?"

"Outgoing messages, personnel evaluations, and ship's maintenance reports, Captain."

David smiled inwardly as Brown never called him Skipper, like many crew members. It was always Captain or Commander. As he opened his eyes, he knew he'd see Brown in a perfectly tailored uniform, even if it was dungarees. He'd have his cover on and look sharp in all respects. He knew Brown would pass him the ubiquitous clipboards in a specific order, with the messages and correspondence carefully organized. There would be a paperclip wherever his signature was needed. First was the incoming message board that the Communications Officer and the XO had already chopped. Second would be the outgoing message board with each message signed by the originating Department Head and the XO. Third would be Personnel Evaluations, all-important for each man's advancement and service reputation. David looked at every grade and read each required comment as he thought about the 820 quarterly evaluations that Selfridge sent to BuNav every year.

Next were the outgoing letters, reports, and memorandums, followed by incoming correspondence. The last clipboard had today's Plan Of the Day, the next day's draft POD, the ship's newspaper, a cartoon or two, and other miscellaneous material. It took him an hour to work his way through all of it as First Class Brown patiently waited. Some days, Brown would track him down two or three times with a new stack of paperwork on the multiple clipboards. David estimated that he signed or initialed 100 to 200 documents a day. He had once thought the Navy ran on coffee, but as CO, he could see that paperwork was a close second. Petty Officer Brown was crucial in establishing and maintaining Selfridge's fleet reputation. Reports, correspondence, and messages were sent on time, the spelling and grammar were correct, and the distribution of paperwork up and down the ship's chain of command was prompt and efficient.

"Petty Officer Brown."

"Yes, Captain."

"You look very sharp today. I want you to know that I greatly appreciate how well you organize the ship's paperwork. Well done."

"Thank you, sir. See you later tonight."

As Brown departed the Bridge, the ship's Engineer entered the Bridge and stood by his chair.

"Good Morning, Mr. Haldane; what can I do you for?"

"Skipper, there is a rumor floating about a high-speed run, longer than any we've done."

"Damn, I must be talking in my sleep. Yes, I want to crank her up to forty knots for ten hours on our way to the Midway operating area."

"Captain, I don't recommend that. We have been out for a week. Playing Plane Guard burned a lot of fuel. A run like that may require us to ballast the fuel tanks with seawater, which gives my gang a lot of extra work in the long run. The Porter destroyer class is top-heavy, and we must maintain certain weights to keep her safe."

"Duly noted, but we need to push the power plants now when it doesn't matter. We can pick up some fuel at Midway after we shell Kure Island."
"Aye sir. I'll touch base with the navigators and go run the numbers."
"Outstanding, thank you, Engineer."

David took his meals on the Bridge until finishing the high-speed run. He passed his standing orders for cruising to Kure and conked out in his cramped bunk. About 0600, he was aware of someone moving about his at-sea cabin and guessed it had to be Jackson.
"Is that you, Steward Jackson?"
"Yass suh. Just picking up a little. OOD sez we be at Kure in an hour. Thought you might want to greet the Lord's new day before that."
"Thank you, Jackson. I'll take working khaki down to my other cabin and do that."
"Yas sir, yas sir. There be a hot pot of coffee there and some fresh donuts."

David smiled and thought, *Who's in charge here? Much of my day is organized by the tag team of Brown and Jackson. Bet they look at the POD, figure out where I'll be, and call each other about my location.* He used his more spacious Inport Cabin head, dressed, and returned to the Bridge as the JOOD called out,
"Land ho, Kure Island dead ahead about eight miles."
David answered, "Very well, OOD, call away the whaleboat launch. Once it's launched, make turns for twelve knots, and call away Dumbo flight ops. Then, announce General Quarters will be set in twenty minutes."
"Aye, aye, sir."

Pelham knew that Ops had briefed Smitty on the Kure Island targets that morning. They had picked the two western edges of the reef opening into the atoll lagoon, the northernmost tip of the beach, and the two sand spits on the east and west ends of the Island. Each had numbers, and the order of the attack was unknown to the Dumbo pilot. The island proper was ruled out as a target to spare the thousands of birds that nested there. They had also briefed Bosun MacGregor to pick a crew for the whaleboat, including the gunnery officer, to observe the shell hits from a mile south of the lagoon. Once Dumbo was flying, and the whaleboat launched, David instructed the OOD to pass within a half-mile of the island. They made sure there were no people or boats visible there and anywhere on the atoll area.
"2MC Talker."
"Sir."
"Instruct Dumbo to strike target 3, first with the forward guns, and finally with aft guns."
"Aye, aye, sir."
The 2 MC Talker responded, "Sir, Dumbo acknowledges strike on the north reef and gun sequence."
"Very well. Keep us just in sight of the atoll and tell Dumbo not to hit the whaleboat. OOD, take your course corrections from Dumbo. If I don't like something, I'll let you know."

Selfridge shelled the north atoll reef from the edge of her visual horizon with each of the ship's main guns. Then Pelham instructed Smitty to bombard the northwest reef surf line with the big forward five-inchers and the southern edge with the aft five-inch guns. Pelham then had Smitty attack the Island's west sand spit with all eight guns in salvo mode, and the east sand spit. Dumbo was secured, and Ops directed the OOD to call away the sniper and mortar crews. The shooting platforms were manned and raised; the M1 and M2 mortars were assembled and ready to fire. Two large oil drums were placed near the lagoon entrance as targets by the ship's whaleboat. David had the OOD position Selfridge about 500 yards south of the drums. Upon hearing the weapons-free command, the Gunnery Officer first directed the mortar fire at the drums and then cleared the snipers to shoot. After the drums sank, each mortar crew fired a bursting smoke shell and an illuminating shell for practice. Finally, David had a rocket fired to indicate that the firing exercise was over. In the whaleboat, MacGregor had orders to reconnoiter a route to the southwest tip of the island and to pick up dead fish for supper.

Once MacGregor returned to the ship, he attended a briefing from Ops on the whaleboat's nighttime mission. Lt. (jg) Simpson, MacGregor, two Seamen, and two Amjaps were selected to go ashore on the south spit of the island and stand guard. Simpson carried a .45-caliber pistol, and MacGregor a Tommy Gun. Lt. Wolfe briefed them that they were on their own and not expect the ship back until after sunrise. He said that no Jap activity was expected, but it was always possible. They had an hour to provision the whaleboat before Selfridge sailed for Midway.

After the whaleboat departed for the Kure atoll island, Pelham, Ops, the 1st Lieutenant, and the XO planned the Mid Watch events when Selfridge returned to stage a mock attack against the Kure Atoll.

One day blended into another until the two weeks of drills, watches, and ship handling was coming to an end. On the sail back to Hawaii, Selfridge shelled the reefs at French Frigate Shoals during the day and at night. DD-357 spent the last two days at sea in live day and night gunnery practice at Kahoolawe. Selfridge was firing at targets just on the edge of the five-inch guns' effective range. Low-lying targets, such as shorelines, were best seen from the gyro glider. Smitty was flying from 200 to 300 feet above the ship's wake while calling in different gun patterns and shell types.

For the final night gunnery exercise, Pelham, the Gunnery Officer, and Ops positioned themselves in the ship's whaleboat to see the shells hitting the island. They had given Smitty a script similar to what they expected to do at Wake. He was directing fire at three bonfires on the beach, separated by a quarter-mile, as the ship sailed on a parallel course heading west. The officers could see Smitty walking the shell hits from offshore to the beach. Then he called for a concentrated fire that obliterated the first bonfire. He followed the script, and laid down shell fire along the shoreline and then a mixture of star shells and flares before wiping out the next blaze. Captain Pelham and Lt. Wolfe applauded the performance as the shells crept along the coastline, and a tight salvo extinguished the last fire.

They knew this was how they would attack the Wake Island airfield with deadly accuracy, and the ship nearly out of sight while laying smoke.

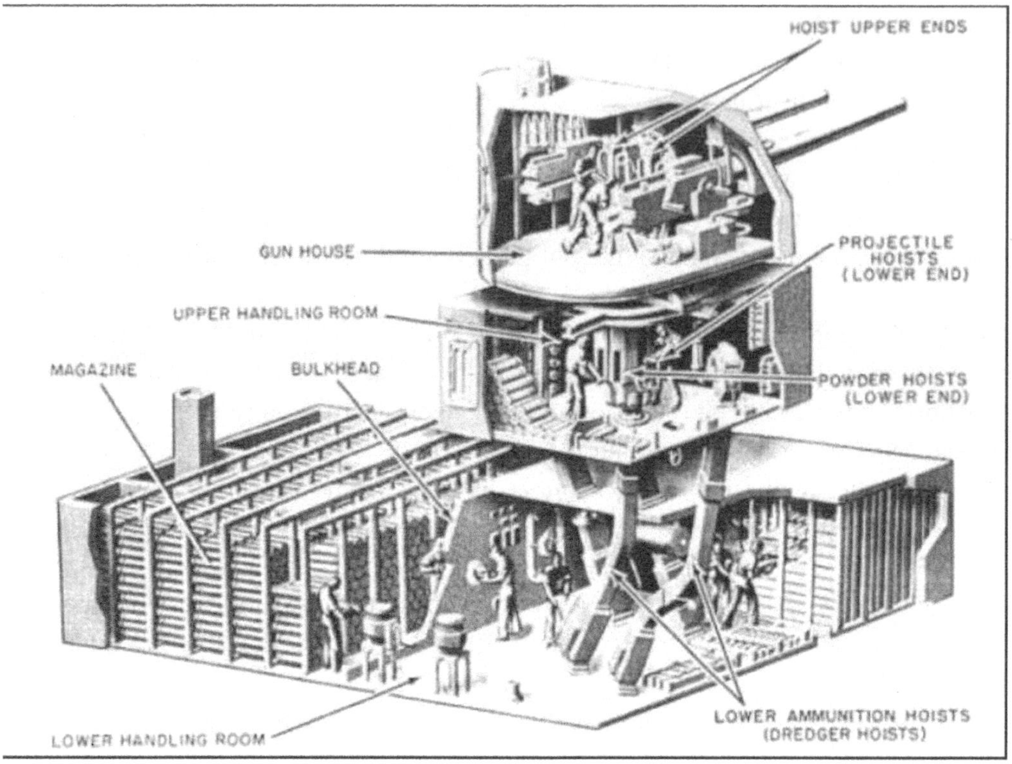

Typical 5-inch gun components

Source: Navy training documents

Chapter Twenty-nine

REMEMBER WAKE – Pearl Harbor Replenishment

May 1942

USS Selfridge completed her exercise training and returned to the ammunition dock at Pearl Harbor during the first week of May. The crew was elated to hear the first details of Colonel Jimmy Doolittle's B-25 raid on Tokyo. Unfortunately, their celebration was reduced when Captain Pelham had his XO announce some bad news. There would be no off-base liberty during the short inport period, and Selfridge would execute the Strip Ship Bill the following day. They needed to offload hazardous and flammable materials not required for the combat mission of the ship. Non-vital wood, paper, flooring, paint, solvents, gasoline, kerosene, furniture, acoustic panels, false overheads, and personal articles of the men would go over the side. All items of glass or materials that could splinter and wound must go if they weren't critical to the ship's operational performance.

After the ship was tied fast to the pier, Pelham and Yeoman Brown went down the gangway and drove to CinCPac HQ. David was buzzed through the lobby door to meet with Lt. Comdr. Layton in his office.

Layton welcomed him with a firm handshake. "Welcome back, sailor. How did your workup go?"

"Very good. The ship and crew are ready for Slingshot even if most of them don't know what it is."

"And your new Signalman, how did he do?" Layton inquired.

"Better than I expected. Akemi learned many words in English, the daily routine, and how to move around the ship to discharge his duties. He's drawn diagrams of how we should paint the Jap markings on the side and stern of the ship and any changes we'll need to flag masts. We're making the changes to the bow, stern, and forward stack the last two days of this inport period."

"Pelham, Kimicho has sent us word that the double-carrier Battle Group will sail in a few days. After they take care of your mythical carrier, they plan to join the Midway attack force."

"Do we know what ships are in company with the two flattops?" asked David.

"Two cruisers, six or seven destroyers, two oilers, two troopships, two freighters, and a seaplane tender."

"Damn," exclaimed David, "I'm ready for everything but that tender. With calm seas, she could be sending search planes out 2,000 miles or more!"

"The boss is ahead of you," Layton replied. "He has a couple of subs that will be astride the battlegroup path. The Skippers have "Eyes Only" orders to kill the tender first if they tangle with Orange Force. By the way, how did our new civilian employee do at sea?"

"He was great. It turns out that Petty Officer Akemi wouldn't board the plane from Hickam to Johnston Island without him. Michael Nagy is responsible for most of Akemi's training success. Now that we're going to strike Wake, can you tell me the background on how we lost it? Why did we not shove the Japs off of the atoll right away?"

Layton opened a locked desk drawer and pulled out a briefing folder marked SECRET. "I can tell you that a relief force sortied but hit bad weather, refueling delays, and concerns about mixing it up with Jap carriers before we were ready. So the first thing our new boss wanted to know was what happened at Wake. This synopsis is what I gave Nimitz."

David read, **"Report on the Surrender of Wake Island:**

Wake Island is a coral atoll over 4,000 nautical miles southwest of Los Angeles and 2,000 miles west of Hawaii. The group of three islets (Wake, Peale, and Wilkes) was defended by the Marine First Defense Battalion, which was at half-strength with less than 500 men. Since August 1941, they had been emplacing 5-inch and 3-inch guns for ship and antiaircraft defense, respectively. Work was also proceeding on .50-caliber and .30-caliber machine gun nests for beach defense.

During 1940 and '41, the Wake complex was under major construction by eleven hundred civilian workers to become a capable forward base for submarines and aircraft.

The Battle for Wake Island commenced when Admiral Shigeyoshi Inoue unleashed his Marshall Island bomber forces on the atoll - seven hours after the attack on Pearl Harbor. Due to heavy cloud cover and a lack of radar on the island, the Japanese planes were unopposed. They destroyed eight of the Marine Corps' twelve Wildcat F4F fighters that had just arrived three days earlier. The Jap bombers returned the next day and nearly every day for two weeks.

The bombers returned on 11 December after a Jap Destroyer Squadron attempted amphibious landings on two of the three Wake islets. Wake shore batteries sank one ship and three other smaller craft. A Marine F4F fighter destroyed another tin can, and another F4F shot down one of their bombers. Gunners on Wake scored hits on the light cruiser Yubari. Wake suffered two bombing attacks on the fifth day without significant damage, and Fighter Squadron 211 claimed a possible sinking of an enemy submarine.

Roughly 1,600 Military and civilian construction personnel on Wake were subjected to multiple bombing attacks during the next week while a U.S. Navy Task Force sailed from Hawaii to reinforce the atoll. On 21 and 22 December, fighters and bombers from two Jap aircraft carriers, Soryu and Hiryu, repeatedly attacked Wake personnel and facilities. They were returning to Japan after attacking Oahu targets on 7 December. Their aircraft shot down the second-to-last Marine fighter aircraft and set the stage for a pre-dawn invasion by over 1,000 troops on 23 December. The Marine and Navy defenders bravely fought off the larger force of attackers with devastating air cover from the two carriers. It became clear that defeat was only a matter of time. Commander William Cunningham, USN, who had recently reported to Wake as Commanding Officer, and Marine Major James Devereux in command of the nearly 450 Marines, conferred and decided to surrender. Over 400 Marines and Navy men, a few Army radio operators, and 1,000-plus construction workers became prisoners of war. When the Jap carrier aircraft attacked Wake, Admiral Frank Fletcher commanding Task Force 11 (USS Saratoga), and Admiral Wilson Brown commanding Task Force 14 (USS Lexington), were recalled by Admiral Pye, acting CinCPac Fleet.

Intelligence estimates are that the Wake defenders killed and wounded 1,000 Japanese. Wake forces suffered 120 personnel killed in combat (military and civilian) and about the same number injured. In addition, many construction workers assisted in the Island's defense by manning guns, firing rifles, and providing logistic support. The remaining men on the base are now POWs. Jap losses included two destroyers, a damaged cruiser, a probable submarine kill, fourteen to twenty enemy combat aircraft downed, and several smaller boats or landing craft sunk.

The last physical U.S. contacts with Wake defenders were made on 20 December by a Navy PBY and 22 December by a Marine C-46 cargo aircraft. The Catalina landed in the lagoon to deliver mail and orders; she returned to Midway with a high-priority passenger. The C-46 Commando ferried critical supplies on 22 December to Wake and took off full of wounded Marines and older construction workers the following morning. The first Japanese landing on Wake was the only one repulsed of many enemy amphibious attacks in December 1941."

David handed back the report and whistled. "They gave it their all, and we let them down."

"Yes and no," Layton answered. "Black Jack Fletcher in Saratoga was a victim of Kimmel's plan that instructed him to keep his forces fueled; that and the weather slowed his speed of advance. Wilson Brown in Lexington was charging down on the Marshalls on a hit-and-run mission to take the heat off Wake. Vice Admiral Bill Halsey in Enterprise was south of Midway in reserve. With luck, we might have resupplied Wake Island and done some damage to a Jap landing. Except each of them had orders to be aggressive, but not to the point of losing a carrier.

"When Admiral Pye relieved Kimmel, the game was over. Truth be known, we couldn't provide consistent air cover at Wake in the short run. Also, Saratoga was taking them obsolete Brewster Buffalo fighters that could not stand up to the Zeros. The Japs could bomb Wake at will from the Marshalls despite our resupply efforts. It was only a matter of time until they controlled the island."

"Our outer bases were in disarray. Guam fell on 10 December, Philippine forces were falling back, and Wake was more or less a goner. Nevertheless, we desperately needed to hold the line at Midway, Hawaii, Johnston Island, Howland, Palmyra, Samoa, Australia, and Port Moresby, New Guinea. Currently, we have MacArthur retreating to Australia, and his main force captured – shit, 70,000 or more!"

"Do you have any more intel on Orange Force?" asked David, changing the subject.

"We know they are forming up in Tokyo Bay, and we expect them to sail on the 7th of May."

"What about the submarine surveillance?"

Layton replied, "We have one sub along their route 600 miles east of Tokyo, but detection is a long shot. A boat going out on patrol will try and pick them up and shadow them 100 miles from where they plan to launch aircraft to support Wake. How about your PBY's?"

"We have three assigned. Two will operate out of Midway, and the other Catalina will rendezvous with Blue Force at Johnston Island. They will be searching for the Orange battlegroup the morning of our twilight attack on Wake."

David thought, *We aren't going to talk about the forces you don't know about - the C-46 makeshift Commando bomber out of Midway or the hospital ship sailing with the tanker.*

"Why Johnston?" Layton asked.

"Johnston puts us about 700 miles closer to Wake, where we top off the PBY and Selfridge from the tanker Neosho. That gives us two and a half days to position Neosho 300 miles southwest of Wake, top off Selfridge again, and possibly a PBY with calm seas. We then position Selfridge 125 miles south of the atoll for a final run-in at sunset."

"What time zone will you be using?" Layton questioned.

"We're setting up the whole operation, all event times, on Midway's time zone. It's too confusing to add a day and subtract an hour because Wake is on the other side of the Date Line."

"Agreed. Is there anything else that you need? Anything else the staff can do for you?"
"No sir. I think we're good to go. Our new Assistant Ops, Lt. (jg) Simpson, will be returning here in a PBY. He's our step-by-step planner and will have our latest intentions. Do you still plan to set up a Slingshot Command Center here in Pearl on the night of the attack?"
"Affirmative, the Admiral has his aide organizing a set of specialty desks like Destroyer Ops, Aircraft Carrier Desk, Engineering, Communications, etc., etc. Comdr. Rochefort will have his radio intercept crews double-manned and focusing on the area around Wake. I have radio frequency plans for you. Will you be on the shortwave with our friend in San Francisco?"
Pelham replied, "We're going to give it a try. Kimicho may have something at the last minute that could make a difference. Perhaps you can have Michael Nagy monitor the two freqs here?"
Layton, "Consider it done. I'll have Rochefort set it up. How will we know what you're doing? How much progress are you making?"
"Lt. (jg) Simpson has defined a series of two-letter Morse transmissions using baseball terms. We'll be radio silent until after striking Wake. We will report that event in code since they already know we have forces in the area. I think H-R, H-R, H-R, H-R is Home Run, excellent attack, no damage to the ship. We go radio silent again until engaged by the Japs, or we have a report on the carrier attack."
"Excellent. We need those code words and planned frequencies before you sail. Lt. (jg) Simpson could have aircraft problems and not make it here."
"Then I'd better get back. We've restricted the crew to base liberty and are doing Strip Ship today. I need to find a way to raise their morale," David replied.
"How about a band?" Layton shot back.
"A band, what about a band?"
"Yes, a band concert. Everyone likes some music. We have a good Navy band, an outstanding Marine Corps band, and an amateur jazz band on the base."
"Great idea. Could you please arrange for one of those tonight at 1700 at the ammo pier? Tell the band we'll feed them and buy them a beer!"

David returned to the ship and directed the XO to have "S" Division arrange a Bar B Q with beer on the pier and expect a band concert at 1700. He climbed up to the Bridge and was surprised to see Boats MacGregor and two of his carpenters building something out of wood.
"Good morning, Boats. What do we have here? Should we be doing this when we're executing Strip Ship?"
"My question exactly, Skipper, but Ops and Lt. (jg) Simpson say it's mission essential. It's a light table that Mr. Simpson plans to put maps and overlays on during our next op period. It will have red or dim white lighting and a curtain to shield the Bridge crew from the lights."
David put out a call for Ops and Lt. (jg) Simpson to come to the Bridge. When they arrived, he said, "Good morning, gents. Please explain how we'll use this table that Boats is building."
Simpson jumped in. "Good morning, Captain. I've been laying out Slingshot as a series of Phases and Phase Steps. Then I plot them on tracing paper with diagrams, forces in play, and objectives. For example, we're in Phase 4, Final Preparation and Deployment. I have placed the Slingshot units on maps at their locations. As we move to other steps or a new Phase, I update the positions and actions expected.
David looked meaningfully at Lt. Wolfe. "Do you think it's time to tell him the whole plan?"

"Yes sir, he needs it for his charts."
"Mr. Simpson, I'm going to brief you on some of the more sensitive parts of Slingshot. When I'm done, you're restricted to the ship, except for the beer bust this afternoon."
"Yes sir."
"First of all, there are more units involved than we've told you. When we get to Johnston Island, Point Delta, we'll meet with a hospital ship, the USS Solace, AH-5. She'll sail with our fleet oiler, the USS Neosho, AO-23, from Pearl to Johnston. If we need her for a later Phase, we can have her close Wake from their station 300 miles southwest."
Simpson took notes, "Yes sir, got it."
"Secondly, there is a submarine, the USS Gudgeon, SS-211. She'll sortie from Midway as we sail from Pearl. Her mission is to identify Orange Force when they're northwest of Wake Island and report when they're 100 miles from their aircraft launch point. Also, there will be two PBY Catalina long-range search aircraft flying out of Midway to detect Orange Force during the two days before we attack Wake."
"Yes sir. Is there another sub out west?"
"Yes, I don't have her name yet, but a returning sub off of patrol will attempt to intercept and shadow Orange Force 600 miles out from their current location in Tokyo Bay."
"Thank you, sir."
"Thirdly, an airplane will be flying from Midway to Wake to drop bombs on the main runway. It will be a modified transport aircraft. They will arrive at dusk and start their bomb drop exactly at 2050 Midway time, or Zulu minus 11. All event times will be Midway time, is that understood?"
"Yes sir, understood."
"Good. Solace and Neosho will leave a day before us. We will meet them at Point Delta. We top off our fuel and the avgas for one PBY. That PBY will conduct a search pattern west of Wake, while another will arrive later from Midway to search the northwest. The third PBY will be in reserve at Midway. Selfridge will accompany Neosho and Solace to Point Charlie, 300 miles southeast of Wake, then Selfridge sails to Point Bravo, 125 miles due south of Wake. We must arrive in a firing position on the southern shore of Wake Island, Point Alpha, at precisely 2030, 14 May, Midway time. Sunset is at 2032. We launch two of the fast boats and Dumbo just after sunset. At 2040, we turn on our aircraft ADF radio beacon to guide the C-46. The first bomb explosion from the C-46 will be our weapons-free signal."
"Yes sir. What will we be shooting at?"
"We'll have to train Smitty on this; the first priority is any aircraft that has a spinning prop or is ready to launch. Second, are any gun emplacements that are shooting at us with accuracy? The third is other aircraft, especially seaplanes in the lagoon and the gas dump. Fourth, are the shorter runway and taxiways. Fifth, are subs or ships at the harbor pier and the command center. Sixth, are buildings, gun emplacements, and fortifications. You two Ops guys need to drill Smitty on the latest photos of the field, the buildings, Avgas storage, and the gun emplacements."
"Yes sir, copied all of that. How long will we be firing? Simpson asked.
"Good question. We keep firing if we feel secure and the shore guns aren't dropping shells near us. We'll have one fast boat leading us at a quarter mile and one off the stern at the same distance. They will be making smoke and setting off fireworks. It will look like a line of ships

a half-mile long firing away from shore. Our SOA will be ten to twelve knots to keep Dumbo airborne, so that gives us about ten to fifteen minutes on the first pass at the southern Island, which is Wake Island. If we're undamaged, we may make a second pass."

"Understand, sir. Will we be using the mortars?"

"Another darn good question. What do you think, Ops?"

Wolfe answered, "The larger M1 can lob a lighter shell roughly two statute miles, and the M2 less than that. We'll be six or seven miles from the island, so mortars don't apply."

"Mr. Simpson, any more questions?" David asked.

"Yes sir. How fast will the transport aircraft fly? I want to arrange a signal when they get airborne and estimate if they will be on time."

"Figure on a ground speed of 165 nautical knots. Tell me again how this lightbox will work."

"Yes sir. Once we have it up and running, we can put maps on it, or my tracings that will have each event and the location of the units, their radio frequencies, call signs, and their role in each step."

"You and Lt. Wolfe have done an outstanding job organizing Slingshot and moving it along. Well done. I'd better find Boats and see how Strip Ship is going. See you at the Band concert."

The rest of the day went quickly. He found MacGregor, and they toured the ship. Men were hauling out all kinds of glassware, wood objects, and flammable liquids. There was a storage plan for personal gear ashore, and a slim chance the crew could reclaim some of the items if they returned to the States via Oahu. In the late afternoon, Lt. Comdr. Layton delivered the Marine Corps band that planned to practice that night anyway. They rehearsed their music next to the ship. All hands enjoyed the beer, burgers, and dogs. The band was surprised and delighted when they saw their brother Marines come ashore from Selfridge. They finished with Anchors Aweigh and the Marine Corps Hymn.

Selfridge's crew rotated on and off the pier for supper and a couple of beers, except the Amjaps. They were kept out of sight. David was at the pier-side party talking to the XO when Lehman winked at him and nodded toward the ship. David saw two mop handles taped together with a cup holder on one end. The apparatus extended out to the pier from a porthole. A bottle of cold beer was placed in the cup holder, and the prize was carefully drawn back into the ship. The cup holder was extended and refilled again in no time at all.

"XO, being in charge of morale, I don't see anything, do you?"

Lehman smiled. "No, Skipper, the Amjaps deserve to have a little fun. They've been helping out all over the ship. I wish we could keep them after this mission."

"Me too. But I doubt we'll fool the Nips with our new look more than once or twice. Do you think we're ready for Slingshot, XO?"

"Yes sir, more than ready. We finish Strip Ship tomorrow. We're loaded for bear once we have more food and ammo!"

"Good. We have two more full days and nights before we pull out. The second high tide is at eleven hundred. Warn the men tonight that there will be a dungaree personnel and space inspection at 0900 that morning before we set sail. I want the crew and the ship looking sharp as a tack."

"Aye, aye, Skipper."

Captain Pelham thanked each band member as they were packing up their gear. He gave Yeoman Brown the high sign, and they drove to the BOQ. His cramped single room was still empty, and David reserved it for the next three nights. He had Brown take him to the Navy Base Officer's Club to check the hours and the menu. Then he placed a payphone call to Betty to see if she could join him for dinner the next night. The XO took Eight o'clock Reports on the Quarterdeck and reported the results to David in his Inport Cabin. Before Taps, Captain Pelham fell asleep reading a book.

Steward Jackson had him up, drinking his first cup of coffee, and dressed in service dress whites by 0600. He went out on the Main Deck and toured the gun mounts. The guns were loaded and at fifty percent manning. *Let the Nips attack again*, he thought. *Selfridge is ready*. Captain Pelham had breakfast in the Wardroom that no longer had linoleum on the deck or a false overhead. The glassware was gone, and all liquids were served in metal coffee mugs. Food was served on aluminum mess trays similar to those used in the crew's mess instead of china plates. The wooden serving doors had been removed, and half of the Wardroom faced directly into the open galley. David asked Wolfe and Simpson to show him the completed light table in Bridge Plot as breakfast finished.

Once on the Bridge, Lt. (jg) Simpson showed him a traced map of the Western Pacific on the light table and the current Phase 4 locations of the Slingshot units. David was impressed. The operation encompassed 4,000 miles from Hawaii to Tokyo Bay.

"Sir, are there more phases that we've not discussed, or any more details on Phases 8, 9, and 10?" requested Lt. (jg) Simpson

"Yes Grant. My thoughts on Phase 8, Exploiting the Tactical Situation, are shifting to how we can confuse and attack the service ships following the carriers. We have two troopships, two tankers, two freighters, and maybe a seaplane tender as targets. Let's assume we've put the carrier down and have most combatants running for the hills. How can we damage the service ship group that's probably in two columns and has three destroyers in company?"

Lt. Wolfe took up the challenge. "Let's see; it's 0200 or 0300 in the morning and pitch black. No doubt they saw the explosions and flashes but may not know what occurred. Say we have the heavies running away; how about ordering the destroyers to detach and cover their rear quarters and warn them that there are subs ahead. Then we assume the lead of the service ships and start turning them away from the combatants, making the gap bigger and bigger."

"I like it. I have it in my mind's eye to head them toward Wake and use them in a later Phase," David said.

"Sir, would that Phase be number 9, Withdrawal, or something new?" Simpson questioned.

"It can be part of Withdrawal. We will be heading out of the area. Perhaps we can get the sub to support us and sink some service ships."

Wolfe shook his head no. "Skipper, after we've creamed Wake and put down a Jap carrier, we must get out of town before dawn. The remaining carrier will have planes up, Wake may be sending planes after us by then, and we can expect aircraft from the Marshall Islands in the area by noon. The cruisers may launch their scout planes earlier than that. So we need to retire fast, maybe after a swipe at the support ships."

David paused, "You may be right. I thought maybe we could swap one of their supply ships for some of the prisoners on Wake."

"A noble endeavor Commander, but without air cover, we'll be sitting ducks," Wolfe insisted.

"What do you recommend?"

"Sir, we should clear to the east about 100 miles north of Wake at thirty to forty knots. If we start at 0400, we can be over 100 miles away, and hard to find an hour after dawn. Each hour they're looking for us, we move further away. Nothing from the Marshalls can reach us except those damn flying boats."
Simpson queried, "Skipper, was the hospital ship part of this prisoner release scheme?"
"Yes, it was. I could see us trading one of the Jap supply ships for prisoners. The Japs on Wake are hungry for food, fuel, and water. I thought maybe we could take on two or three hundred POWs, especially the sick, wounded, and older civilians. We would then transfer them to the Solace."
"Captain, are we agreed that's a non-starter, and we need to start running toward Midway as soon as possible after engaging Orange Force?" Wolfe persisted.
David gathered his thoughts. "Yes, you're right, but we'll take a swipe at the service ships. How do we do that?"
"We try and get their forward destroyers to clear the area, and then we run down the center of the two columns firing port and starboard. The rear destroyer will come at us. We have to take it out before she gets our range," replied Lt. Wolfe.
"As good a plan as any. Our ace in the hole, Signalman, should come in handy for spreading fear of subs and rebuffing challenges once we've penetrated their formation," David answered.
"Captain, can you expand on using the Fireflies during Phase 6, Search and Approach?" questioned Simpson.
"Yes, once we've penetrated the destroyer screen of the main body, our next challenge is to pass behind the starboard cruiser escort. We will use the bubble machine to suppress our glowing bow wave and wake. Dumbo will be flying, so we'll make turns for ten to twelve knots. Hopefully, we ID the cruiser from a mile out. We reel in Dumbo and launch our three fast boats. Make sure they're briefed to go in a line abreast of the cruiser about a half-mile east. They start Firefly mode and slowly bring the cruiser's attention to the randomly blinking blue lights – first the lookouts and then the enemy's Bridge crew. When we see the blue lights, that's our cue to approach and cross the cruiser's wake a few miles astern. Once across, we approach the starboard carrier in the formation."
Simpson responded, "Sir, I need the important checkpoints before the attack so we can diagram them. Could we back to finishing the shelling of Wake? What happens next?"
"OK, we're making good progress here. Grant, your diagrams are forcing me to put in missing details. Remember, Phase 5 is the Confirmation of Enemy Intentions and Neutralization of Wake Island. Through the subs and Catalina PBY surveillance, we hope to confirm they're sailing east on their PIM. The worst case is that we have no reports and use their planned aircraft launch position as a starting point. Sunrise is 0720 Midway time. They will be at that position by 0650. They will turn into the wind at 0630 to 0640 - probably within five to ten miles of the launch point."
"Let's call the launch position Point Lima. Skipper, why should we count on them being in that position? How did we get this information?"
"Just consider it good intel. They could get delayed by bad weather or equipment failures, but if they can get to Point Lima at dawn on the 14th of May, they will. Professor Nagy has taught me that the Japs have a sort of groupthink. Once a plan is approved, it must be followed. Like

us, they no doubt keep logs on all of their ship movements, so during a later reconstruction, there better be a darn good reason for not following the plan. "Back to Phase 5. Assume we have confirmation that Orange Force left Tokyo Bay, and we have a sighting or two along their route. Once we finish blasting the crap out of Wake Island, any Jap seaplanes off Peale Island, and gun emplacements on Wilkes Island, we clear to the west and start our high-speed run north. I expect that'll be 2130."

Simpson nodded as he took notes. "Excuse me Skipper, what will our speed be heading north?"

"If the sea state allows forty knots."

"Which brings me to my next question, sir. Why are there Marines on the crew list for the fast boats?"

"Glad you asked. Two boats will have Marine bazooka teams, and the third will have a .50-caliber machine gun manned by our crew. When we attack the carrier, they will pounce on the cruiser while her crew looks at the burning carrier. First, they will attempt to shoot up her catapult aircraft and anyone on deck. Then, the bazooka boats will attack up her wake and try to jam her stern steering gears. They also have a smaller version of the blinding rockets to degrade the Bridge crew, plus grenades and submachine guns."

Wolfe was visibly alarmed. "Sir, that sounds like a suicide mission. Do we want to take that risk?"

"Mr. Wolfe, this whole operation is a suicide mission. A Porter Class destroyer against two carriers, two cruisers, and six destroyers? The only reason it may work is that it's improbable and unexpected. About a couple of hours after we attack Wake, the Neosho will transmit signals as if she is an aircraft carrier recovering aircraft 300 miles southeast of Wake. The Japs will think they have a 500-mile buffer, secure flight ops, and plan pre-dawn search launches - we hope. We attack the carrier, then the service group, and then retire to the north and east to recover the fast boat crews. Plan to save time by taking on the men and scuttling the fast boats."

Simpson interjected, "Sir, I have some recommendations on sequencing to the attack position."

"Very well, go ahead."

Simpson looked at his notes, "I'm thinking of naming these events and locations after fruit. For example, Wake is watermelon, the first contact with Orange Force is grapefruit and confirmed visual contact with the Jap Battle Group is orange. Launching the fast boats could be lemon, and safely past the cruiser's wake is lime. Reaching a good firing position on the carrier could be banana, and getting to our quarter-mile attack position is strawberry."

"Good ideas, something the whole crew could follow while we're sneaking in at GQ. Change strawberry to peach. Everything will be peachy keen if we get to peach."

"Yes sir. Could you define Point Banana for my diagram?"

"OK, Point Banana would be after we safely cross the cruiser's wake and are within a mile of the carrier."

"Yes sir, what happens if we're discovered between Banana and Peach?"

David thought and stroked his chin. "We fire the blinding rockets, and all guns are free. Then we increase our speed, close the carrier firing from their port or starboard and have our torpedoes ready."

"Thank you, sir. Who will call out the attack sequence if we make it to Peach?"
David looked confused and reluctant to answer, "I'd not thought about that. I guess it would be Lt. Wolfe with you as a backup."
David excused himself and used the head in his At-sea Cabin thinking, *Mr. Simpson, you've brought up another good point. You don't know it yet, but you'll fly to Hawaii before we attack Wake. When should that be? At Johnston Island? At Point Charlie? We can't count on a benign sea state to land and refuel a PBY flying boat three hundred miles southeast of Wake. So it has to be a PBY at Johnston going back to Pearl or AH-5 Solace as a backup.*

When David returned to the Bridge Plot, the three officers thoroughly reviewed the Point Peach attack sequence. Simpson took copious notes and promised to produce a script and diagrams on tracing paper. David wrote instructions for the USS Solace and asked Lt. Wolfe to deliver it to the hospital ship's Captain. Then he toured the weather deck and talked to the gun crews. Captain Pelham went down to the crew's mess deck, grabbed a tray, and got in line. He now knew every man on the ship, their name, rate or specialty, and hometown.

Pelham found it easy to banter with the men in line and asked them if they would be ready for inspection in two days. They all offered for him to go to the head of the line, which he turned down. He said waiting in line reminded him of when he was a Petty Officer. This sparked a lot of interest, and the men around him wanted to know his rank and specialty. David told them about being a Navy Reserve Gunner's Mate Second Class on a destroyer and a battleship. He knew the talk would spread that the Skipper was a Mustang who once slept in bunks stacked four high, pulled mess duty, and stood lookout watches before commencing officer training. Dinner was beef stew with cornbread, washed down with fresh milk. David found the supply chief in charge of the mess and complimented him on the cleanliness of the foodservice line and the quality of the food. Back in his Inport Cabin, he discovered a fresh stack of paperwork compliments of Yeoman Brown.

After processing the messages, letters, memos, and evaluations, David started reading his current book and took a nap on top of his well-made rack. Yeoman First Class Brown woke him at 1630, retrieved the ship's paperwork, and asked David if he wanted a ride to the Officer's Club. David said no and asked Brown for the keys to the car. Next, Pelham showered, shaved, and donned a fresh service dress white uniform. He put his shaving kit in his briefcase and was ready to go ashore after providing Lt. Comdr. Lehman his BOQ room number in case of emergencies.

David told his XO to expect him back onboard for breakfast. At the Base Exchange, he bought flowers, a bottle of bourbon, and some crackers and snacks that he smuggled into his room using the briefcase. He was waiting at the Officer's Club entrance when Captain Betty White arrived at 1800.
David shook Betty's hand. "Miss White, allow me to introduce myself, Lieutenant Commander David Pelham, USNR. Thank you for joining me for dinner."
"Mr. Pelham, so nice to meet you. Captain Mueller has told me all about you. You must be great friends."
David escorted her into the dining room. "We don't see each other so much with the war on, but I'm glad to help with the surprise birthday party. Let's get a table and order some drinks. The assistant caterer is going to show us the banquet rooms for the party."

After being seated and ordering wine, they were joined by a woman in her thirties who showed them the available meeting rooms and provided a price list for catered items. They returned to their table, and David started writing notes in a spiral binder. He talked to her about the mythical birthday party while he perused her appearance. She was wearing the same khaki skirt as when he first met her. In addition, she wore a long-sleeve khaki shirt with a matching tie and a dark brown tunic with her Captain's insignia on the shirt's lapels.

"Aren't you taking this a little far? I don't think anyone can hear us," she chided.

"You can't be too careful. Say, is that an engagement ring? Who's the lucky guy?"

"A naval officer, would you believe? He dashes around the ocean in one of those destroyer boats."

David frowned and corrected her. "Ship, not boat. A submarine is called a boat. A destroyer is a ship. So, is he handsome?"

"Fair to middlin'. I'm concerned he might be a vampire; he has an albino look about him."

David laughed in a low voice and sipped his drink. "You look ravishing. Tell me, how does one take off that cute little skirt you're wearing?"

She gave him a naughty look. "One undoes the button on the side and unzips the zipper, and it falls right off. So tell me, sailor, how does one take off your lily-white trousers?"

David spoke in a whisper. "First, you have to release the belt buckle; then there's a button and a zipper. Then, of course, the shoes have to go also."

"Too complicated, too much work for this gal," she replied saucily.

David spoke louder and handed Betty the notebook. "I've written several things down for the birthday party. Take a look."

She turned the page. "This looks like a map. What does it mean?"

"It is a map and a diagram of where you should park your car and the route to my BOQ room through a side door."

"My roommate is working tonight; wouldn't it be easier to go to my apartment?" she replied.

"Can't. I restricted the ship's company to the base. I can't violate my own order unless it's official business. Besides, I have bourbon, flowers, and snacks for you. We do need another water glass. Think you could swipe one of these from the club?"

She held up a brown purse. "I can take care of that. This purse is a bit larger than the regulation one. But, it needs to be since it contains my toothbrush, a make-up kit, and a very naughty nightie."

"Music to my ears, my dear. Let's have a very leisurely dinner and repair to my penthouse after sunset."

Betty was surprised at the cramped room but happy to be with him. David woke up at 0500 and found her sleeping in the easy chair with her feet on the foot of the single bed. He used the bathroom and brushed his teeth. Betty got up from the chair and went to the bathroom. Then, she came over to him, leaned down, and gave him a long wet kiss.

"Ummm," she said. "You taste good. Is that a pistol in your shorts, or are you just happy to see me?"

"Very happy to see all of you. Climb up on top and take your time."

"Yes my Captain, your wish is my command. Do you want the nightie on or off?"

"Let's start with it on and take it off when you want to drive me crazy."

Betty slipped out of the BOQ while it was still dark in the parking lot. David showered and shaved. He tidied up the room and drove to the ship. Coming down the gangplank, Captain Pelham put his finger to his lips so the Quarterdeck crew didn't pipe him onboard. It was an open secret that he had been out all night since some of his officers saw him dining in the Club, and the XO passed his BOQ room number to the OODs. David went to his Inport Cabin and attacked the stack of correspondence left by Yeoman Brown. He had breakfast in the Wardroom with the XO, Ops, the Engineer, and several other officers. The conversations were lively and spiked with anticipation of getting back to sea. Pelham could tell that morale was up, and the crew was ready to set sail the next morning.

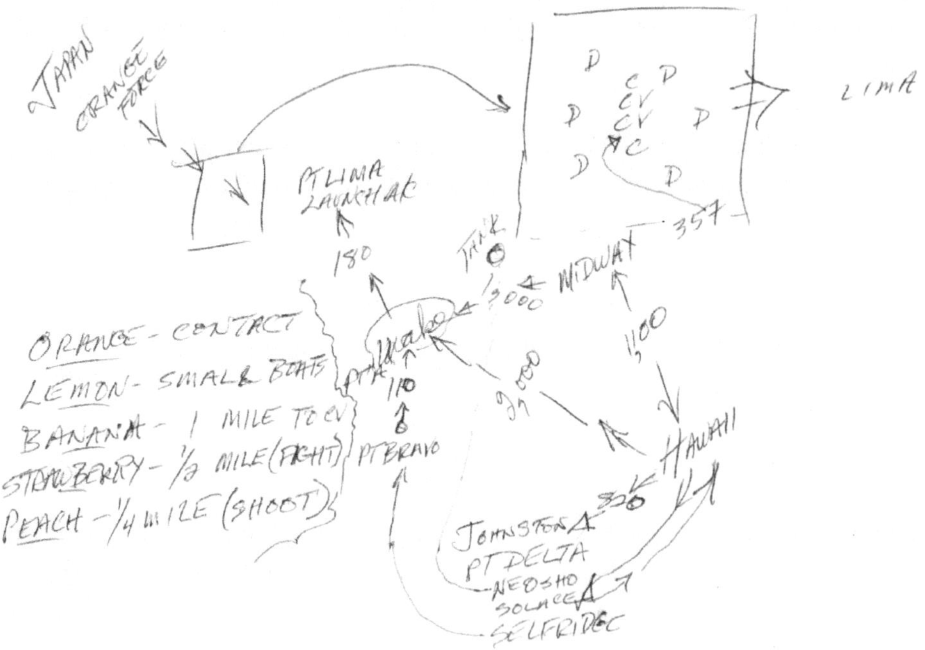

Lt. (jg) Simpson's rough diagram

Chapter Thirty

TRANSIT TO JOHNSTON ISLAND – USS Selfridge

May 1942

David took Lt. Wolfe aside and gave him new instructions for the PBY crew meeting them at Johnston Island. He called Michael Nagy using the Quarterdeck landline and arranged the transfer of Petty Officer Akemi back onto the ship. In Bridge Plot, he found Lt. (jg) Simpson and his Gunnery Officer, Lt. Kindall, debating whether a destroyer could sink a carrier.

"Good afternoon, gents. Believe me, and our five-inchers are up to the task. Layton has informed me to expect two inches of armor above the waterline on the Shokaku carriers and six inches protecting their magazines. We know our armor-piercing shells can penetrate five inches of armor at two miles. At a quarter-mile, we might even get at the magazines. In the briefs to CinCPac, Lt. Wolfe and I've been overly conservative on the number of hits in five minutes. Our best gun crews are firing twelve shells a minute. If you take an average of ten per minute times eight guns, we can fire eighty shells per minute. If ninety percent of them hit the carrier, that's over 200 hits in three minutes. With low sea states, I think the Marine mortars will make mincemeat out of the airplanes on the Flight Deck. Otherwise, they've got a severe aiming challenge with the ship pitching and rolling."

Lt. Kindall raised his hand, "Skipper, how do you want the shells set?"

"It's up to you, Guns, but my thought is half set for contact and half set for a minimum delay. We open up 200 one-foot to two-foot wide holes along the waterline and blast out several interior bulkheads giving the Japs a damage control nightmare. We have several problems with how to start our attack. Our target carrier has four, five-inch, dual-barrel guns on the starboard side that can blow us out of the water and six triple mount AA guns firing one-inch diameter shells. We have to assume those guns are manned and loaded. Do we try and knock them out, or just concentrate on sinking the carrier? How much damage can our Bofors AA guns do to their lighter and heavier guns? How bad will the visibility be for both sides after one minute of firing? We will have planes burning, shell hits on the carrier's side, and smoke from our guns. It won't be hard for them to fire back at our gun flashes. It seems a real Hail Mary pass to think we can damage all their guns. It may be better to destroy their power plants and generators first. No electricity, no way to aim heavy guns. That line of logic would have us concentrate all of our firepower on her generators, but we don't know exactly where they are installed or where the backup diesel generators are."

Lt. (jg) Simpson spoke up, "Skipper, are you thinking out loud here, or should we be taking notes?"

"Just weighing things out with Guns here and posing questions. If we concentrate on sinking the carrier right from the start, can we induce a list that jams her gun shell hoists and maybe her elevators before the carrier can shell us? Will either carrier have aircraft at the ready to launch and attack us at night, or the cruisers for that matter? Hold it; I just remembered that Layton told me that these carriers have upper and lower Hangar Decks. The lower one is barely above the waterline. That means our waterline shots will impact just below the lower Hangar Deck and into it. Scratch hitting boilers, engine rooms, and generators. They are probably all below the waterline.

"Let's go back to the first minute of firing. We need half the shells to explode in the lower Hangar Deck and the other half to blast holes at the waterline to get maximum flooding below the Hangar Decks. Our medium AA weapons concentrate on their five-inch guns from the start, and our light AA guns fire on their six triple mounts.

"Now, let's take minute two. We keep firing four guns into the upper Hangar Deck and shooting at the waterline with the other four - expect ninety percent hits or seventy-two solid hits.

"At minute three, we've suppressed their fire, or we're dead. The carrier should be taking on a starboard list. We put all eight guns on the waterline, which should cause flooding of the lower Hangar Deck and a rapid list to five or ten degrees unless they initiate massive counterflooding and close off every Lower Hangar Deck partition. We get seventy-two more hits.

"By minute four, the mortars will have fired about eighteen rounds times two tubes per minute for the heavier M1, and the same for the lighter M2. That would be seventy-two rounds per minute times three minutes equals 216. We have fifty-four heavy delayed-fuse shells and fifty-four lighter fragmenting shells hitting the Flight Deck at a fifty percent hit rate. We might get lucky and drop a couple through an open elevator, maybe punch through an elevator, or make a hole that allows burning avgas to drop into the Upper Hangar Deck. Next, we need to decide if we want the mortars to put up some smoke shells or parachute flares."

Lt. (jg) Simpson interrupted. "I have two questions. Why do they have two Hangar Decks? And why would we want mortar flares or smoke?"

"Grant, they've not mastered the engineering technology for folding all of their aircraft wings yet. Only their wingtips can fold on some planes, so they need a lot of parking space. On question two, we need to fine-tune our attack objectives. We may want to blind them with mortar smoke and flares while we pound away and hole her with over two hundred five-inch shells and 108 mortar hits. If she's not exploding and progressively listing at that point, we should fire all of our torpedoes and haul ass out of there."

Lt. Wolfe arrived on the Bridge and started listening to the discussion.

Simpson said, "I agree there will be a lot of smoke. Could even more smoke obscure the target?"

Pelham answered, "Maybe so, but they present a much bigger target to hit than we do. Perhaps some extra smoke and flares in their eyes will degrade their shooting more than it does ours. Also, we need to practice with the mortars and get them adjusted for a half-mile out and Point Peach at a quarter-mile. Through all of this, we have Marine snipers taking out any Jap on the starboard side of the carrier."

"Skipper, some of your thinking out loud seems to be taking us in a new direction," theorized Simpson.

"I know. I've been oversimplifying the results of the attack. I thought we would be shelling the power plant and their essential services, especially electricity. I was wrong not to expect a rapid counter-gun response. I now think the first thing we have to do is take out their five-inch guns. They're in two batteries in two locations, about a hundred to 200 feet apart. One set of two turrets is forward of the island. The other is halfway between the island and the stern. Shells that miss their big guns will penetrate the Upper Hangar Deck and wreak havoc. Hopefully, we damage the big guns, shell hoists, and cook off some ammo. Getting some

planes burning in the Hangar Deck will make it hard for additional personnel to man the guns."
"Skipper, can you give us the sequence, so I can put it on tracing paper?" Simpson requested.
"Lt. Kindall, as an old Gunner's Mate, I'm going to take a stab at how to employ our guns. I expect you to improve on my ideas."
"Yes sir," Guns answered.
"Layton gave me this top and side view drawing of the Shokaku. Let's put it on the light table. Assume we've made it to Point Banana unopposed, and the Jap battle group is idling along at less than five knots. Selfridge is at General Quarters; the snipers are in place but not raised yet. Our gunners are ready, mortar crews are ready, our Gun Director is starting to acquire the carrier guns, and our rockets are ready. We have half a mile until we're committed to the attack. We're showing no lights, closing at ten knots to reduce the visible wake, and the oil/soap bubble machine is pumping to squelch the plankton buggers."
"Skipper, it sounds like half a mile is important. So let's put Point Strawberry back in the plan." Simpson asserted.
"OK, from Point Banana to Strawberry at half a mile out, if we're fired on or challenged, we may turn tail and run at flank speed. Lt. Wolfe, I'll want your recommendation, but it will be my call."
"Yes sir," Wolfe replied.
"From Strawberry to Peach, we're committed. If the carrier fires on us, we increase our closure speed, launch the rockets, raise the sniper platforms, and fire back with everything we have. If we're just challenged by searchlight or blinking signal light, we continue to Peach at a quarter-mile from the carrier with our Signalman buying us time."
"Excuse me sir, how does he do that?" asked Kindall.
"I talked it over with him using Professor Nagy as an interpreter. Akemi will be watching any light signals from the Jap ships that might indicate if they've detected us. If they have, they would first challenge using a two-letter Kanji code. He has two courses of action. He quickly blinks the challenge at another ship, perhaps the cruiser. If they reply, he blinks that code back to the challenging ship. Or two, he starts a dialog acting like a confused escort unit. He signals we aren't sure about the day or the time because of the proximity of the Date Line. He's buying us time. The next thing we can expect is a quick sweep of the ship with a searchlight. They will see a Japanese destroyer with the proper flags and some of our Amjaps on deck. We'll have one of them up here on the Bridge dressed as an officer. He'll step out on the Fly Bridge and salute the light."
"Skipper, what about the mortar crews? How do we hide them?" Simpson challenged.
"There will be mattresses strung along the railings where they are. The Marines hit the deck, and the Amjaps look like they're cleaning the area. If we're nearing Point Peach and get challenged again, our Signalman will report that we have a sub contact ahead of the carrier, and we're engaging it. In rethinking this, we should create a scenario for each type of approach we may have to make to the carrier. Dropping the Army gun means we can go at either carrier from either side, but her starboard side is preferred. I'm digressing, Mr. Simpson. We need a diagram for attacking the formation from the front, either flank or dead astern."
"Aye, aye, sir."

Pelham continued. "OK, we've made it to Peach without being engaged or taking fire. We swing parallel to the carrier to unmask all of our guns. We raise the snipers then, or when we cross Strawberry. Snipers report target acquisition. We fire the blinding rockets. The snipers start plinking aircraft and fuel bowsers.

"Ops, you time the rockets and transmit on the 1MC five seconds before the rockets burst, 'All hands cover eyes, cover eyes.' After the two air blasts, you transmit, 'Weapons free, weapons free.' Mr. Kindall, have the mortars first aim for the forward area of the Flight Deck and work aft to stop any aircraft from taking off. Direct our forward guns at the carrier's forward five-inch guns, and our aft guns first at the carrier's aft five-inch guns using armor-piercing shells. Then have the aft guns sweep along their triple-A gun mounts.

"Missed five-inch shots will blow up planes in the Upper Hangar Deck. When you're sure we've damaged the carrier's heavy forward guns, direct our forward guns at her waterline. Same for the aft guns, when you feel they've smashed the five-inch and triple-A weapons, aim them at the waterline. Have half the guns firing contact-fused shells and the other half firing delay fuses."

"Aye, aye, Captain, I concur," said Lt. Kindall.

"Ops and Guns, make sure that the snipers know to increase their rate of fire with guns free, and the mortars and Bofors open up also. The forward Bofors should first fire into the carrier's bridge, then sweep the big guns and the forward catwalks. They should immediately engage any gun firing at us from the bow to the rear of the island. The rear Bofors should sweep all the triple-A gun mounts and the rear catwalks. They should immediately engage any gun firing at us from the stern to the rear of the island.

If our lighter guns have time to blast a few parked planes, all well and good. The Bofors and our .50 calibers should shoot down any aircraft entering our area. With several aircraft starting to burn on the Flight Deck, the snipers should shift to shooting Jap sailors on the Fly Bridge or the Signal Bridge, in the catwalks, or on the flight deck."

"Yes sir. Simpson is capturing all of this," Wolfe advised.

"Good, now we may be so lucky to reach minute three. Our five-inchers are punching holes in the carrier's hull at an alarming rate. After that, she should start to list toward us. Even if they start counterflooding on the opposite side, they're helping us make their ship heavier and less seaworthy.

We keep punching holes and flood the Lower Hangar Bay. Our bow deck crew should have reloaded the blinding rockets if we need them. We should start showing Freddy the dragon randomly for a few seconds and broadcasting his soundtrack at the cruiser. The Fireflies will be closing on the cruiser to harass her and attempt to jam her steering gears."

"Yes sir," said Wolfe.

"Now men, we come to the moment of truth. If we have the Flight Deck and Hangar Decks burning and are still slugging away with our big guns at the five-minute mark, we need to give it one more minute and retire at high speed. We fire rocket signals for the Fireflies, telling them where to rendezvous.

"If possible, we'll locate the service group ships and sink some of them. We may have to shoot our way through a destroyer or two. The first plan is for our Signalman to order them to detach and support the carriers and cruisers. He may report enemy subs ahead to scare them. I think we should break for lunch; any questions?"

Simpson held up his hand. "Sir, you've given us a ton of work. Should we still plan on an inspection in the morning?"
"Yes, Mr. Simpson, the men have been getting themselves and their uniforms ready for three days. They will have their spaces clean and neat. The officers and chiefs will be alert for Strip Ship items that must be removed. The XO is working on the POD. We will get underway at 1100. The oiler and the Hospital Ship have an eighteen-hour head start on us. We should catch them as they're nearing Johnston Island - then top off our fuel and learn how to refuel a Catalina PBY."
"Pardon me Captain, I thought we dropped the Hospital Ship?" questioned Ops.
"Solace wanted some sea time and may also try to refuel for practice. She'll be coming back to Pearl from Johnston."
"Thank you, sir. When do we engage Orange Force?"
"In five nights, including tonight, 14 May at 0200 Midway Island time. Mr. Simpson, I'll need those new overlays finished by 2000 tomorrow night. That is all, dismissed."

Pelham had lunch in his Inport Cabin while he processed a ten-inch pile of paperwork left by Yeoman Brown. He had Brown drive him over to CinCPac HQ for a final meeting with Lt. Comdr. Layton. Layton had Comdr. Rochefort joined them, and they worked out a few more details on the Communications Plan for Slingshot. Rochefort reported that the first sub looking for Orange Force did not make visual contact but had plenty of Sonar contacts of a large force heading west at fifteen to twenty knots at the expected intercept time. David smiled and thought, *I know what that's like. So we slowly trailed the Lexington Battle Group for several miles when she headed west at the end of the fleet exercise using our new sonar gear. We were still getting multiple contacts when the Battle Group was sixty miles away. I remember listening to the carrier props that sounded a little like an agitator washing machine.*
David feigned mock anger. "You both had destroyer duty. Why didn't you tell me that I needed to worry about the carrier's big guns first?"
"We figured you would come to that conclusion sooner or later," said Lt. Comdr. Layton.
"We go after her five-inch guns first, damage them, injure personnel, and set fires around the mounts – same for the triple-A," David replied.
"Sounds like a plan. Are you still sending Mr. Simpson back for the Command Center drill?" Layton asked.
"Yes, it will be Lt. (jg) Simpson, our Assistant Ops. He'll fly back on a PBY day after tomorrow. Who should he report to?"
"Have him come and see me. I'll keep him busy until the 14th," replied Layton.

David shook hands with Layton and Rochefort as they wished him luck and a safe return. He went back to the ship and took a nap. Brown woke him up with paperwork at 1600. During supper, he asked the Engineer and XO to be sure and see him on the Bridge after Eight o'clock Reports. Later on the Bridge, he stepped through the latest battle plan. David was particularly interested in Lt. Comdr. Haldane's reaction since it was his first detailed exposure to Slingshot.
David looked at Haldane. "Well, what do you think?"
"I think it's a high-risk, high payoff plan. No dual-carrier Jap Battle Group will expect an attack by a single destroyer. They may get confused and hold back, thinking there must be a

larger force nearby. Why would a single destroyer commit suicide against a far superior force? There's one thing that bothers me, though."
"What is that?"
"You plan to go after their guns first; nothing wrong with that, but how will you see them on a pitch-black night?"
"Good point. We may need the mortars to put up some flares," David answered.
"That might work, except you also illuminate yourself to the carrier and its supporting ships."
"What do you suggest?"
"You might need the searchlights we took down from the aft mast. I bet Chief Bennet knows where they are. Cover them with metal or something similar and shine a pencil beam out a hole in the middle. Put the beams of light on their big guns for our gun directors."
"How big a hole?"
"I don't know; start testing with three or four inches and work outward."
"XO, any comments or questions? David asked.
"No sir. It's good to see more detail added as we go along. If you take a hit on the Bridge, how am I going to know the plan in Secondary Conn?"
"A very good point, Gary. We will draw up a second set of tracings and make you a smaller light table. Let's get Bennet and MacGregor working on the searchlights and the light table ASAP."
"Aye, aye, sir."

Before Taps, Captain Pelham went down to his Inport Cabin and found Michael Nagy there with Petty Officer Akemi. They weren't restricted to the base, had dinner at a Japanese restaurant, and consumed a liberal amount of sake. After shaking hands with Pelham, Michael announced that he needed to sail with the ship and continue Akemi's English training. He wanted to make sure Akemi understood every detail of Slingshot. David said it was impossible to take Michael into combat. Michael's response was to flop down on David's rack and pass out. *Oh well,* David thought, *I can send him back with Simpson from Johnston Island either on the PBY or USS Solace.*

David assured Akemi that everything was OK, and Steward Jackson magically showed up with coffee and a snack for the Petty Officer. Jackson woke David in his At Sea Cabin the following morning at 0600. Pelham was sipping his first cup of coffee and putting on his short-sleeved summer whites while he read the Plan of the Day. The XO scheduled the ship's inspection for 0830 and the Special Sea and Anchor Detail at 0930. They had agreed that the ship would sail without a pilot. Pelham inspected the Pilot House and the Bridge watchstanders while his officers and CPO's examined all the other working and berthing spaces on Selfridge.

First Division stowed the mooring lines at 1100. The duty OOD and his Bridge crew guided the ship away from the dock and the Navy Yard. As they rounded Hospital Point and could see the channel out to the ocean, the OOD stepped out on the Port Bridge Wing and used his glasses.
"Skipper, I think you better see this."

David went outside and focused the binoculars. He could see an Admiral's car next to the quay wall with four-star flags on the front fenders about a quarter-mile away. There were three naval officers outside the vehicle; he was sure that one was Nimitz.

David stepped back inside. "OOD, I have the conn. Helmsman, have Engineering cut our turns in half, turn twenty degrees to port." Pelham punched the 1MC ALL button, "Good morning, Selfridge!. I trust everyone had a good inspection. Every man on the ship looked extra sharp to me, and the spaces were very shipshape. Well done. The Bridge crew has a special treat for you this morning. Now hear this. Man the port rail with your cover on for a memory that will last your lifetime. Working uniforms are OK, but don't forget your cover. Captain out."

The OOD looked concerned, "Captain, veering out of the channel center is making me nervous."

"You are right to speak up, OOD. Please check the chart depths along that large ship dry dock ahead and monitor the depth gauge. Sing out if it goes below twenty-five feet. Helmsman, add twenty turns."

The port rail filled with sailors, CPOs, and officers as they pulled abreast of the Navy Hospital. A couple of cooks with their chef hats on, and David could see Gunny Giles lead his Marines to the center of the ship's rail, line them up, and put them at parade rest. When Selfridge reached the dry dock, the men on the bow could see Admiral Nimitz ahead and started to wave at him. Nimitz had a big grin, and he waved back at them. Several men waved their hats, and he did the same. Pelham could see Lt. Comdr. Layton and Comdr. Rochefort standing behind the Admiral.

"Quartermaster, add twenty turns, turn starboard twenty degrees. OOD, how's our depth?"

"Depth is good sir, about thirty feet.

David punched the 1MC transmit paddle, "Selfridge crew manning the rail, attenshun! Selfridge crew, parade rest! He released the 1MC switch, "OOD, you have the conn, add twenty turns in one minute and steer for the center of the channel."

David put on his cover, stepped out on the port wing, and yelled, "Bosun MacGregor, you out there?"

"Aye sir, right below you on the weather deck!"

"Good, I want you to call the salute."

"Aye, aye, sir!"

As the ship's bow neared Nimitz, he was standing at attention with his cover on, as were Layton and Rochefort.

MacGregor thundered, "Men of Selfridge, attenshun! Hand salute!"

Nimitz and his two subordinates returned and held the salute as the destroyer passed and added some speed. When the stern of the ship came even with Nimitz's party, the Bosun bellowed, "Ready, two!"

Pelham completed his salute and saw Nimitz's party do the same. The sailors resumed waving at Nimitz, and he again waved back. The four-star looked up at David, pointed to him, and gave him a thumbs-up. David went back to his Captain's chair and marveled at how far the ship had come in two months. He monitored the Bridge crew but didn't have to interfere for any reason. They took the destroyer out into the open ocean and set a base course for Johnston Island. They zigzagged for an hour at twenty-two knots to throw off any Jap sub that might be lying in wait off of Oahu. David thought, *The ship is in great shape. The men are prepared. I'm ready for a fight. Morale is outstanding, and everyone knows their job. All the drills and practice exercises are paying off. We have a high state of material readiness and the fastest destroyer in the fleet. Send us more Japs.*

Selfridge transited to the Johnston Atoll at an average speed of twenty knots. While en route, she received a short-wave message from Boss Kimicho in San Francisco. It was decoded and delivered to Captain Pelham on the Bridge.

He read: "from k to dp – shokaku class built with prefab bow, 25-foot length top, 12-foot length bottom, ship is 32,000 tons full load, happy punching, congrats on coral sea. Stop"

David did a quick calculation and thought, *32,000 tons or sixty-four million pounds, fifteen times the size of my ship. What is the message? What kind of punching? Artillery shells or something else?* He called the Engineer and the XO to the Bridge to read the note. Lt. (jg) Simpson came over from his tracing table. They all reread Kimicho's words. They read it slowly out loud. None of the more experienced officers had a clue. Simpson started imitating a prizefighter's footwork while jabbing and punching. Haldane was laughing until a stunned look came on his face. He asked Simpson to please stop, and looked nauseous.

Haldane was pale and asked, "What does a prizefighter punch?"

David said, "He punches the opponent's body, his chin, his kidneys, and his nose. Hmmm, is that it?"

"Think of it skipper, a bow twenty feet long and maybe fifty feet wide at the top, and twelve feet long and ten or fifteen feet wide at the bottom. It has steel stringers, vertical ribs, several bulkheads, water-tight doors, and hatches. It weighs tens of tons."

"How do they put it on, fit it to the ship?" David asked.

"I imagine that they mate it up with the main hull and bolt it on, rivet it on, and weld it on. The point that K, whoever K is, is making is this bow, this nose of the ship is never going to be as strong as one that's an extension of the main hull."

David replied excitedly, "So if we fire away at that joint, probably fifteen to twenty feet back above the waterline, we can rip the bow off and make Shokaku stop dead in the water."

Haldane looked stunned. "No, I don't think so. How heavy is she?"

Pelham replied, "Thirty-two thousand tons, or sixty-four million pounds, I figured."

Haldane gave his Captain a distressed look. "Let's say you rip off a big chunk of the nose, a ten-by-ten-foot chunk starting at the waterline, and she's cruising at five knots."

Lt. (jg) Simpson started bobbing and weaving and punching at the air again. "I've got it. I've got it. Sixty-four million pounds pushing against 100 square feet is 640,000-foot pounds of pressure per square foot. Ougah, Ougah, water is incompressible. No bulkhead or dogged door, or hatch can withstand that pressure. Dive. Dive! Dive! Shokaku or a sister ship has just become a submarine!"

Captain Pelham looked at Haldane. "Engineer, is this true?"

Haldane nodded, yes. "It's pure physics, Captain. Every door and bulkhead in the waters' path will collapse like dominoes. There will be terrible shrieking, grinding, and exploding sounds. She'll go down by the nose in just a matter of seconds."

David looked at the other two officers with a dazed expression. "You're saying, nearly 2,000 men and 32,000 tons of ship gone in seconds, not even minutes? Right to the bottom?"

"I'm afraid so. The faster she's moving, the greater the probability it all collapses. What's this part about the Coral Sea?" Haldane inquired.

"Layton told me that the Japs were thrusting south toward Port Moresby to threaten the East Coast of Australia. We sent our carriers and an Allied surface force to stop them. We lost on a ship count basis but kicked their ass in killing their aircrew and destroying planes."

David thanked his XO and Engineer and told the OOD he'd be taking a nap in his At Sea Cabin. He had to do five Radar relative motion problems before he conked out. Yeoman Brown was outside his cabin with the usual paperwork when Pelham woke up. David told Brown it would have to wait until morning. He got up, used the head, and had Wolfe, Simpson, and MacGregor report to the Bridge. He sat in his Captain's chair and discussed the earlier conversation with Haldane and Lehman.

"Here's the moral question, men. If you're at war and planning to kill a man, does it matter whether you shoot him, burn him to death in a fire, drown him, or blow him apart? Does the time frame matter? Is it immoral to kill a lot of men in a short amount of time rather than kill the same number over a longer period?"

Simpson was first to speak. "Commander, we're trying to sink a capital ship. The more men that die, the less we'll have to fight in the next battle. Certainly, there are many things in war that are immoral. I think it's immoral to machine-gun survivors of a sinking ship or shoot a pilot who has bailed out. I think the bombing of noncombatants that's happening in Europe is despicable. But if you're warship against warship, anything goes. Our job is to destroy the enemy's capability to fight while preserving ours. The faster we do the former, the better."

"Boats, what do you think?" asked Pelham

"Captain, it's a sobering thought that we can kill so many so fast. I ask myself if the Japs could do the same thing, would they hold back? Would they be worried about morality? Or would they happily kill our asses? I think we all know the answer."

"Ops?"

"My answer is that I think we've found what we do in minute five. We encourage the target to pick up speed, and then we rip her God Damn nose off as we start to retire."

"Boats, you'll find out what this is all about in two days. In the meantime, forget what was said here today. Dismissed," David commanded.

Boats raised two fingers to his brow. "Aye sir."

"Mr. Simpson put this new tactic in for minute five. Change your tracings by tomorrow morning."

"Aye, aye, sir."

Selfridge caught up with USS Neosho and Solace the next day at 0900, about fifty miles from the Johnston Atoll. David had Lt. (jg) Simpson report to the Bridge. When Simpson arrived, David motioned for him to enter the cramped At Sea cabin and sit on the rack while he stood by the closed door.

"Lt. (jg) Simpson, you've spent the least time onboard and made the greatest contribution to Slingshot of any of our officers. Thank you. When we attack Wake, three nights from now, I need you to be ashore at the CinCPac Fleet Command Center."

Simpson started to protest, "Sir…"

"Hear me out. I have to send someone ashore who knows all of our plans and the rationale for our operational decisions. I can't send Ops or the XO. You're my logical choice, the only one qualified. Admiral Nimitz will set up a Slingshot Command Center; you will be Selfridge's voice. You'll take a set of your tracings that I think will be very impressive to the Admiral and his staff. The men will count on you to also deliver their letters to the censors in Pearl - perhaps their last letter to their loved ones. You have three hours to get your gear together and board the PBY that will land, refuel, and return to Hawaii. Dr. Nagy will be going with you. He's done as much as he can to prepare our Signalman. It's time for him to return to his family."

Simpson cried out, "Permission to speak frankly, Commander."

"Go ahead."

"This is not fair! The men will think I haven't done my job. Some will think I'm a coward. I'd rather die than be thought a coward. I've given everything to this plan since I came aboard. I have rarely been off the ship. I know this plan better than anyone. I should be on the Bridge calling the shots. I think I can do better at that than Lt. Wolfe. Send him ashore."

"Simpson, Grant, I know it hurts. I'll make sure everyone knows you did your job, and you aren't a coward. Go pack up and get ready. What you do at CinCPac Fleet may save our backsides. You're part of this mission, just in a different location. Dismissed."

David walked back onto the Bridge, selected the ALL button on his 1MC unit, and held down the transmit paddle. "Now hear this, this is the Captain speaking. In less than three hours, we will be pulling into Johnston Island. We will be topping off our fuel with the tanker you see on our starboard side. The Hospital Ship off to our port will also take on some oil and return to Pearl. A PBY Catalina will land in about four hours. We will be launching all of our boats to help with refueling the aircraft. After that, we will have a GQ and live mortar firing and gun practice. Three nights from now, we'll be in combat. CinCPac Fleet HQ wants a representative from our ship to explain our tactics. Only three men know all of our attack plans, myself, Lt. Wolfe, and Lt. (jg) Simpson.

"Mr. Simpson has done a great job refining our operational procedures and documenting every step. If we're successful, it will be mainly due to his efforts. He does not want to go, but he must. It's an order.

"Listen up! Mr. Simpson will be taking your letters with him and delivering them to the censors at Pearl. This may be your last letter to your loved ones. You have three hours to write them. You can say we're going into combat. You can say you are on Selfridge. You can't mention our location when we left Pearl or anything about the ships in company. You cannot say anything about Selfridge's capabilities or manning. Thank Mr. Simpson for taking your letters. Also, say a friendly goodbye to Professor Nagy. He's been training Signalman Akemi, who may be the key to our survival in combat. That is all."

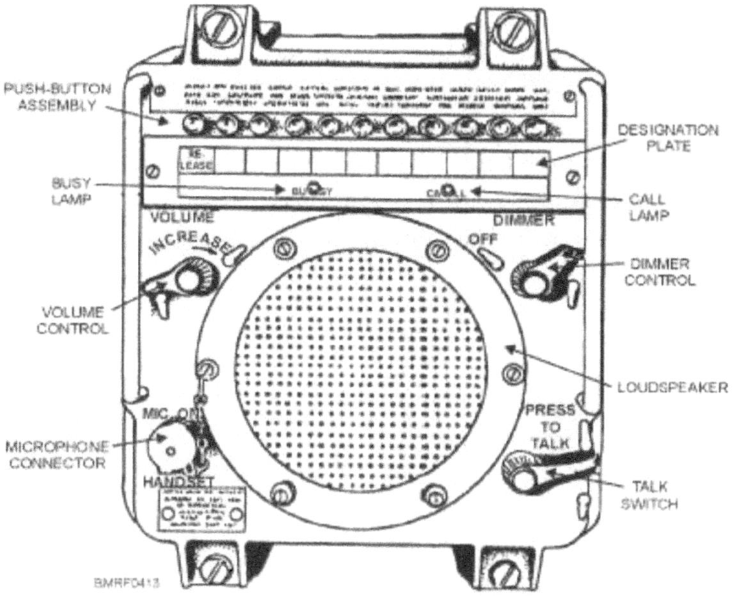

Navy 1 MC receiver-transmitter

Source: Navy training manual

USS Selfridge before removal of torpedo storage lockers, aft gun director, and searchlights

Source: Google Images

Chapter Thirty-one

TRANSIT TO WAKE ISLAND – DD-357

May 1942

Prior to reaching Johnston Island, Captain David Pelham left the Bridge and descended to his Inport Cabin. He found Professor Michael Nagy writing a letter at his desk and Signalman Goro Akemi resting in the stuffed chair. David indicated that they both should remain seated.

"Michael, pack your gear. You leave on the PBY or the Hospital Ship in a few hours," David directed.

"Nope, not packing my gear, and I'm not leaving. Instead, I'm writing a letter to Rachael and the boys just like you said."

David became irritated. "You are leaving under my orders. I'll not have a civilian on the ship in combat!"

"Civilian, pamillion, big deal. I've thought it all through. I must stay."

"Give me one reason why," David demanded.

"David-san, I'm going to give you a dozen. Akemi!"

PO Akemi jumped to his feet. "Sir!"

"You heard the Captain speak a few minutes ago, right?"

Akemi touched his ear. "Yes, I hear."

"Tell the Captain in English what you heard."

Signalman Goro Akemi looked worried and scared. "Captain speak to silver box."

"What did he say to the silver box?"

"Captain say we go Johnston Island.

"What else?" Nagy demanded.

"We have GQ and drop little boats."

"What will the little boats do?"

"I not know. Captain speak fast."

"What did he say about Mr. Simpson?"

"Mr. Simpson wrote letter."

"What did he say about me?"

"He say you his friendly."

Nagy pointed to David's bathroom. "Please Akemi, you go to head. Shut door. Me talk to Captain. Now David, do you see the problem? He knows some English words, but his understanding of complex sentences and slang is minimal. When we're working together, we use a mixture of English and Japanese words; he reacts instantly."

"I'll put an Amjap with him. You must go back," David insisted.

"No, you're stuck with me. I've reviewed your plans in detail with Wolfe and Simpson. Akemi's quick and correct response is critical to the ship's mission and possibly its survival. I'm familiar with his Hokkaido dialect. I have to translate all numbers and your crazy Navy phonetic words. Once I do that, he responds perfectly. None of your Amjaps speak his dialect. We're a team. If you want to win, you can't separate us."

"I can't do that to Rachael and the boys," David replied with a distressed look.

"It's not your choice. You must do the best you can for the ship and your crew, your excellent crew. Besides, Rachael already knows I'm staying, and she concurs. Better I fight for my country than sit in a cushy staff job – I picked that attitude up from you."

"Do you know what you might be doing to your family?"

"Captain, do you know that sixty-five Hawaiian civilians died in the Pearl Harbor attack, and many were wounded? Some from our falling AA shrapnel, but most from being strafed and bombed by the Japanese. My family now lives in Hawaii, and you're the only one I can work with to stop the enemy from coming back. Don't talk to me about my family. Do you know what you might be doing to Betty White? How many men does she have to lose before she's permanently damaged?"

David ignored his comment. "You're saying that Akemi won't be effective on the Signal Bridge without your presence. You're saying that I can't count on him to understand our tactical situation and do the right thing - when we need him to buy us some more time, he might get us all killed?"

"Exactly, that's the tactical situation, my good Captain. Like I say, you're stuck with me. Learn to enjoy it."

David was exasperated. "Michael, you win, but if you get yourself killed, I'll never forgive you."

Johnston Atoll barely qualified as an air route stepping-stone to Australia. There was a small airfield, very few buildings of little consequence, and a handful of tiny islets. A north-south barrier reef created an open lagoon to the west, but not the whole circle reef found at Midway, Wake, and many other volcanic coral atolls in the Pacific. Once the three ships arrived at Johnston, they carefully positioned themselves in the lee of the reef and east of Johnston Island. Neosho anchored and fueled Selfridge with hoses over her stern and assistance from Selfridge's whaleboat.

Shortly after Selfridge's fuel was topped off, a Navy PBY circled the lagoon and landed into the gentle waves just east of the ships. David had the OOD put the rest of Selfridge's boats in the water while MacGregor assisted Neosho refuel the aircraft. MacGregor had the whaleboat crew put a mooring line on the PBY and pass the line to the tanker. Then he ferried out two gas lines to the crewmen waiting on the aircraft wings. In the meantime, the three fast boats were skimming across the lagoon and skirmishing with the islets. One fired .50 caliber rounds into the shoreline while the other two formed a line ahead formation and attacked an islet coral outcropping with bazookas. Hundreds of birds rose from the islets as the firing disturbed their morning nesting.

When the token refueling of the PBY was complete, Selfridge fired one white flare and got underway at five knots. This was the signal for the fast boats to return and their Marine and Navy crews to board the ship by climbing up cargo nets thrown over the port side. One Coxswain fell into the ocean while descending the rope ladders and was picked up by the whaleboat. He and two other crewmen in the whaleboat were transferred to the empty fast boats to be hoisted before the planned firing exercise.

Lt. (jg) Simpson lowered himself down the cargo net and was taken by MacGregor to the PBY along with his seabag and a pouch of mail. With the boats back in their divots, the OOD called away General Quarters. The snipers manned their platforms, and the four Marine-led mortar crews quickly positioned their tubes. Nagy and Akemi took up their positions on the

Signal Bridge. No one on the Bridge noticed when MacGregor and a crewman escorted a man in civilian clothes and his luggage across the deck and down to the Captain's Inport Cabin.

Selfridge maneuvered to attack the reef area to the west of Ahau Island with mortar rounds from a distance of half a mile and a quarter-mile. The five-inch guns were aimed using Gun Director commands, but didn't fire. Instead, the Bofors and .50 caliber machine guns swept the area with rapid fire as if a carrier existed. The firing was first from the port side of the ship and then from the starboard after the Selfridge reversed course. Finally, they were ready for a full-scale rehearsal.

Captain Pelham had the OOD reposition the ship one mile southeast of the northernmost shoals of the Johnston reef. The Bridge 2MC talker informed all circuits that the ship was at Point Banana, heading for Strawberry. The wake was least noticeable at nine knots with the bubbler machines on. They raised the sniper platforms as they reached Peach; David ordered a thirty-degree starboard turn and had the sound-powered phone talker announce, "Standby rockets, snipers weapons-free."

In twenty seconds, the blinding rockets were launched. Ops was standing at the OOD's 1MC with a stopwatch. After ten seconds of rocket flight, he broadcast throughout the ship, "Cover eyes! Cover eyes! Cover eyes!" All hands on the Bridge, gun crews, on deck, lookouts, searchlight crews, men in the Signals Bridge, and the sniper crews closed their eyes and covered them tightly with a forearm. A second after they all heard the double bang, Wolfe on the 1MC and the 2MC Talker transmitted, "Guns free! Guns free! Continuous fire!"

The mortar crews were first to fire using the last quarter-mile drill settings. Next, the Bofors and .50-caliber guns joined in as the five-inchers sent a roaring broadside on each side of the small shoal area. The firing continued for a full three minutes when Ops and the Talker transmitted, "Cease firing! cease firing!" David was impressed by the big guns' rate of fire and how the entire area around the shoal erupted in a hail of explosions. The first blasts at the shoals were from the mortars since the big guns were firing higher, simulating disabling the carrier's five-inch weapons. They then fired at the waterline during minute two. After ten minutes, David had the OOD reverse course, and they repeated the firing exercise from the starboard side. This time the big guns waited ten seconds before commencing fire, allowing the mortars to fire two ranging shots. Again, the noise was deafening, and a large cloud of smoke trailed Selfridge. Hundreds of birds took back to the air from the islets and Johnston Island. Selfridge was ready!

After watching the gun exercise from a mile east, Solace gave one long blast on her horn and headed for Pearl Harbor. Selfridge secured from GQ and set condition YOKE. On deck, men were policing shell casings and throwing them over the side. David's ears were ringing as he tried to make out what MacGregor was saying to him on the Bridge. He took the cotton out of his ears, and it helped.

"Skipper, there was a visitor on the PBY who insisted on coming onboard, says he's an Army doctor."

David was still stunned by the ferocity of his guns. "A visitor. We can't have visitors. Why did you let him board the ship?"

"Sir, he has orders to Selfridge. I put him in your Inport Cabin. He's an Army volunteer who heard we needed a doctor."

David looked at Boats narrowly and thought, S*ure, an Army doctor hears Selfridge needs a doc and just can't wait to do some combat sea duty. I smell Rachael Nagy or Betty White behind this or both.*

He went down to his cabin and found Nagy talking to a crewcut young man in civilian shorts and a Hawaiian shirt. Akemi was sleeping on the cot.

Nagy provided the introductions. "Captain Pelham, allow me to introduce Captain Floyd Lawson, United States Army Medical Corps."

"Captain Lawson, Floyd, welcome to Selfridge. Do you know what you've got yourself into? We're starting a hazardous combat mission. You could get yourself killed or maimed," David admonished.

"Yes sir, that's why I'm here."

"To get yourself killed or maimed?"

"No sir, to keep other men alive. I heard you would be in a shoot-up with no doctor. I'm tired of being in a secure hospital well behind the front lines."

"Who did you hear this from?" David demanded.

"I heard it from Captain Betty White, and she introduced me to Michael's wife, Rachael." He reached into a satchel. "Here, they sent you both presents."

"Captain Lawson, what's your specialty? What can we expect you to do?" asked Pelham

"I'm a trained M.D. I have a lot of experience with trauma cases and triage. I am also a board-certified surgeon and have treated hundreds of wounded men from Pearl Harbor."

David looked at Nagy, who was nodding yes. "I'm sold, but I have to know you understand the risks. You saw the Solace hospital ship when you landed. We can catch up with them and transfer you if you have any doubts."

"No sir, I have orders as your ship's doctor. I brought a bunch of supplies, including new sulfa drugs."

"How did you get orders? Did you bring uniforms?"

"Yes sir, three or four working uniforms."

"Get one of them on. We need to sail. Michael will bring you up to the Bridge."

Pelham found Yeoman Brown waiting outside the cabin with an envelope that had arrived on the PBY. David carried it up to the Bridge, and told the OOD to get underway at one-third with a Sonar sweep along their westward PIM. Selfridge sent Neosho a blinking signal to pull anchor and follow. Captain Pelham perched in his chair, opened the envelope, and read:

SECRET

10 May 1941

From: CinCPacFlt

To: Commanding Officer USS Selfridge

Subj: Operational Orders for Operation Slingshot

1. (S) Commanding Officer USS Selfridge is directed to proceed to Wake Island and attack the enemy facilities there at a time of his choosing.
2. (S) Commanding Officer USS Selfridge is authorized to conduct combat operations against all enemy forces in the Wake Island area.
3. (C) Commanding Officer USS Selfridge shall use aircraft, submarine and tanker units to facilitate and complete the above-assigned missions.

4. (S) Slingshot forces shall be well clear of Wake Island by 2300 Midway time on the night of the attack.
5. (S) Slingshot forces shall retire to the Midway Atoll when the above missions are completed or aborted.
6. (U) Commanding Officer USS Selfridge, Lt. Comdr. David A. Pelham, USNR has been frocked as Commander David A. Pelham, USNR.
7. (U) Executive Officer USS Selfridge, Lt. Gary P. Lehman, USN has been frocked to Lt. Comdr. Gary P. Lehman, USN
8. (U) Operations Officer USS Selfridge, Lt. Donald K. Wolfe, USN has been frocked to Lt. Comdr. Donald K. Wolfe, USN
9. (U) Insignia for these deserving officers sent by separate cover.
Congratulations, happy hunting, and good luck,
Chester W. Nimitz
Admiral USN
SECRET

David finished reading the orders. He noticed Petty Officer Brown had eased into the Bridge and whispered to the OOD. They both had big smiles as the OOD announced that the Skipper was promoted to Commander. There were some yells and a round of applause from the Bridge crew. Brown brought him a small box with silver oak leaves and shoulder boards with three broad stripes.

"These also arrived on the Catalina. Congratulations Captain," Brown said as he shook David's hand.

David admired the insignia. "Thank you, Yeoman Brown; I'll need a new cover."

Brown replied in his low, quiet voice. "Sir, I have a cover frame for you with scrambled eggs on the bill. I picked it up a couple of months ago."

"Brown, thank you. I'll need to pay you for that."

"No sir, it would give me great pride if you wear the cover that I bought for you. Myself and Steward Jackson, he insisted on paying half."

When he shook Brown's hand, Pelham felt a catch in his throat and thanked him again. Nagy, with Lawson in tow, arrived on the Bridge. David showed Nagy the orders, and Michael pounded on his back, shook his hand, and insisted on pinning the silver insignia on his khaki shirt. Lawson gave David the typical Army uncovered salute and congratulated him. David returned the salute.

"Captain Lawson, we have two days to teach you about the ship," David pointed to the 1MC. "This receiver-transmitter connects me to specific areas of the ship or to all of the ship. And it allows critical operational personnel to call me. Watch this."

He selected a button and depressed the transmit paddle, "XO, are you in Secondary Conn?"

"XO here. That's affirmative."

"Good. We have a surprise visitor, a doctor. Please come forward and meet him. Also, put out a call for Ops to report to the Bridge. Captain out." Pelham turned to face the new ship's doctor. "Mr. Lawson, the XO is in the Secondary Control space aft in the ship. He has communications similar to what you see here and would control the ship if this Bridge is wiped out. I need to know, will you be able to take your first sick call at 1900 tonight?"

"Yes sir," Lawson replied.

The XO arrived on the Bridge; Pelham had him read Nimitz's orders. David took off Lehman's railroad track emblems, pinned on his new gold oak leaves, and congratulated him. The XO received handshakes and congratulations on the Bridge when Lt. Wolfe arrived, and the short ceremony was repeated. Wolfe was introduced to Lawson and briefly explained his duties as Operations Officer.

David said, "Captain Lawson, the Officer of the Deck, or OOD is my representative for operating the ship. He's assisted on his four-hour watch by the Junior Officer of the Deck or JOOD. The other men you see up here are their crew. OOD!"

"Yes sir!"

"If our Sonar sweep for subs is negative, set the course for Point Charlie at fifteen knots. Signal Neosho to stay close aboard and that we'll be zig-zagging for the first hour."

"Aye, aye, sir.

"JOOD!"

"Yes sir."

"Announce on the 1MC that the Captain will make an important announcement at 1800."

"Aye, aye, sir!"

"Doc Lawson, Michael will take you down to our very cramped dispensary while the XO and I figure out where you're going to sleep. We have fifteen officers onboard, counting you, and 191 enlisted men. Supper is at 1730 in the Wardroom. There is always a snack there if you're hungry before supper."

"Thank you sir, and congratulations again," Lawson replied with a big smile.

Nagy chuckled and pointed at the gift bag next to Pelham's Bridge chair, "I'd wait to open that if I was you."

As the ship was picking up speed and Nagy departed with Lawson, David declared Lawson would mess in the Wardroom. Lt. Comdr. Lehman recommended Nagy join the Officer's Mess. They decided to put another cot in David's Inport Cabin for Lawson. It would be crowded, but the three of them, Lawson, Nagy, and Akemi would have the luxury of their private head and shower. The XO left to prowl the ship and talk to the men. David jotted down a few notes for his 1800 announcement of their mission, Nimitz's orders, and Admiral King's words of encouragement.

When Selfridge's Captain got a moment, he went into his At Sea Cabin, used the head, and opened the gift bag with the red ribbon tied around it. There was a tin of chocolate chip cookies, a silver necklace with a cross that Betty wore on Sundays, a note, and another smaller bag. The message said to please wear the cross for good luck, enjoy the cookies, and think about your favorite fiancé when you wear the other gift. It ended with love and kisses.

Pelham opened the smaller sack and exploded with laughter. They heard him laughing in Bridge Plot and on the Bridge proper. He laughed until tears started flowing as he held up the unexpected gift. It was an athletic supporter, a jockstrap with a thick sheet metal formed over the pouch. Etched in the metal were the words:

"RESTRICTED AREA!

U.S. GOVERNMENT

PROPERTY"

David wondered what Michael's supporter said as he flopped on the rack and fell asleep.

At 1730, David had a quick supper with the XO, the new ship's doctor, Lt. Comdr. Wolfe, and several other officers in the Wardroom. Lawson had a good supply of medical jokes, and everyone was in good spirits.

At 1759, David was on the Bridge and had the JOOD announce, "Standby for an announcement from the Captain."

David stepped up to the 1MC and transmitted, "Good evening, men of Selfridge. As some of you have guessed, we're on the way to attack the Japs on Wake Island." He could hear cheers throughout the ship; the Bridge crew was yelling and pounding each other on the back. He waited until things quieted down on the Bridge. "I told you we were going to raise our ship, get it back in fighting trim, and go kill some Japs. Your response has been magnificent. Our ship is ready to fight, and our crew is prepared for any challenge. The firing exercise today was beyond impressive. Lord, would I hate to be on the receiving end of those thousands of bullets, mortar rounds, and dozens of five-inch shells. Two nights from now, we'll be shelling Wake at sunset from over the horizon while Dumbo directs our fire. We will destroy the aviation facilities, blow up the planes, and make the runways unusable so they can't come after us. Get all the rest you can on the 14th, suspend all nonessential ship's work that day.

"After we pulverize Wake, we'll sail north at high speed to engage an enemy surface force coming to reinforce the Wake Atoll. We plan to infiltrate that formation in the dead of night and destroy or cripple a Jap aircraft carrier. When we're executing that plan, we'll be at darken ship. The smoking lamp will be out, period. If you try to sneak a smoke or look out a porthole, you can kill yourself and everyone on this ship.

"I'll not minimize the danger to all of us. Some consider this a suicide mission. I don't. If every man does his job and no man gives us away to the enemy, we have a fighting chance. We will go over the details of this plan after striking Wake hard. Tonight, we'll escort Neosho until dawn. Then we will separate and assume the identity of a new Jap destroyer named Akizuki. We then head for a point 125 miles south of Wake – about 1,200 miles west of our current position. I told you about Petty Officer Akemi, who joined our crew. He's painted the Akizuki markings and numbers on our ship covered in canvas. Boats MacGregor and his carpenters will build up our prow and make our stern look rounded. The deck crew will stretch gray canvas to blend our forward stack into our superstructure and erect a fake mast where we'll fly the Japanese flag.

"Petty Officer Akemi is training our Amjaps to go out on deck in Japanese Navy garb and make our ship look like it's the real McCoy. If we're spotted by Jap airplanes, we will attempt to entice them down for a close fly-by so our Amjaps can take their picture and vice versa. When they're abreast of the ship, our Marine snipers, forward Bofors, and .50-caliber guns will shoot them down. This will be called away as Operation Nippon, repeat Operation Nippon. Only the rifle and gun crews hid under camouflage, and the Amjaps can be on deck.

"During Nippon-modified GQ, if you stick your head out of a hatch to see what is going on, or come strolling down my deck, I'll personally come out on a Bridge Wing, shoot you with my .45, and have the Amjaps shove your sorry body over the side. Selfridge crew, your officers have stuck their necks way out on this one. We've briefed this operation to Admiral Nimitz and even Admiral King. We're striking back at the Japs, and I expect this crew to back its officers 110 percent. Every man does his duty. Every man uses his head and thinks about the consequences of his actions. And every man helps his shipmate.

“Here’s more good news. We now have an actual ship’s doctor, Captain Floyd Lawson, United States Army. He heard we would kill some Japs and signed on as ship’s company. His first sick call is at 1900 tonight. See him if you have a medical problem. Professor Michael Nagy and Petty Officer Akemi work together on the Signals Bridge. I asked Mr. Nagy to return to Oahu from Johnston Island, but he wants to assist Akemi during air attacks and the most dangerous phases of attacking the Jap carrier.

“By order of Admiral Nimitz, Lieutenants Lehman, and Wolfe are now Lieutenant Commanders, and your Captain promoted to Commander. Admiral Nimitz also ordered us to attack Wake Island and any enemy forces around Wake. He ends his letter with congratulations, good hunting, and good luck. When Lt. Comdr. Wolfe briefed this operation called Slingshot to Admiral King in San Francisco, the Admiral said, quote, if Slingshot is a go, tell the crew from me, give them hell, Selfridge. Remember Pearl Harbor, unquote. That is all.”

Pelham could hear more cheering echoing throughout the ship.

The following day, Selfridge bid the tanker farewell by blinking light and semaphore, turned sixty degrees to port, and headed for Point Bravo. There was a light breeze from the west. The seas were benign with three-foot swells, and the deep blue sky was full of whimsical, puffy white clouds. Before the OOD increased speed from All Ahead Slow to All Ahead Two Thirds, the deck crew installed the new prow, the rounded stern, and canvas masking the forward funnel. Another group of men painted each side of the Bridge structure in blue and white camouflage to make it seem smaller.

Boats and three crewmen erected a twenty-foot wooden mast in front of the aft guns, and Akemi raised the rising sun flag to its peak. They moved the stern ensign jack forward a few feet, and a smaller Japanese flag flew from it. The Amjaps shifted into their enemy sailor uniforms and prepared to go on deck at a moment’s notice.

On 14 May, Selfridge arrived at Point Bravo, 125 miles south of Wake Island at 1500 Midway Island time. All the clocks had been set to Midway time at 2400 the night before. The OOD slowed the ship and deployed Sonar in compliance with the Captain’s standing orders. David was sleeping, as was much of the crew. Before lunch, Captain David Pelham realized he had been on the Bridge for seventeen hours and better get some rest. During this lull, the XO and his Assistant Navigator shot the sun with a sextant, confirming their position slightly north of Bravo as they slowly sailed north. Lt. Comdr. Wolfe planned to charge north at thirty-five to forty knots starting at 1800. An early supper of steak, mashed potatoes and gravy, green peas, the last fresh milk, and apple pie a la mode was on the menu in the Wardroom, the Chief’s Mess, and on the Mess Deck.

Lookouts were double-posted and searching for aircraft while the Radarmen searched the skies with three sweeps every five minutes. Quiet hours were in force, and the Bridge crew suspended ringing the bell and routine announcements. The OOD gradually increased speed to All Ahead Two Thirds in the calm seas. At 1700, they rang two bells and announced that an early supper was served throughout the ship. Steward Jackson roused David in his At Sea Cabin, provided him with a fresh khaki uniform, and asked if he wanted supper on the Bridge. David said no and took the khakis down to his other cabin. Nagy, Lawson, and Akemi had already left for supper. He took a shower, shaved, and put on a clean uniform.

David took a long look at himself in the head's mirror and thought it was time for a haircut. He'd have to see if Yeoman Brown was free after supper. They had a routine. Brown would cut his hair and refuse payment. David would insist on paying fifty cents, which Brown would claim was too much. He knew Brown cut enlisted crewmembers' hair for two bits, so he'd pull rank and tell him that an officer should pay more. Pelham thought about the larger ships, the cruisers, battleships, and carriers that had actual barbershops - sometimes operating around the clock. He smartly put his cover on while looking into the mirror and wondered if a carrier like Shokaku had a barbershop.

Captain Pelham entered the Wardroom and removed his cover. A junior officer shouted, "Attention on deck," and ten officers snapped to attention. David was surprised since he had said this was not required during meals. He thought, *They know this is a big night in our lives. They want some words from their Captain.*

"Good evening, gentlemen. This is a momentous night, an exceptional occasion, something we all will remember for the rest of our lives. Rarely do we get steak on a Wednesday night! As you were."

His words broke the ice, and there was a round of laughter. The Wardroom served steak every other Thursday, if at all. Supper was served cafeteria-style so that more stewards could enjoy their steak dinner on the mess deck with their Division. David went to the galley serving window, got a tray, and selected a steak. Each steak had a flat toothpick that said Rare, Medium, or Well. It didn't matter which one you picked since there was no correlation between the toothpick and the time the steak was cooked. David loaded his tray and filled a coffee mug with almost fresh milk. He hated powdered milk and powdered eggs; such was the life of a sailor at sea. He sat at the head of the table with Lehman at his right. The XO served as the Wardroom President. A junior officer was the Mess Treasurer, who collected monthly payments from the officer members of the mess. The supper discussions were lively that night - nothing was said about the battle ahead.

At 1800, Captain Pelham was pacing the Bridge. The Nav plot showed them ninety-five miles from Point Alpha. Lt. Comdr. Wolfe directed the OOD to increase speed to All Ahead Full. The game was on. Selfridge was in condition ZED heading for war. They had run for an hour at forty knots when Radar reported an aircraft target dead ahead at twenty-five miles and closing fast.

David quickly issued orders, "OOD reduce speed to two thirds, then one third when the stern wave permits. JOOD call away Operation Nippon. 2MC talker, report enemy aircraft at twenty miles closing from the north on all circuits. Radarman report air target speed."

From back in the Bridge Plot, "Estimate 1-5-0 knots sir."

"Very well. Ops, that's 2.5 miles a minute. We have less than two minutes before they see our wake and eight minutes before they're overhead."

"Concur sir. Setting Nippon takes five to six minutes. Radarman, any composition on the air target?"

"Sir it's either two planes or a large plane. I'm getting a big return."

David went out on a Bridge Wing and saw the aft snipers man their platforms and pull gray-shaded camouflage netting over themselves. He saw Nagy and Akemi climbing to the Signals Bridge wearing Jap uniforms. Amjaps in enemy uniforms were pouring out on the deck with cleaning equipment, and the forward Bofors crew was in place under a gray netting framework. He felt the ship slowing to One-Third and remembered to have the gun crews completely lower their guns when the "friendly" aircraft was identified. He stepped back into the Bridge and saw a breathless Amjap youngster dressed as an officer rush into the Bridge armed with an old-style Japanese camera.

"OOD take off thirty turns. Radarman, say range of air target," David directed.

"Ten miles, sir. Dead ahead, closing fast.

"Very well, secure Radar."

We have to fool them, make them think we're a Jap destroyer, or have a fight on our hands and more planes attacking from Wake. Thank God the ship was fitted with newer five-inch 38 mounts in San Diego before transferring to Pearl. Now they can elevate to 85 degrees, making them lethal against aircraft.

The 2MC Talker reported, "Signals reports two aircraft bearing 360, closing fast. Standby, forward lookouts report two aircraft with twin engines."

"Very well. Ops start looking up Jap twin-engine planes."

2MC Talker spoke again, "Signals reports that Akemi says it's Mitsubishi G4M, two G4Ms at 1,000 feet. He's gonna signal them for a picture."

"Very well. Ops, what do you have?"

Wolfe responded after a short pause, "G4M, Navy twin-engine bomber, our designation is Betty, crew of seven, range 1,800 miles, bombs and guns, twenty mike-mike on the tail, one light machine gun on the nose, two in the waist."

The 2MC Talker spoke loudly, "Signals says put Amjap officer on the port wing now with a camera."

The Amjap put his hat on firmly and ran out on the Port Bridge Wing. David could see three Amjaps on the bow drop their mops and point at the G4Ms. Then they started waving and excitedly jumping up and down at the sight of their pretend airborne comrades. Ops ordered all guns lowered and stowed fore and aft. David could see the Bridge Amjap salute the planes and hold up his camera. He knew a group of Amjaps on the stern would suspend their make-believe maintenance of the Bofors guns and wave to the aircraft. Another group having their picture taken at the railing, would wave wildly, and the cameraman would gesture for his airborne comrades to descend so he could take their picture. The two Jap flags were snapping in the breeze, and the ship's markings were Japanese. What could go wrong?

"Ops, what's our best outcome?" David requested.

"Sir, we talk them into a low flyby, shoot them down, and none of their crews report our position."

"To do that, we have to annihilate them. No one can survive once we start shooting."

"Yes sir, the gun crews will keep firing until the aircraft sink."

"Very well. Give me the worst case."

"They report our position and decide to bomb us. Aft lookouts will be watching for the bomb bay doors opening. Next worse, one plane stays at altitude, reports our position, and the other does a low pass to photograph us. We pass ourselves off as Japs or take out the low plane."

The 2MC Talker sang out, "Signals says they agreed to photo; they're coming down."

David replied calmly, "Very well. Ops, what can we expect?"

"Skipper, ninety percent chance of flying up our wake and passing down our starboard side," Wolfe replied.

"Flying in formation?" Pelham queried.
"Yes sir, the wingman will be tucked in for a nice picture. He'll be lower and on the starboard side of the leader."
"Show me with your hands," David said. "Thank you. 2MC Talker, tell the forward guns and snipers to expect planes from aft along our starboard side." Once that was transmitted, David continued, "2MC talker, advise aft sniper to shoot wingman pilot when he's alongside our ship. Advise forward sniper to shoot aircraft leader pilot when aft sniper calls his shot. OOD, take off twenty turns."
Lt. Comdr. Wolfe said, "The Amjap pretend maintenance crew will man the aft Bofors and the starboard aft .50 calibers as soon as the planes pass them. The five-inchers will be getting gun director aiming and training on the aircraft when the snipers fire."
"Very well, we're at ten knots. OOD, continue ahead for two minutes after we splash the aircraft, then set your rudder twenty degrees starboard."
The 2MC Talker stated, "Signals reports two aircraft low at one mile should pass us on our starboard side. Signals recommends putting the Amjap officer on starboard wing. Aft lookout reports two aircraft bearing 180 relative, bomb bay doors closed."
"Very well, talker. Tell snipers weapons are free," Pelham responded.
David walked over and looked out the window of the starboard wing door. It had tinted acetate taped to it, as did the other Bridge windows, so the aircrew wouldn't see the Caucasians on the Bridge. He heard the planes coming. He looked aft, and saw the lead plane's inside windscreen covered with blood from the sniper's headshot. The wingman veered away as the forward Bofors and .50 calibers blasted the port wing off the leader's plane. It spun into the ocean, throwing up a huge spray of white water. David opened the door and pulled the Amjap pretend officer back into the Bridge. He told him to cover his ears as the big guns started to swing toward the wingman, who was now turning away, trailing yellow and orange flames. He could hear the aft Bofors and .50 caliber guns swing into action and see their shells tracking in the water up and into the remaining bomber until it exploded into several pieces.
"2MC Talker, tell guns to keep firing until all the aircraft parts sink," David ordered.
Pelham hesitated and fixated on the shots striking the second aircraft debris. Wolfe called a halt to the guns firing and secured the ship from GQ and Operation Nippon.
He questioned, "Skipper, skipper, you OK?"
"Yes, I guess I got lost in my thoughts there for a minute. Do you think they could have transmitted our position?"
"Yes sir, I think they could have, but chances are they decided to wait until they had a positive ID. I bet their usual tactics would be to have one stay high, and the other comes close enough to make an ID. Our Amjaps made the Jap flyers throw caution to the wind."
David thought, *I should feel something right now, exultation, victory, regret, revenge, sorrow, shame. I feel nothing. We just killed fourteen Japs that would've gladly done the same to us. I hate their guts for starting this war. I'm going to kill every last one of the stinking yellow bastards!*

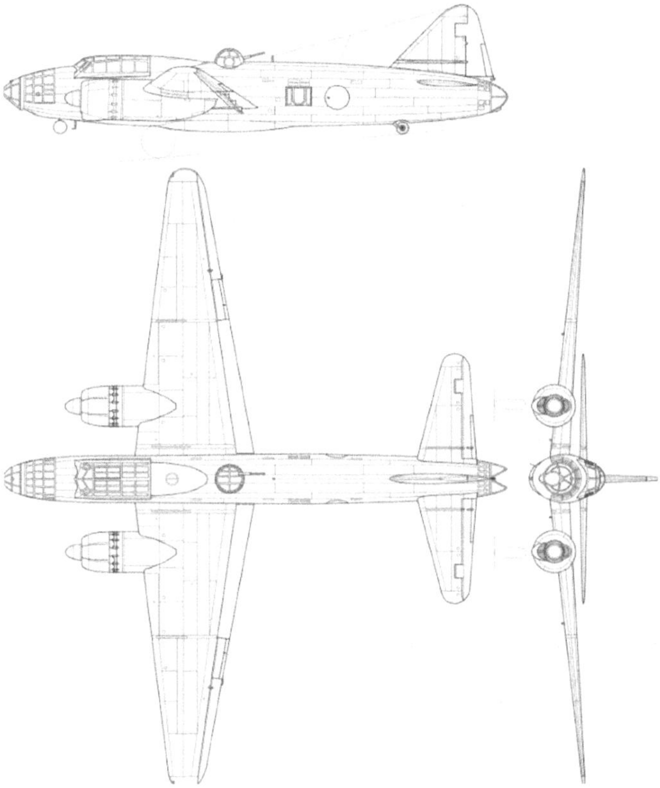

Mitsubishi G4M "Betty"land-based bomber

Japanese war strategy was to use submarines and land-based bombers to attrite American surface forces when they advanced west from Hawaii

Source: Google Images

Akizuki Class destroyer

Source: Wikipedia

Chapter Thirty-two

ATTACK ON WAKE – DD-357 and Ginger Snap

May 1942

David Pelham returned to the Bridge after using his At Sea Cabin head. His body had calmed down from the adrenaline rush of downing the two planes that could have destroyed his ship.

"Ops, give me an update on Point Alpha."

Wolfe looked at his watch. "We have forty-one miles to run and less than an hour to get there. We'll be about five minutes late but in time to light off the homing transmitter for the C-46."

"Very well. OOD, get us back at All Ahead Full. You have the conn."

"Aye, aye, sir!"

"JOOD, enter in the Ship's Log that Captain's opinion is no crewmember in the A4M aircraft survived the impacts with water, and we saw no survivors in the water. Ops, wake me when we hear from the C-46, or we're ten miles from Alpha."

Captain Pelham went into his At Sea Cabin and lay down on his bunk. He took deep breaths to calm his pulse and realized his right thumb was twitching. David could mentally command it to stop, but it started moving again when he relaxed. It seemed like he had only been asleep a couple of minutes when Wolfe shook him. "Skipper, Golf Sierra reports visual on Wake."

David was groggy and confused, "Golf Sierra?"

"Yes sir our call sign for Ginger Snap, the C-46."

"Yes, yes. Man, I was deep asleep. How is our position? What time is it?"

"We're approaching ten miles out from Alpha. Midway time is 2020, and sunset has started."

Pelham replied, "Good, that gives us twenty minutes to launch two fast boats, put Smitty up in Dumbo, and get to our firing position seven miles southeast of the field. At 2045 we turn on the ADF for the C-46 final approach. I'm hitting the head. See you on the Bridge."

After being announced on the Bridge, David settled into his captain's chair. The nap had helped; his thumb was not shaking. He shook the cobwebs out of his head as he remembered the split-second violence that had destroyed two medium bombers and fourteen enemy aircrewmen. *War is hell. Now we'll kill more men when we blast away at Wake. We could be killing our own Americans being held there as slave labor. Don't think about that. Don't think about that!*

Wolfe came closer to his chair and reported, "Captain, we're approaching Alpha, the sun is just down, recommend launching the boats, and then Dumbo."

"Very well, make it so Ops."

Pelham thought about the training sessions that Lt. Kindall and the Gunnery Division had conducted with Yeoman Smith. At first, they tried to teach him every command and gunnery term. It was too much for Smitty, who complained that he had to fly Dumbo and remember all the different ways the five-inch guns could be employed.

Lt. Kindall had come to David and asked for his advice as a former Gunner's Mate. They worked out a simple method for Petty Officer Smith to control the two-forward twin-gun turrets or the two-aft twin-gun turrets. He could call for ranging shots separately from the forward four guns or the aft four guns by providing a relative bearing from the ship and an estimated range. Then Smith could correct the train of those guns left or right in degrees and ladder the range up or down in 50, 100, or 200-yard increments. Once he reported "On Target," all four guns would slow fire while compensating for the ship's forward speed and movement from the ocean waves. He could ask for rapid-fire, parallel fire that did not correct for the ship's movement, or parallax fire that trained and elevated all eight guns on the same target area. Kindall taught Smitty how to use parallel shots to walk the shell explosions along a line of targets like a runway, using four guns or all eight.

"Mr. Wolfe, time check and status report."

"Skipper, time 2040, C-46 ready to attack, ADF beacon on now, small boats deployed, Dumbo deploying, up to about a hundred feet, all guns manned and ready, torpedoes not manned."

"Very well, OOD, sound General Quarters on 2MC and internal 1MC circuits only. Ops, take the conn and position us for a gunfire mission."

"Aye sir. Helmsman, steer course 330, make your speed Ahead One Third plus twenty turns."

Pelham said, "JOOD report on Dumbo."

JOOD after stepping out on the Starboard Bridge Wing, "Dumbo looks about 200 feet up, sir."

"Very well, 2MC Talker. Check communications with Dumbo."

"Comms check good with Dumbo, Gun Captains hearing Dumbo," Reported the 2MC Talker.

They could all hear a very low aircraft passing overhead and saw the C-46 Commando slowly heading for the runway with its flaps and gear fully down.

David spoke calmly, "Ops, we have about five minutes. 2MC Talker, transmit weapons-free under Dumbo command, have Comms strangle the ADF."

Lt. Wolfe issued orders, "Helmsman, add twenty turns to your speed. 2MC talker, have Dumbo report field in sight.".

"Ops, state your intentions." Captain Pelham instructed.

"Captain, when Dumbo sees the field, we engage any ready aircraft with the forward guns. We're coming up on the eight-mile mark; then we turn to port on the seven-mile mark. Forward guns are loaded with fragmenting anti-aircraft shells and contact fuses - initial range is 14,000 yards."

"Dumbo reports field in sight bearing right thirty degrees," reported the Talker.

"Very well, Pelham replied. "Put Dumbo's circuit on the speaker and tell him we're at 16,000 yards. OOD, broadcast on 1MC circuits, standby for gun firing. Ops, you still have the conn."

"Ops aye."

They heard on the speaker, "Forward guns, this is Dumbo. The target is two aircraft near the runway, props not turning, bearing right thirty-five degrees, range 16,000, fire single shot now. Break, bombs are exploding on runway."

"Forward guns, shot away." After a short pause, "Forward guns, Dumbo, one-degree right, ladder up 100 yards, fire single shot now."

"Forward guns, shot away."

After several seconds, "Forward guns, ladder up 50 yards, slow fire now."

"Forward guns, slow fire."

Another long pause. "This is Dumbo, Targets destroyed, forward guns check fire.

"Helmsman, in two minutes, set your rudder thirty degrees left, steady on course 280," directed Ops.

Before turning port, the Bridge crew heard, "Forward guns, this is Dumbo, target runway, bearing right eighty degrees, range 15,000 yards, fire single shot."
"Forward guns, shot away."
"Rear guns, this is Dumbo. Target aircraft starting up, bearing right ninety degrees, range 16,000 yards, fire single shot. Forward guns ladder up fifty yards each four shots and one degree right, commence slow fire now. Rear guns, left one degree, ladder down 100 yards, fire single shot. Forward guns, check fire, parallax on rear gun's target. Enemy aircraft taxiing. Rear guns one-degree right on each shot, continue slow fire. Forward guns, rapid-fire now."
David saw a geyser of white water about a mile toward the island. "Ops, we're taking incoming fire about a mile short."
"Roger that, Skipper. We should see more action soon from our fast boats. They'll be making smoke and firing off Roman candles.
They could hear Smitty on the speaker, "This is Dumbo, aircraft destroyed, all guns check fire. Rear guns target airplanes and gas tanks bearing 100 degrees right and range 16,000 yards, fire a single shot. Forward guns target runway, eighty degrees right, range 14,500, fire single shot now.
"Forward guns, shot away. Rear guns, shot away."
"This is Dumbo, Rear guns, one degree left, ladder up 100 yards, parallel slow fire."
"Rear guns, Wilco."
"This is Dumbo, forward guns slow fire, parallel fire, ladder up 100 yards every four shots."
David saw a shell hit the ocean about half a mile north and send up a large splash. "Ops, see if Smitty can take out those guns. They may not see us, but they just might get lucky if we ignore them."
Wolfe picked up a 2MC microphone. "Dumbo this is Ops, we're taking some shell fire, engage the guns if you can."
"This is Dumbo. I see gun flashes on the southwest tip of Wake Island. Getting good hits with rear guns right now, burning planes and exploding gas tanks. Wilco. Forward guns check fire. Target gun emplacement bearing right forty degrees, range 14,000 yards, fire single shot now."
David heard his 1MC speaker squeal. "Captain, this is Comms, over."
"Go ahead Comms."
"Captain, you have Golf Sierra calling you on the short-range radio."
David transmitted on a radio microphone, "Golf Sierra, Alpha Sierra, over."
He recognized Rats' voice. "Alpha Sierra, we still have twenty eggs for delivery and plan on another pass."
"Golf Sierra, that's a negative. Your signal is return to base, over."
"This is Golf Sierra. Having trouble reading you, first egg in four minutes, out."
David threw his hat down on the deck. "Damn it, Rats! Talker, tell Dumbo to put all guns on the latest target. Ops, show me our current position on the plotting board." David picked up the 2MC microphone, "All guns this is the Captain, prepare to fire star shells in two minutes."
David transmitted again on the short-range radio, "Golf Sierra, give us a two-minute mark and prepare to cover eyes."
"This is Golf Sierra. Reading you now, Wilco."

David gave Smitty another minute to destroy the enemy gun emplacement and spoke on the 2MC circuit. "This is the Captain, all guns load star shells ASAP. On my command, forward bow turret fire bearing right sixty degrees, rear bow turret fire bearing right eighty degrees, forward turrets reload flare shells, repeat flares, and fire twenty seconds later above the island. Break, aft forward turret fire star shells bearing right 100 degrees, aft rear turret fire bearing right 120 degrees, all guns use eighteen-thousand-yard range, stern turrets add twenty degrees right and fire a second round of star shells. All guns wait for my command."

On the radio speaker they heard: "Alpha Sierra, this is Golf Sierra, two-minute mark."

David transmitted, "Golf Sierra, roger. Firing star shells in thirty seconds. Cover eyes for two rounds, out."

David ordered his guns to fire on the thirty-second mark. A 120-degree wedge to the northeast lit up like daylight as the star shells streaked across Wake Island. In seven seconds, the rear battery barked again, illuminating the whole southern portion of Wake Island. The forward battery placed flares high above the lagoon thirty seconds before the C-46 started its second bombing run. Any Jap that looked at the star shells was blind for several minutes.

They could hear Smitty on the 2MC speaker. "Bridge, this is Dumbo. I see two large flying boats at the seaplane base area."

Wolfe replied, "Dumbo, engage targets with all guns."

The Bridge crew heard Captain Mueller's voice on the radio speaker. "Alpha Sierra, this is Golf Sierra. Ain't nobody gonna use that runway for a day or two. Nice show you swabbies put on. Don't forget the taxiways. We gone, out."

Captain Pelham looked at his watch. They had been shelling the island for almost fifteen minutes. It seemed like two hours. It was now nearly dark. Dumbo was going to need flares soon to mark the targets. David's body was fatigued, and he wanted to use the head. The ship was halfway up the western side of Wake. They needed to strike the aircraft facilities and taxiways again, or they would have to make a second pass. He ordered Ops to shift the gunfire once the seaplanes were damaged. Then he went into the Bridge Plot to the red-lit light table to review the latest Simpson tracing for attacking Wake.

He remembered there were two major gun emplacements further north on Wake and the southern shore of Wilkes Island to the west. Their plan had them shelling the small harbor that had been blasted in the coral. They hoped to catch a supply ship or a sub unawares at the piers. Before making a quick trip to his head, he noticed that the suspected Wake command center and the northern Wilkes Island guns were also on the target list.

David returned to the Bridge, "Ops report."

"Skipper, we're working over the aircraft parking area and the taxiways. We've got a couple of secondaries, but it's getting harder for Dumbo to see the targets."

"Very well, let's have a review. Simpson's drop-dead time to leave the Wake area is 2130. We still can attack the harbor, the command center, and three-gun emplacements on our way north. Then we reel in Dumbo and recover the fast boats before sailing north at high speed. Smitty is going to need flares from now on. Coordinate Guns and Dumbo to engage the remaining targets. Comms just handed me a PBY position report on Orange Force that's four hours old. I need to work that with the Navigator and figure our intercept course. Call me if you need me."

"Aye, aye, sir!"

David gestured to the Assistant Navigator; they went over to the light table and placed a chart on it. Lt. Gehres plotted the carrier position at 1710 and extended a line along the reported easterly course of 100. It passed close by Point Lima, where they expected the Jap carriers to launch planes in the morning - 160 miles north northwest of Wake. The Orange Force plot placed them 120 miles west of Lima, traveling at a reported speed of fifteen knots. Selfridge would have to race north at forty knots to intercept Orange Force thirty to forty miles west of Point Lima by 0300. David knew his submarine support, USS Gudgeon would be at Point Lima at 2200. He sent a message to Gudgeon via Honolulu radio to move her patrol thirty miles west of Lima.

While the guns were booming away, David heard the Radarman call out a surface target bearing left five degrees, range six miles. He stepped over to the Radar display. "Radarman, what do you think it is?" David questioned.

"It is not an airplane, sir. Looks like a surface target."

"Yes, coming right at us, at our smoke and gun flashes. It probably came out of the harbor entrance, could be a patrol boat, a destroyer escort, or even a sub." Pelham said, "Ops, lob a couple of flares over a surface target dead ahead, range 12,000 yards. See if Dumbo can engage."

"Copy, Wilco." Lt. Wolfe answered.

David moved back onto the Bridge next to his chair. Two flares ignited under their small parachutes several miles in front of Selfridge. Smitty reported no contact. "Ops, tell Dumbo that's our highest priority target. It could be a sub. I have the conn. Helmsman, put your rudder left thirty degrees, steady on course 300.

The Helmsman responded, "Aye sir, rudder is left thirty degrees, new course three-zero-zero."

"Very well. JOOD, go out on the wing and use your glasses on surface contact seventy right at 11,000 yards. David transmitted on the 2MC microphone, "This is the Captain, all guns check fire, forward battery fire two more flares at bearing right seventy degrees, range 10,000 yards when we steady on course 300. All other guns load a mixture of contact and delayed projectiles."

Wolfe asked his Captain, "Skipper, do you want more speed? We're at one-third plus twenty."

"No, this is fine. Radarman, report your contact!"

"Sir, contact bearing 025, range five miles, course 230, speed twenty knots."

"Very well, it still could be a sub on the surface. Ops, rapid salvo fire forward battery, bearing right thirty degrees, 9,500 yards, ladder up 100 yards for three salvos, except one mount fires a flare every thirty seconds."

On the 2MC speaker, they heard, "This is Dumbo. I think I see a bow wave bearing right thirty degrees at five or six miles. I see splashes in front of it. Lots of splashes right on top of it. Check fire, check fire!"

"Radarman report," David commanded.

"Captain, we got blossoming when the shots were hitting. Just a second for the sweep. No, there is no contact, surface contact sunk, sir!"

There was cheering on the Bridge from the crew until Captain Pelham told them to settle down.

"Helmsman, set your rudder thirty degrees left, steady on course 260, add fifty turns to our speed. Ops, we're running in case it was a sub that dived. He's between us and our next

targets, but he can only go nine knots max on batteries. We'll circle around him to the west and work our way back in to fire at the harbor. He can hear us, but he can't catch us."
"Are you sure it was a sub, sir?" Wolf questioned.
"Pretty sure, it's just too pat that we fire a few shells, and a ship, even a small ship, completely disappears. JOOD, did you see anything other than our shell flashes? Any secondary fires or explosions?"
"Negative, sir."
"Ops call Dumbo and ask him."
Wolfe reported after a short delay, "He says no. He only saw our shots hit the water. Dumbo reports that this higher speed is uncomfortable. I think the waves whipsaw him on the end of his tether."
"Noted. Helmsman, reduce speed by twenty turns, set your rudder ten degrees starboard, steady on new course 350."
"Skipper, we're running out of time; it's 2115. We can't attack the northern targets, reel Dumbo in, and recover the boats before 2200," Lt. Comdr. Wolfe warned.
"Ops, you're correct. Reel Dumbo in until we get in firing position, then relaunch him. Increase speed to All Ahead Full once Dumbo is secured. You have the conn, Ops. The fast boats can keep us with us. Attack the north targets, recover the fast boats, and secure from GQ. I'm going to rerun the carrier intercept numbers."
Wolfe responded, "Understood, I have the conn."

Selfridge returned to six miles offshore. She launched Dumbo, bombarded Wake Harbor, and the northern gun emplacements while cruising north. After reeling in Dumbo a second time and recovering the fast boats, she sprang forward at flank speed. David had Ops, and Lt. Gehres put together an after-action report for approval and release by the XO. He went to his At Sea Cabin, used the head, and sat on his bunk for a few minutes collecting his thoughts.
Back on the Bridge, David took a deep breath and spoke to the crew on the 1MC. "Men of Selfridge, we took the fight to the enemy tonight. We must have knocked out the airfield and several planes, or they would be harassing us right now. Kudos to our gunners, the Bridge crews today, and our fast boat crews. We had a few shells lobbed at us, but they never knew where we were thanks to Dumbo and her superb pilot, Yeoman Smitty Smith. Now is the time to grab some chow and get some rest. We're going to be racing north for five hours to intercept an enemy battlegroup.

"Our goal is simple yet highly dangerous. We plan to get into an ideal firing position on a Jap aircraft carrier and sink her. We're going to hit her with sniper fire, mortar rounds, and all of our ship's guns simultaneously. We will set her planes on fire and blast away at her waterline until she sinks. Then, while it's still dark, we'll run for the hills and sail back to Midway. I see Steward Jackson has brought me some sandwiches here on the Bridge. Thank you, Jackson. I plan to have a bite and take a nap. I recommend that you men do the same. Well done, Selfridge. Captain out."

While he was finishing his sandwich, Lt. Comdr. Haldane joined him in the Bridge. He nervously asked David what their chances were to transit to Midway safely. David told him they had a good plan and showed him Simpson's overlays for the carrier attack. They both reviewed the tracings for the Transit Phase at High Speed; Intercept and Shadowing Phase;

Screen Penetration and Masking; Attack Approach and Commitment; Attack Timing and Weapon Sequencing; Disengage and Retire.

David told the Engineer that Simpson would be providing the same details to an ad hoc Command Center at CinCPac Fleet as Selfridge prosecuted the attack. He said Nimitz and his staff would evaluate the ship's coded radio report sent on the Wake Island raid and have Lt. (jg) Simpson step them through the phases as they were happening. Selfridge would transmit nother message after the carrier attack. Simpson's overlays for the carrier attack identified the units involved, their call signs, and functions. There was no PBY aircraft support during the final phases to prevent Orange Force from believing it was under surveillance. It would be up to the submarine scout to report the last position for the carriers before Selfridge's Intercept Phase.

The Engineer said all was well with the ship's power plant before bidding Pelham good luck and goodnight. David once again reviewed the overlays. Before hitting his rack, he memorized the units involved, their call signs, and functions.

Steward Jackson woke Pelham at 0200. The ship was in material condition modified Zed. The Bridge crew gave as many crewmen as possible time to rest and sleep. The watch bells were silent until 0230. David used the head, brushed his teeth, and noted he needed a shave. *No time for that now*, he thought as he sat on the bunk and sipped the hot coffee provided by Jackson. *Men will be waking up all over the ship, and breakfast is at 0230. GQ is scheduled at 0300 unless we make contact earlier. We'll use lookouts, Radar, and Sonar to find the Japs. Funny, the watch Radarman challenged my evening standing orders to sweep twice every ten minutes. They taught him how to use the equipment in A School, but not that it can be detected at more than double its range against large targets! Our big decision is whether to launch Dumbo before contact or not. Smitty should see the bow waves and wakes several miles out and direct our screen penetration, but then we would have to reel him in to slow down and launch the Firefly fast boats. Once we get Dumbo flying, all the practice with the Lexington Battle Group should help us analyze the formation and find the carriers. Akemi is our wild card. He talked those two Bettys into a fly-by with his flashing light. Hope he can save our skins if we're challenged.*

There was a knock on his door. The XO informed him that Gudgeon had reported Sonar contact with Orange Force. Gehres, Ops, and the XO had just checked and refined the intercept solution. Selfridge should see the lead Jap destroyers just after 0300

David was sitting in his skivvies, sipping his coffee. "Outstanding, XO! How are you holding up? I feel guilty not talking to you in person for twenty-four hours."

"Skipper, I'm doing fine. I put a mattress in Secondary Conn and had a fine rest until Jackson brought me my coffee before yours."

"Good. I'm sitting here puzzling whether to launch Dumbo before we make contact or not."

Lehman said, "Ops and I just went over that on Simpson's plan. The consensus is that we can penetrate their destroyer screen without Dumbo and his flying speed requirement."

"Understood XO, thank you. It strikes me that Ops and I took the conn away from the OOD much of the day and missed some of the drills. When we slow down, expect a shift to Secondary Conn. You can do a shift steering drill using the Helmsman for a few minutes in the steering engine room. Carry on, XO. We have a carrier to sink!"

"Yes sir, the crew is with you come hell or high water!" Lt. Comdr. Lehman exclaimed.

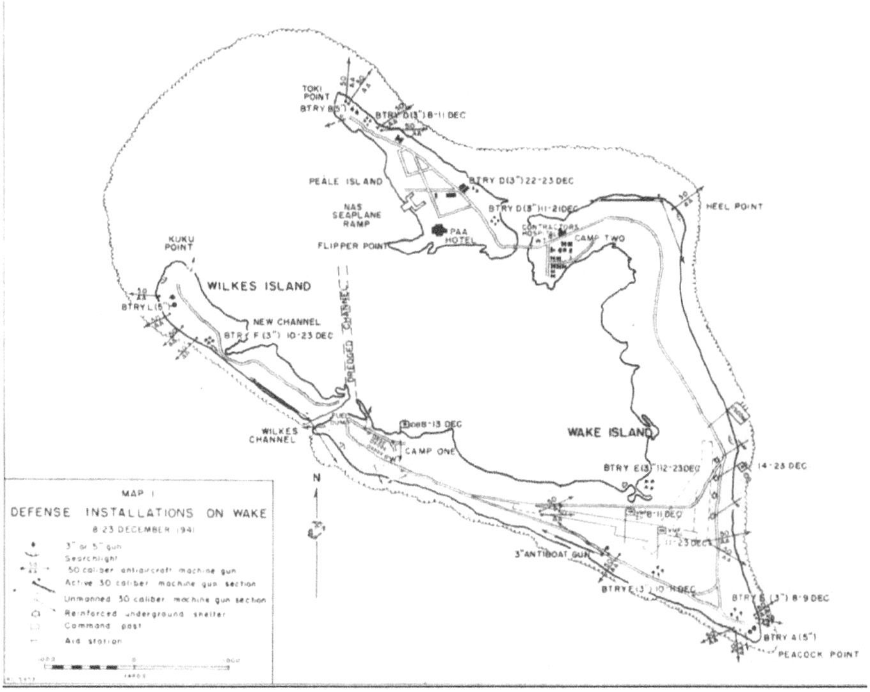

American installations and defenses taken over by Japanese 23 December 1941. Larger guns were destroyed before capture, airfield on lower section of Wake is oriented 280 and 100 degrees (unfinished portion in dotted lines)
Source: Marine Corps history of Wake Battle

Japanese aircraft carrier Shokaku

Source: Wikipedia

Chapter Thirty-three

PELHAM'S REVENGE – "Fighting 357"

May 1942

Captain David Pelham put on a working khaki uniform and retrieved his shoulder-holstered Colt M1911 .45-caliber pistol from the desk safe in his Inport Cabin. He briefed Nagy and Akemi on what he needed from the Signals Bridge during the attack. Dr. Lawson was already below and arranging an operating area in the Wardroom to receive wounded men.

David decided to have breakfast on the enlisted mess deck. He caused quite a stir as he talked to many of the men and sat eating with several of the gunner's mates. He knew the word of the Captain wearing his .45 pistol would travel around the ship. Each Division had the fruit checkpoints diagram leading to the Point Peach firing position. Captain Pelham saw a copy posted on the Mess Deck as he headed for the Bridge.

ORANGE - first contact with enemy battle group
LEMON – planned launch of Dumbo
LIME – cross wake of protecting cruiser
BANANA - last mile to engaging Jap carrier
STRAWBERRY – last half mile, game is on
PEACH – quarter-mile to carrier target, sink 'em!

The Midwatch Bridge crew was conning the ship at forty knots into calm seas and inky blackness. Lt. Comdr. Lehman, Lt. Comdr. Wolfe and Lt. Kindall were in Bridge Plot reviewing Simpson's diagrams for attacking Orange Force. They had a tracing for every possible approach to the enemy formation, from dead ahead, the left or right flanks, or astern. At 0245, Wolfe estimated that they were twenty miles east of the carrier group and had a closing speed of forty miles per hour. They all agreed to hold off on launching Dumbo until they were inside the enemy destroyer screen.

Simpson had penned that one penetration tactic was barely making way and letting the Jap formation pass through them. They would see the bow waves and wakes of the destroyers but be nearly invisible to the enemy as Selfridge threaded their way between them. David made sure that everyone understood he'd only use the Radar as a last resort. The Japs didn't have Radar, but nothing stopped them from receiving Selfridge's transmissions, raising the alarm, and determining the bearing to Selfridge. He wished each of his officers good luck, shook their hands, and went into the Bridge to his Captain's chair.

"OOD, decrease speed to All Ahead Full, in five minutes to Two-Thirds, and another five minutes to One-Third."

"Aye, aye, sir."

"JOOD, double up the forward lookouts at 0250 and personally brief them on reporting bow waves and wakes. Completely darken the Bridge at 0300 or when we make first contact. Call away General Quarters at 0255."

"Yes sir!"

Here we go, he thought, *For better or for worse. We need a bit of luck to pull this off. No, we need a ton of luck. The Japs aren't expecting us, that helps. We need to be the Apache scout that Layton talked about, sneaky, and hard to detect. If they see us, it will be nip and tuck to escape at high-speed. Their cruisers can light us up with flares and shoot us out to fifteen miles. The cruisers or carriers can put planes in the air to track us and attack for two hundred miles. Our high-speed wake is a dead giveaway to the Jap ships and aircraft. It might be better to retire at low speed and play the Operation Nippon card if required. Damn, I'd like to go over these choices with Simpson. Wolfe is talented, but Simpson can visualize these tactical situations better than us.*

At that very moment, Lt. (jg) Grant Simpson was entering the first floor of the Pearl Harbor Administration Building, accompanied by Lt. Comdr. Edwin Layton. Hawaiian time was 0340. They had set up a jury-rigged Command Center the night before in the administration's open bullpen area. There was a Command Desk for the Admiral, a Carrier Desk, a Destroyer Desk, an Intelligence Desk, and a Communications Desk. Several blackboards were along the outside edges of the desks, and Simpson had a projector to display his tracings. Nimitz and the other officers were expected at 0400 when Selfridge should be making the first contact with Orange Force. The Admiral's steward and valet arrived with a tray of sweet rolls and made coffee. Layton went over the format of the initial brief with Simpson:

OPERATION SLINGSHOT INTRODUCTION (Admiral Nimitz)

March briefing to Admiral King.

April briefing to CinCPac Fleet Admirals.

Actual Blue Forces, USS Selfridge DD-357, USS Neosho, AO-23, USS Gudgeon, SS-211, PBY Catalinas (3).

Notional Blue Forces, USS Lexington (CV-2), two cruisers, USS Selfridge, and 4 destroyers, USS Neosho (fleet tanker), 1 troop carrier, and 1 supply ship.

Estimated Orange Combat Force, 2 aircraft carriers, 2 cruisers, 5 destroyers.

Estimated Orange Service Group, 2 tankers, 2 troopships, 2 freighters, and 3 destroyers.

MISSION OBJECTIVES (Lt. Comdr. Layton)

Surface gun Attack on Wake Island to disrupt flight ops.

Infiltration of Orange Battle Group Force and surface attack on an aircraft carrier.

Confuse Orange Forces on the safety of operating in Wake area.

COVER AND DECEPTION PHASE (Lt. Comdr. Layton)

Leak plan to reoccupy Wake using a single carrier.

Ships in the leaked plan known to be at sea.

Let troops in April exercise near Johnston Island think it's a rehearsal for landing on Wake Island.

ATTACK ON WAKE AIRFIELD (Lt. Comdr. Layton) NOTE: All times are Hawaiian Time Zone

Completed by Selfridge and modified C-46 Commando aircraft last night at 2240 Hawaii time.

Selfridge radio report at 2245 claimed destruction of two medium bombers eighty miles south of Wake at 2025 our time.

Selfridge claims Wake airfield was temporarily disabled with many runway bomb craters by 2210. Several aircraft destroyed and much of the gasoline storage was burned.

Selfridge claims possible sunk patrol boat or damage to sub that submerged on the west side of Wake at 2230.

Selfridge claims destruction of one H8K flying boat and damage to another while shelling the Wake Seaplane Base.

Selfridge claims destruction of two-gun sites and damage to two others.

Selfridge shelled the Wake port and command center, damage unknown.

TRANSIT TO INTERCEPT ORANGE FORCE (Lt. Comdr. Layton)

2255 - Selfridge starts high-speed transit at forty+ knots for five hours.

0030 - USS Neosho transmitted fake Carrier air operations recordings 300 miles southeast of Wake.

0130 - USS Gudgeon reported Sonar contact on Orange Force heading east at latitude north 22 degrees 45 minutes and longitude east 164 degrees 53 minutes.

INTERCEPTION AND PENETRATION OF ORANGE FORCE (Lt. (jg) Simpson)

Expected Orange Force formation – two carriers side-by-side in the center of the force. One cruiser outboard of each carrier. Three destroyers in the forward screen and one aft on each flank. All ships are about one-half to one mile apart.

Penetrating the destroyer screen.

Launching Fireflies and Dumbo.

Crossing the cruiser's wake.

Identification of carrier target.

Attack approach to carrier target.

Detected – signal subs ahead, attacking subs, press the attack.

Detected and engaged – dash through Service Group ships.

Not detected, reaching Point Lime, reel in Dumbo. Proceed to Point Banana (one mile to target)

Point Strawberry - Selfridge committed to the attack, range a half-mile from target carrier.

Point Peach - Ideal firing position, range a quarter-mile from target aircraft carrier.

SEQUENCE OF CARRIER ATTACK (Lt. (jg) Simpson)

Sniper platforms raised at Point Strawberry.

Sniper guns free after reaching Point Peach– plinking aircraft and fuel bowsers on flight deck

Fire blinding rockets, cover eyes, cover eyes, warning to the crew.

Four mortars fire at flight deck after the double flash. Rate of fire is thirty-six rounds per minute times 3 minutes and 104 hits using a fifty percent hit rate.

Selfridge's eight five-inch guns engage carrier's heavy guns (4 double 5-inch mounts, two forward, two aft).

Bofors and .50-caliber guns rake carrier's AA guns and gun directors.

Carrier's five-inchers spiked, Selfridge's aft five-inchers fire on AA guns and catwalks.

Selfridge's forward five-inchers fire on carrier waterline, fifty percent contact shells, fifty percent quarter-second fuze delay.

End of minute 1 - Carrier's AA guns spiked, Selfridge's aft five-inchers fire on waterline, same fusing.

End of minute 2 – 100+ shells into side of carrier, half punch through two or three bulkheads before exploding

End of minute 3 – 200 shells into side of carrier, she should be listing with her hangars and flight deck on fire.
End of minute 4 – seventy-two more shells into the side of the carrier, she should be listing and burning.
Minute 5 – Complete destruction, retire toward Jap service group. Show Freddy to cruiser, play sound effects, and get away from the sinking carrier.
RETIRE AND ATTACK SERVICE GROUP (Lt. (jg) Simpson)
PLAN A – Possible attack on the cruiser harassed by Fireflies. Then attack service group. Retire northward and recover Firefly crews.
PLAN B – High speed run to west, attack service group, retire to north and recover Firefly crews
PLAN C – High speed run to north and then east and recover Firefly crews
If still afloat at dawn, continue east and set condition Nippon.
QUESTIONS AND DISCUSSION (Lt. Comdr. Layton, Lt. (jg) Simpson)

Over 2,000 miles west of Oahu, Commander David Pelham was mulling over the least dangerous path to penetrate the Jap destroyer screen. Selfridge was at All Ahead Slow, and Sonar was deployed. Both arrays had loud rhythmic ship engines and prop thrashing sounds to the west. The Bridge Plot Sonarman said it sounded like the whole Jap Navy was coming at them and were about ten to fifteen miles in range.
"Yes," David said with a big smile, "They're sailing right into our trap, the poor buggers."
He went back to his Bridge chair and thought for a moment. "OOD, I have the conn. Raise Sonar. Helmsman, set your rudder left forty degrees, steady on course 170, All Ahead One Third. JOOD, tell Engineering to be ready for emergency speed and keep the stack smoke at a minimum. Ops, Assistant Nav, OOD, we will come at them from the south. We either slip between the two starboard flank destroyers or fall behind the aft one.
Wolfe selected the relevant attack tracing and placed it on the red-lit light table. "Ops copy, Skipper."
"OOD, use the 1MC, remind the entire crew that the smoking lamp is out, and we're at darken ship. Helmsman, All Ahead Two Thirds."
David picked up his phone and punched the SEC CONN button. "Secondary Conn, XO speaking."
"XO, Sonar had them on our nose at ten or fifteen miles. We're going to penetrate their starboard flank, between the two destroyers, or circle aft of the rear one. If anything happens to me, remember that. If we don't get to Strawberry, we run. We don't slug it out with any unit except an attacking aircraft. Firing our guns gives them an excellent target. Good luck to us all, Captain out."
After running a couple of miles south, David brought the speed down to All Ahead Slow and turned west. He ordered the bubble machine tested. The JOOD reported that their wake was dark and hard to identify. After securing the anti-plankton pumps, he thought, *This must be how a bullfighter feels when he steps out in the ring to face a belligerent young bull. He knows what the beast is capable of, but not exactly what he'll do. We don't know what we'll see when the first ships are sighted. They could be coming right at us or zig-zagging away and out of sight. The ship is ready. The crew is spoiling for a fight. I'm prepared. No more*

coffee, don't want to be too keyed up. We need to deploy the sonar again and see if the bearing is drifting starboard as they pass north of us.

David stationed himself at the Sonarman's plot. The sounds were very similar but not as loud. What could that mean? Pelham decided the ships in Orange Force were slowing down because they had over three hours to reach Point Lima and launch aircraft at dawn. *A mixed bag,* he thought. *It makes them harder to see but easier to intercept without detection.* The Sonar Petty Officer couldn't give him a precise direction on the carrier screw sounds. There were just too many ships plowing through the water. The bearing to the loudest sounds was drifting starboard to the east. Akemi on the Signals Bridge saw the faint bow waves of the lead ships that were probably destroyers.

The 2MC Talker sounded off, "Signals reports two ships, bearing right seventy degrees, range estimated 8,000 yards and 10,000 yards. Course estimated easterly, speed estimated at twelve knots."

"Very well. OOD, turn on the bubble machine. If we can see them, they can see us. Talker, tell the crew we're at Point Orange."

Ops brought over the tactical plotting board showing Selfridge and the two contacts. They both knew that there should be a third lead destroyer, either on the other side of Orange Force or nearer to Selfridge – perhaps coming right at them.

"Talker, ask Signals for a recommended course to breach the destroyer screen," David directed.

After a brief pause, the 2MC Talker said, "Signals recommends three-fiver-zero, sir!"

"Very well. Helmsman, right standard rudder, steady on 350, add twenty turns to our speed."

Ops was out on the Port Bridge Wing scanning for the missing destroyer that should be covering the rear right flank of Orange Force. Selfridge was either about to pierce the formation or stumble directly into the force's core, consisting of two cruisers and two carriers. Sonar had retracted their sensors when he increased the ship's speed. Radar was ready at the standby - if he wanted to chance Jap detection of the transmissions. David went over to the unlit Radar Plot.

"Radarman, get ready to make three sweeps. I'll mark the large targets ahead. You will concentrate on any contacts astern that might be a destroyer. Transmit now."

Pelham plotted two contacts thirty left at five miles, two more beyond those, and the two van destroyers at ninety right about three and four miles. The Radar tech reported no ship contacts behind Selfridge for ten miles or more. In three rotations of the Radar antenna, they could not estimate the course and speed of the ships. Ops came over and copied the relative locations of the ship echos onto the tactical plot board. David took the board back to his Bridge chair to ponder the surface contacts and his ship's next move.

After a couple of minutes, Wolfe joined him. "Skipper, what do you think?"

"It's puzzling, Ops. There's no destroyer guarding the right rear flank of the formation." Pointing at the two targets that were close together at five miles, "I think for whatever reason, the missing destroyer is alongside the formation's starboard cruiser. When she peels off to resume station, her course will take her right through our current position. We need to move further south and hope she doesn't spot us. Helmsman, left full rudder, steady on course 190, decrease speed to All Ahead Slow."

"Skipper we're burning through our darkness time allotment. We need to launch the Fireflies and Dumbo," Wolfe warned.

"I hear you, Ops, but what if the cruiser is so preoccupied with the destroyer that we can cross her wake without being seen? It could be our golden opportunity to pass behind her. We were going to put Dumbo up to find the nearest carrier. I think we painted it on the Radar about a mile north of the cruiser. No alarm was raised so that we can radiate again. We can't launch the fast boats with that destroyer near the cruiser. Their job was to distract the cruiser while we cross her wake and start with the spooky stuff to put the Japs off their game. Our boats will close on the cruiser and harass her when we attack the carrier. We can cross the wake now or wait for the destroyer to clear, launch the Fireflies, and then cross the wake. The Radar showed no ships to the south or out west. Who knows where the Service Group and her destroyers are? They should be further west, behind the carriers and cruisers. We head further south to get out of the way of that destroyer. Then head north to cross the wake at three or four miles behind the cruiser that's heading east. A cruiser lookout with binoculars will see a dark shape pass through the wake, or not. We could use a little luck on that one."

The 2MC Talker reported, "Captain, Signal Bridge reports bright bow wave and wake bearing right 170 degrees relative, range estimated 10,000 yards. Signals estimates the contact is turning starboard and picking up speed."

"Very well. Helmsman, right fifteen degrees rudder, new course 210, reduce speed to just making way, tell Engineering to make the least amount of smoke possible while keeping the steam up. Shut down the evaporators and all nonessential equipment. OOD and JOOD, go out on the wings and confirm the bubble machine is working on the wake. Ops, get on the 2MC and tell the crew that a Jap destroyer is closing our stern, set silent ship, darken ship and extinguish all red lighting, no one moves about the ship, no one talks, secure all fans and equipment now."

After Wolfe passed the word on the 2MC circuits around the ship, David punched the Guns button on his phone, and Lt. Kindall answered at the Gun Director's Station.

"Guns, this is the Captain. Lower all guns and get the small gun crews off the deck and into hiding. You have about five minutes to make us as invisible as possible. Turn your red-lit instruments off and cover the director lenses with canvas. Even if this destroyer shoots at us, we won't fire back. Captain out."

"2MC Talker. Tell the lookouts to sit down and not move. No one uses binoculars. The Bridge will take our contact reports from the Signal Bridge. No one moves, no one talks."

David called the Signal Bridge on his phone and talked to Nagy. Michael reported that the closing Jap destroyer had completed her turn to the southeast and came right at Selfridge at twenty knots. David reminded Nagy to have all lights out, including instruments, not use binoculars, and report the contact every minute. David knew that the gun crews would be chafing under his order to hold their fire. He called Guns again and told him if Selfridge was on fire, the gun crews could shoot back. Pelham reminded Lt. Kindall that the Jap cruiser five miles away could make mincemeat out of Selfridge if she spotted her gun flashes. As he hung up, Signals reported the destroyer at 8,000 yards closing rapidly.

David thought, *This is it. Ten years of training in the Navy has brought me to this do-or-die moment. Just like throwing dice for an outcome in a wargame - the destroyer will see us or not. Shit! She may already know we're here! They don't have Radar, but we may have shone a light, or some obscenely glowing plankton may have alerted them. They may even be expecting us. Too many people knew about our plan before we left the base. Too many men go into bars and tell secrets. The Jap Captain could be closing on us right now with all of his guns aimed and ready.*

David felt sweat trickle down his spine and knew his armpits were stained with uncontrolled perspiration.

The 2MC Talker relayed, "Signals reports destroyer at 6,000 yards, closing rapidly. They estimate it will pass a half-mile to port."

"Very well. Helmsman, add ten turns on the port engine, keep your rudder amidships. We want to present the smallest possible stern silhouette to her.

Yes, he thought, *And we would all like to live through the next five minutes, too. Thank the Lord for a moonless night and no thunderclouds that would give away Selfridge's position. She was a black shape, slowly moving through a black ocean with a black sky as a backdrop. Lord, give us some low clouds on the horizon so we aren't blocking the stars. When would I fire if I was the Jap Captain? Probably at two miles. Fire off a couple of star shells to ID the enemy and pump him full of H.E. before he can fight back. We've no credible armor anywhere on the ship. They can shoot right through us like in a Popeye cartoon. Destroyers are a dime a dozen.*

"Signals reports destroyer at 5,000 yards and closing," reported the Talker.

David could feel his mouth going dry and his body tensing. He prayed to God and Jesus for his ship and her crew to survive. The tension in the Bridge crew was stifling. No one spoke except the Talker. Without ventilation, the temperature was rising rapidly.

Lt. Comdr. Wolfe came over to David, saluted him, and shook his hand. "It has been a pure pleasure serving with you, Skipper. In for a penny, in for a pound."

David looked at Wolfe and nodded his head. They both knew that it would be soon if the screaming shells were coming to destroy Selfridge. The Talker called 4,000 yards and that the contact was turning slowly to port. David slapped his chair arm in relief until he had a fearful stabbing thought. The other Captain could be turning to unmask his aft guns and fire the most effective torpedoes in any Navy at Selfridge. David desperately wanted to unleash Kindall and his weapons. He almost pushed the Guns button on his phone.

Instead, he calmly directed the Helmsman to add ten turns on each engine. "Ops, he's going to be a mile away. If he doesn't shoot us now, he didn't see us. We stay in this hiding mode for one more minute. Then we need to get back on our horse and push on to Point Lemon. What do you recommend?"

"Sir, swing back north, aiming for four miles aft of the cruiser, launch the Fireflies when we're halfway there, use the Radar again to ID the target carrier."

"Make it so. Ops has the conn. OOD, secure the bubble machine until we're within five miles of an enemy ship. I'm hitting the head." In a louder voice, "When I return, all hands on the Bridge are authorized to use the Captain's head! Good job men. Well done!"

Pelham went into his At Sea Cabin, closed the door, and sat on his bunk for a few seconds. Selfridge's Captain felt drained and exhausted. He went into the head, used the toilet, and splashed water on his face. David cleaned up his armpits and put on a fresh T-shirt and khaki shirt. Upon his return to the Bridge, he could see that Wolfe had them heading north at One-

Third power. The three fast boat crews would be ready to launch. The XO and Ops had briefed the crews on their roles as the ship departed from Wake.

David had talked to a couple of the men on the mess deck earlier. Giles and his Marines had blackened their skin and wore dark blue jumpsuits. Other than the Amjaps, every man on deck who manned the light guns, the sniper pads, the mortars, the boat crews, searchlights, and lookouts was also camouflaged. They were itching for a fight and walked about the ship with a swagger in their dark clothes.

David took the Plotting Board over to the Radar Plot and ordered four Radar sweeps. Ops joined him, and they could see Radar returns on the center and starboard flank destroyers, the starboard cruiser in the formation, two large targets north of the cruiser, and the rear destroyer five miles north of the port flank. There was no sign of the port cruiser or the lead port destroyer. They decided they were masked from the Radar by the two carriers. Lt. Wolfe secured the ship from silent operations. As they approached Lemon, David called the XO.

"XO, how are you faring back there?'

"We're doing fine, Skipper. We were pretty puckered up with that destroyer bearing down on us – too close for comfort."

"I couldn't agree more. That tin can is now on Orange Force's starboard rear flank and is about five miles away." David felt Ops slowing the ship, "We will be launching the fast boats shortly, and then cross the cruiser's wake. Orange Force is north and northeast of us, heading east. We just got an update on the Radar and believe we have the starboard carrier in our sights bearing right ten degrees at 10,000 yards. There is no sign of the Service Group ships and their three destroyers. I bet the Battle Group Commander expects them to catch up when he turns west to launch aircraft. Captain out."

Lt. Wolfe spoke loudly, "2MC Talker, call away the fast boats and tell all stations we're at Point Lemon."

Selfridge reached Point Lime and crossed the enemy cruiser's wake without incident. Word was passed around the ship that they were making their approach to the enemy carrier. A new Bridge crew arrived to assume their 0400 watches. Ops quickly whipped them into shape and briefed the current tactical situation. David thanked the outgoing watch, which looked tired and thankful to return to their regular GQ stations. David motioned for Ops to come over to his chair as he called the XO on the 1MC.

"Secondary Conn, XO speaking."

"XO, I've got Ops here. We can only attack the Service Group if they're a handful of miles behind the Orange Force vanguard. I think that if they're lagging by ten miles or more, we drop them from our plan. My thinking is we retire by heading west. We run north at high speed at ten miles separation from the main force and no contact with the Service Group. They may send a destroyer after us, but I doubt it. They've nothing protecting their rear flank, and Gudgeon may get off a spread of torpedoes to confuse their thinking.

The other carrier or the cruisers may send an aircraft at us, but we'll pick them up on Radar and be ready. We make our mad dash to the east when we're about twenty miles north of Orange Force. We pick up the fast boat crews in about twenty miles of eastward sailing and should be ninety miles from the carrier at dawn. The tanker will be waiting for us 200 miles northeast of Wake."

The XO replied on 1MC speaker, "Yes sir, thanks for keeping us in the loop."

Selfridge was three miles from Point Banana - the last mile to the carrier. David left the conn with Ops, who was sailing an intercept course on the closest aircraft carrier. None of Orange Force was zig-zagging, and it had slowed to five knots. The ocean was still dark as the inside of an unlit coal mine and a thin overcast blanked out the stars.

Akemi called the bearing and range to the target's center by keying off her bow wave and glowing wake. Ops took the ship closer to the carrier at All Ahead One Third minus some revs. He knew that the Fireflies were confusing the escorting cruiser with randomly blinking blue lights to the southeast. The lights were swiveling eight-foot masts and had sheet metal shields to shroud them from the right flank destroyers. Akemi was ready to respond to a cruiser challenge during Selfridge's approach to Point Banana. David was in Bridge Plot, going over Simpson's attack tracing that diagrammed the starboard flank approach and the sequence of weapons employment.

At the Pearl Harbor Admin Building, the Slingshot Command Center was manned. Halfway through the Slingshot briefing, Captain Arthur Davis, the CinCPac Aviation Officer, interrupted Lt. (jg) Simpson with a tirade.

Davis was sitting at the Carrier Desk with two pilots. "I can't believe we're all up in the middle of the night to hear such unadulterated BS! A destroyer sinking a carrier? Can't be done. I'm going to applaud Selfridge for her apparent success at Wake. If that Skipper is smart, he better shitcan this pipe dream and haul ass for Midway!"

Simpson responded, "Yes sir, it seemed ill-advised and far-fetched to all of us the first time we heard Skipper Pelham describe the operation. A surprise attack on a Jap carrier by a destroyer gained some acceptance over time because it was such an outrageous gamble - because it would be completely unexpected."

Captain George Stewart spoke up from the Destroyer Desk. "Captain Davis, we all had our doubts at first, and we may lose an outstanding ship. I know you're a betting man. I'm going to bet you your favorite bottle of bourbon that you'll be far less skeptical after this brief."

Captain Davis was a tall, lanky officer with a stern demeanor. He loosened his tie, unbuttoned his shirt, and scratched his neck as he replied with a brief smile, "Captain, you're on."

Simpson briefed that Selfridge was committed to the attack if she reached Point Strawberry, half a mile from the carrier.

Davis butted in again. "Now, how are you gonna do that? Someone is going to see you and start shooting, or at least ask you to ID yourself with a blinking light."

"Yes sir, we have a Signalman from the IJN onboard that will buy us some time," replied Simpson.

"From the Jap Navy?" Captain Davis said with disbelief.

"Yes sir, a Second Class with eight years of experience. A Christian who has decided to fight for us."

Davis seemed confused, "An IJN signals guy who fights for us. How would anyone find such a person?"

Nimitz joined the discussion. "Arthur, you are going to hear a lot of surprising facts. If I hear an untruth, I'll point it out to you. We don't have time to debate every point in the briefing."

"Yes Admiral, thank you."

Simpson continued, "Captain Davis, I want you to envision yourself on the Bridge of this Jap carrier as Selfridge joins a quarter mile on your aft starboard quarter." He pointed at the two officers with the senior aviator. One of you pilots is on the Flight Deck, and the other is in the lower Hangar Bay. The sequence of attack has repeatedly been reviewed and revised. It's now set in stone. At Point Strawberry, we're committed to the attack and raising the sniper platforms."

Captain Davis interrupted, "Excuse me, sniper platforms?"

"Yes sir, two ten-foot platforms on hydraulic lifts, each one with a Marine sniper and spotter team covered in gray camouflage netting. One fore and one aft, thirty feet above the deck. At Point Peach or one-quarter mile from the Jap carrier, the Skipper will call guns free for the snipers who will start plinking aircraft and fuel bowsers on the carrier's Flight Deck. They each should hole eight to ten fuel tanks in twenty seconds, causing the avgas to stream down on the deck in several locations. At that point, Ops fires the blinding rockets."

Davis was perplexed. "What in the hell are blinding rockets?"

Simpson smiled. "They're custom-made five-foot-long rockets with a white phosphorus payload. We launch them up and over the carrier's island - something we've practiced several times. In the middle of a moonless black night, they leave a trail of sparks that invites your Bridge crew to stare at them. Your lookouts are tracking them. Men working on the Flight Deck are looking up, including the pilot sitting next to you."

"And then," he smacked his desk twice, "Bam! Bam! Two blinding explosions occur. Selfridge's crew covered their eyes, but you and your carrier crew now can't see your hand in front of your face. You're blind and won't regain your night vision for twenty minutes or more. We want to do more than blind you on the Bridge. After the two phosphorous bombs, our forward Bofors and .50-caliber guns will fire into your Bridge, turning it into a bloody shambles."

Davis spoke in a low voice, "I'd be blind. If I didn't know what they were, I would be looking right at them."

"Yes sir, but the party is just beginning. The snipers resume plinking aircraft, and four Marine mortars that have been sighted for one-quarter mile start firing continuously at your Flight Deck."

Davis visibly shuddered. "Mortars are firing at the gas leaking onto my deck and at my airplanes filled with Avgas?"

"Yes sir. The Gunnery Officer will let the mortars fire two or three rounds to get the range. Then he'll direct our four forward five-inch guns at the carrier's forward five-inch batteries and our four aft five-inch guns at her aft five-inch batteries. When Lt. Kindall is sure the carrier's forward five-inch batteries are damaged, he will switch to shelling the carrier's waterline. When Guns believes the carrier's big aft guns are out of commission, he'll continue firing at the triple-A guns in the aft catwalk.

"Our aft Bofors and 50-calibers have been raking the catwalk, the flight deck, and the open area of the upper hangar deck since the rocket explosions. Our forward Bofors and .50 caliber guns started by firing at the Bridge and then shifted to raking the forward catwalks and any targets on the flight deck. Once the carrier's AA guns seem damaged and we're taking no fire, our four aft five-inchers also start pounding the carrier's waterline. Each shell opens a two to three-foot hole, except every other shell has a delay fuse, so it penetrates two or three bulkheads, leaving an eight to ten-inch hole, and explodes in her gut."

Captain Davis held up his hand. "How many shots a minute?"

"Seventy to eighty per minute. Ten shots per gun times eight guns. Probably 160 hits in two minutes."
"One hundred shells from one-quarter mile, my God!" Davis exclaimed.
"Yes sir. Many of those shells will explode in the lower hangar deck. Meanwhile, the flight deck is a living hell with gas burning and planes blowing up. In two minutes, the four Marine mortars have fired 140 rounds. A fifty percent hit rate gives us seventy rounds exploding on the flight deck. The two M2 mortars can blast holes in the flight deck, allowing burning avgas to flow into the upper hangar deck, which should already be on fire from our first shells that missed their big guns.

The Marines and light guns are shooting at any enemy on the island, the Flight Deck, or in the catwalks. At the end of the next minute, we have over 200 holes in her side from our five-inchers. She has a starboard list and will roll over without rapid counterflooding. We expect someone will have commanded more speed during these four minutes. We match her speed, secure the mortars, and lower the sniper crews. Our light guns keep firing, and four torpedoes are at the ready. It's time for the coup de grâce."
Davis asserted, "Torpedoes, right?"
"No sir, we're too close. We speed up and aim all the big guns at her bow," Simpson replied.
"Her bow, really? What will that prove?"
"Yes sir, we rip it open roughly fifteen feet back from the bow at the waterline to twenty-two feet back at the flight deck."
"What do you think that'll do?"
"Captain, she has a false bow, a fabricated subassembly. We think if we can open 100 square feet to the sea and she's above five knots, that will be all she wrote."
"Nonsense Lieutenant, every carrier is built to withstand heavy damage."
There was a question from the Engineering Desk. "Mr. Simpson, can you tell us the loaded weight of this ship?"
"It's 32,000 tons."
The Engineer was using a slide rule. "Oh my gosh! Oh my! This is diabolical."
"Engineering, what is so diabolical?" requested Admiral Nimitz.
"Sir, assume the Jap carrier is moving at five knots, and her engines are cranking on more power. Assume Selfridge opens a blunt ten-foot by ten-foot area on the bow or more. Water is incompressible. All the forward bulkheads and those behind them will collapse like dominoes. She's already heavy with water pouring in from the shelling and maybe counterflooding. She noses down and dives in just a few seconds. The seawater will slam through her like shit through a goose. There will be no recovery."
Captain Davis turned pale and scratched his head. "You're telling me a front-line aircraft carrier, an airwing, and nearly 2,000 men will sink in a few seconds!"
"Yes sir, they will go down like a rock, straight to the bottom!" claimed the naval engineer.
Davis made a growling sound, rushed to a sink on a nearby bulkhead, and threw up.
Nimitz called a head break and addressed the Destroyer Desk. "Well, it looks like you destroyer men are getting a bottle of bourbon."

The heat and humidity on the Selfridge's Bridge were insufferable as the ship approached Point Banana. David ordered all fans and ventilation turned back on. He thought, *Less than a mile to go. We can all see the carrier as an enormous dark shape that blots out the stars on the horizon. She can't hide a small bow wave and a glowing wake 850 feet behind the bow wave – jeez, nearly three football fields long. Focus Pelham, what's next? We have reported Banana to the crew and are closing on Strawberry. We're at GQ, all the guns are manned and ready, and Operation Nippon is in progress. All fifteen Amjaps are out in open areas, prepared to give us a few more seconds of closure if we're challenged. They're manning the searchlights, organized to feed rounds to the mortar crews and have asbestos gloves handy to toss hundreds of hot five-inch spent shells over the side. The rear Bofors and .50 caliber crews are all Amjap, and others will feed ammo to the forward light guns. I've got the Amjap officer imposter on the Bridge carrying a spyglass under his arm instead of the camera. Nagy and Akemi are wearing Jap uniforms up on the Signals Bridge.*

As they approached Strawberry and the point of no return, David felt his body go heavy with dread. The carrier flashed a blinking light at them from high up on its Signal Bridge.

"Signals reports flashing light from the carrier. Signals responding," said the Talker.

David saw the carrier light flash again, and shortly after, there was a flashing light from the cruiser. Ops called Point Strawberry in one minute, and the Talker relayed it throughout the ship.

The 2MC Talker relayed, "Signals says to keep closing. They picked off the correct challenge reply from the cruiser. We may be OK."

Lt. Wolfe spoke first, "Very well, but they sure know we're out here, Skipper. OOD, add forty turns to our speed."

David thought, *A quarter-mile to go to Point Peach, only 500 yards. He felt his thumb starting to twitch. Do your job, Akemi, save our bacon.*

"Ops work your script, I have the conn," Pelham ordered.

"Ops aye. Captain has the conn. Standby to raise the sniper platforms."

The carrier's blinking light started again.

"Signals reports that the carrier wants our callsign," said the Talker.

David responded, "Very well. Tell Signals to use story one and then story two. Tell all Nippon units to prepare for a searchlight challenge. Nippon officer, move out on Port Bridge Wing now!"

"Skipper, we need the snipers up now. We're approaching Peach," Ops cautioned.

"Suspend the script, Ops. Akemi is telling them we're confused about today's callsign because of the Dateline. Then he'll tell them we're from the support group and have a badly injured sailor who needs surgery on his mangled leg. There are Amjaps on our bow and another in a Jap stretcher with blood all over his leg."

He finished the last word as the carrier searchlight came on and swept Selfridge from bow to stern. The carrier crew saw an injured Jap sailor being attended to by two Amjap stretcher-bearers on the bow. Amidships they saw two Amjaps manning Selfridge's searchlights and waving, Akemi and Nagy were wearing Jap uniforms and were visible on the Signal Bridge. The Amjap officer on the Bridge Port Wing saluted with his spyglass under the other arm. Other Amjaps were visible on the weather deck near the mortars as the Marines and sailors hit the deck. All gun barrels were down and not pointing at the carrier. The two Jap flags were flying aft, and the rear AA weapons were manned with Amjaps in Jap uniforms. The light switched off.

David spoke loudly, "Now Ops, now! Helmsman, set your rudder twenty degrees right, match the carrier's course and speed."
Wolfe directed calmly, "Talker, raise the sniper platforms, tell searchlights to stand by, transmit to gunners, aim your guns."
The 2MC Talker replied, "Snipers have targets. Snipers have targets."

Selfridge was turning parallel to the carrier at Point Peach. It was now or never.
David ordered, "Talker, sniper guns free, sniper guns free. Ops continue your script."
"Aye, aye, sir. Talker, transmit standby rockets." After thirty seconds, "Talker fire rockets, fire rockets."

Pelham could see the rocket fuses burning forward on the bow. After five seconds, they both blasted upward out of eight-foot tubes used to aim and fire them. Ops started his clock and ten seconds later broadcast on the 1MC, "Standby to cover eyes, cover eyes now, now, now."

Pelham pressed his forearm against his closed eyes. He still felt and saw the blinding double phosphorus flash.
"Talker, mortars free, mortars free. All light guns free, all light guns free. Ops, illuminate," directed Pelham.
Wolfe spoke to the Talker, "Searchlights free, searchlights free. All guns free, all guns free."

David noted that the light guns worked over the catwalks, and two planes burned on the Flight Deck. He knew that the sniper crews were relentless. The Marine spotters would now shift to killing Japs on the open decks and the catwalks. Then all hell broke loose as Kindall aimed Selfridge's big guns, and the shells were exploding against the carrier's four five-inch, double turrets. The booming noise of the destroyer's weapons and the echoes back from the massive carrier were deafening. There was an immediate explosion from the carrier's forward big five-inchers, probably from powder casings igniting. Mortar rounds were raining down on the Flight Deck, igniting multiple planes and pools of Avgas. Selfridge's aft five-inchers had damaged the carrier's aft big guns. They were now methodically walking shells along the catwalk to destroy all the AA mounts and the men trying to man them. Planes were exploding inside the upper flight deck from five-inch shells that hit near the carrier's big gun turrets fore and aft and pierced the thin steel hull.

As his forward five-inch guns started holing the carrier at the waterline, Captain Pelham checked his watch for the elapsed guns-free time - forty seconds! *Jesus,* he thought, *Only forty seconds, and the carrier's starboard side looks like a wreck. Fires everywhere, including new ones breaking out on the lower hangar deck. Better check the cruiser, check the cruiser! She can kill us. He stepped over to the starboard wing door and put his glasses on the cruiser escort. She was turning away, and he could see some small gun flashes. The Fireflies must have got close aboard and have been harassing her. Wonder if they jammed her steering gear. She can't bring most of her guns to bear on them because they're so close. She could shoot at us but risks hitting her own carrier. And our whole area is covered with thick smoke from the guns and the carrier fires. All our big aft guns are now pounding the waterline. Think ahead, think ahead. What do we want minutes from now?*
"Talker, transmit standby Operation Freddy, transmit Freddy sounds at the cruiser. Ops, the cruiser is opening and being harassed by our fast boats. The destroyers are probably turning to close us, but it will take them minutes to get a fix on us."

Wolfe stepped forward to look for the lead Jap destroyer, "Concur, Skipper, I..."

There was a loud, sharp cracking noise. Lt. Comdr. Wolfe slumped to the Bridge deck with blood gushing out of his neck. Selfridge was taking machine-gun fire from the carrier.

David reacted, "Talker, report machine gun fire, a casualty on the Bridge. JOOD and Messenger put Mr. Wolfe out on the starboard wing. Talker, activate Freddy for five seconds."

The 2MC Talker, after carrying out David's commands, "Casualties reported on the port weather deck and the aft Bofors' crew, sir."

"Very well, Lt. Gehres, you will now act as Ops. Review the current attack tracing."

Keep it calm and calculated, Davy. You knew men could die, that you could die. Stick to the plan to save the most lives. No time for an emotional reaction now. Take revenge for Wolfe. Make that work for you.

"JOOD, elapsed time, please."

"Sir, 150 seconds."

"Very well. OOD, start slowly overtaking the carrier to take our nose shot. Radarman, activate Radar and look for closing ships."

The OOD responded, "Aye, aye, sir. The carrier is making smoke out her stacks."

"Good. Continue overtaking her. A bit faster, please. "Radarman, do you have a picture for me?"

"Radar aye sir. Two ships closing at three miles, one opening at two miles."

Damn, they did jam the cruiser's steering and are showing the spooky blue lights while shooting her decks up. The destroyers are coming for us. What are we going to do? Only two choices, run or fight.

"Talker, randomly show Freddy, transmit crowd sounds at the cruiser and Freddy sounds. Mr. Gehres, plot the Radar contacts on the plot board and bring them to me. OOD, your speed is perfect. Keep overtaking no matter what she does. JOOD, time please."

"Nearly four minutes, sir. Captain, she's taking on a list. The carrier is listing!"

"Very well, OOD, increase our overtake speed. Talker, warn Guns, quote standby for nose shots unquote."

Gehres showed him the Radar plot. There were three blips, probably destroyers closing from ahead and the starboard flanks. Pelham thought, *These guys can shoot at us soon. And we don't know what the units on the port side of the formation are doing! Visual and Radar contacts are blocked by the carrier. We need to get out of here, fast! Of course, the damn carrier is listing. We have put over 200 holes in her. The Japs aren't so good at damage control in the first place. If they haven't started major counterflooding, they're capsizing. Let's rip her nose off and run like a rabbit.*

David said, "Helmsman, warn Engineering we'll need full power in two minutes. Talker, tell Guns, nose shots now, nose shots now!" After a short pause, "Talker, lower sniper platforms now, lower sniper platforms now. Secure snipers."

Lt. Gehres directed the 2MC Talker, "Secure from mortar operations, secure from mortar operations. Secure from searchlights, secure from searchlights."

The Radarman reported, "Ship contacts closing to two miles."

"Very well, Talker, tell Signals we have destroyers closing, use stories three and four, I repeat, use stories three and four. Helmsman, report our speed," ordered Pelham.

"Nine knots sir."

"Very well. OOD, All Ahead One Third in fifteen seconds."
David selected the ALL button and transmitted to the ship on the 1MC, "This is the Bridge. Hard starboard turn in thirty seconds. Hard starboard turn coming." The roaring of the guns was continuous. There was no telling how many of the crew heard his warning.
The Talker sang out, "Damage control reports five dead and nine wounded. Two small fires extinguished. The wounded are being treated below. Aft Bofors and .50-caliber guns knocked out."
"Passing All Ahead One Third, sir!" reported the OOD.
"Very well. I have the conn. Helmsman All Ahead Two Thirds on port engine, set your rudder sixty degrees right, hard starboard turn to course 220."
The big guns were finally silent in the hard turn. David picked up his phone and called the 1st Lieutenant, "Captain here. When can we fix those aft guns? We're going to need them to fight off Jap planes."
"Sir, the Bofors is done. We can bring one .50 caliber back online," replied Lt. Jones.
"Very well, get with Boats and Guns, put more firepower back there. Of the five dead, are you including Lt. Comdr. Wolfe?"
"No sir, sorry to…"
David curtly cut him off. "Send a couple of men to clear his body off of the Starboard Bridge Wing, Captain out."
He turned his attention back to the burning carrier and saw smoke belching out of her stack and the bow slightly down. "Helmsman, say heading passing!"
"One seven-fiver, sir."
"Steady on current heading, All Ahead Flank! Talker, tell Signals to blink the destroyer closing to our left that the American cruiser is escaping, watch our gunfire." Pelham selected Guns on the 1MC, "Guns, Captain. We're steadying, fire all aft big guns at the Jap cruiser; estimate bearing left 100 degrees, range 6,000 yards."

As Selfridge's shells fell around the cruiser, David saw the Jap destroyer turn to starboard and open up her forward guns on her hapless sister ship. He secured his weapons as the cruiser returned fire on one of her destroyers. Signals must have blinked the same message to the two other escort ships as they also started firing at their cruiser.

He hoped the fast boats had time to clear the area, but that was not his biggest problem. He yelled at the Helmsman to ring up All Ahead Emergency as he went out on the port wing and looked aft at the stricken carrier. Judging by the billowing black smoke from her stacks, she was listing further starboard and bending on full power. He put his glasses on the carrier's bow as Selfridge accelerated away from the damaged, gigantic ship.

Come on, not now, not now, wait, wait, wait, wait. Let us get further away. Then David could see her bow crumpling and seconds later heard the terrible crunching, crackling, and shrieking metallic sounds of bulkheads collapsing. Her now blunt bow was digging into the unyielding Pacific Ocean. Thousands of tons of seawater were smashing up into the spaces and passageways. The sounds were like a hundred blow torches cutting steel as she went deeper by the bow, and water rampaged aft through the Hangar Decks. David saw burning airplanes on the Flight Deck sliding forward into the ocean as the hungry saltwater rose over her bow and the carrier's stern lifted higher. Then, with the forward third of the enormous ship sloped down into the water, her four props came out of the water and emitted a screaming, high-pitched turbine wail heard for miles.

Then, with aircraft still burning on the Flight Deck and the hangar decks, she slid forward, down deeper and deeper, until nothing was left. Pelham saw some debris on the surface and burning oil through the glasses. Suddenly, a massive bubble of steam shot up through the ocean's surface several hundred feet into the air. It all happened in less than a minute or two. Captain David Pelham released an animalistic, primal scream of triumph on his Bridge Wing and shook his fist at the carrier's last position.

Selfridge had moved far enough away not to get sucked down by the gigantic ship. The horror of what his destroyer had accomplished swept over him. He felt conflicting emotions. Pelham was amazed at what Selfridge had done, and next, sick to his stomach as the initiator of a massive loss of human lives. And finally, he felt the first waves of sorrow from losing several of his crew - especially Lt. Wolfe.

Get back to work, he told himself. *You aren't out of the woods yet.*

Bofors 40 mm twin-mount AA gun

Source: Wikimedia commons

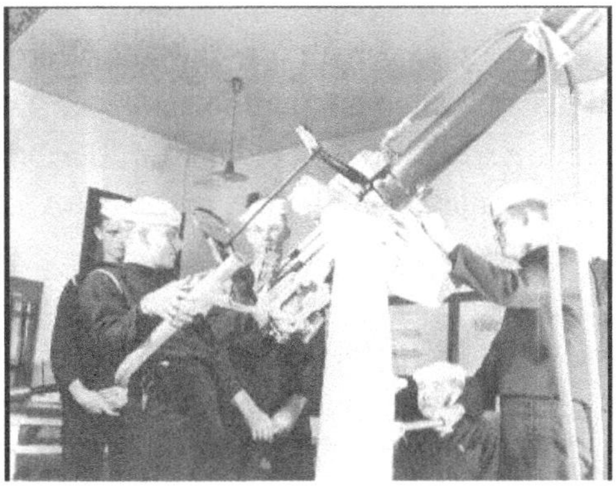

Browning .50 caliber dual-mount machine gun

Source: Navweps.com

Chapter Thirty-four

TRANSIT TO MIDWAY ISLAND - USS Selfridge

May 1942

David Pelham had a cold chill and shuddered as he thought of the Jap carrier sinking. The last thing he saw through his binoculars would stay with him for the rest of his life. As the giant bubble of steam, Avgas, oil, and debris shot to the surface, he saw burning human shapes flung like rag dolls through the night air.

I wonder if this makes me a war criminal. Surely it does in the Japs' books. I'm a goner if they win or get their hands on me.

He re-entered the Bridge, "OOD, dial the speed back to All Ahead Full, you have the conn, come starboard to 270. Radarman, do you have a picture for me?"

"Yes sir, several contacts, none are closing, the carrier is gone."

"Very well. 2MC Talker, tell the crew we sunk our carrier target. Then, Mr. Gehres, plot those ships and bring it to me."

David punched the Signals button on his phone and heard: "Signals, Nagy speaking."

"Michael, this is the Captain. How are you faring up there?"

"Captain, we're holding our own. We see some enemy ships out to the north exchanging signals."

"Yes, I need some advice from Akemi. They all saw the carrier go down, and you got the destroyers and cruiser to fire at each other – good job on that. We're putting up a tremendous wake that we can't hide. I'm thinking of giving them another taste of Freddy to make them run east. What does he think?"

After a few seconds' pause, Michael Nagy responded, "Captain, we've been thinking about it; let us blink at the other carrier and help that along. Then, turn Freddy on in about one minute; make it look like he's coming after them."

"Sounds like a plan, you men OK up there? Did you take any fire?"

"Yes David-san, we had a Signalman striker killed, and I took a metal splinter through my arm that hit an artery. Got a tourniquet on now - hurts like hell."

"Signals, send your message, then get yourself down to the Wardroom and have that wound dressed. That's an order. I'll get an Amjap up there to cover for you. 2MC Talker, standby for Operation Freddy. Direct Freddy sound twenty right from our bow, add the crowd sound in two minutes. Helmsman turn starboard to 020, new speed All Ahead Two Thirds. OOD, you have the conn."

Pelham evaluated the targets on the plotting board with Gehres. Orange Force was continuing east, and no ships were closing their position. There was no scout plane shadowing Selfridge from the cruisers, which did not make sense. Maybe they had maintenance problems. Selfridge had fired two red rockets twice after the carrier went down. Hopefully, one of the fast boat crews saw them, and they were all heading for a rendezvous point twenty miles north.

"2MC Talker, turn Freddy on for five seconds, ten seconds off, four cycles like that. Then, keep the sounds going," David directed.

At the Pearl Harbor Slingshot Command Center, Lt. (jg) Simpson explained who Freddy was and how the dragon, blinking blue lights, huge speakers, and soundtracks were used to unnerve the Japanese.
Captain Davis, now subdued by the thought of a carrier sinking in tens of seconds, was still objecting, "I don't see why you think such tricks will affect the Nips. They train far more than us to fight at night. Some blinking lights and screaming crowd sounds will just give you away and get you killed. And a fire-breathing dragon? Give me a break!"
Simpson responded, "Yes sir, I might agree with you, except these effects were tested on me, and they scared me half to death. During the April exercises, skipper Pelham put me, Chief Bosun MacGregor, two Amjaps, and three sailors ashore on the Kure Atoll sixty miles west of Midway. As the sun was setting, we believed the ship had left for Midway and wouldn't return until sunrise. Our Ops Officer instructed us to land on the island and set up a rotating watch through the night. I had a .45 pistol and two magazines The Chief had a Tommy Gun if there was any Jap activity. We strung some hammocks between the palms and settled in for the night. At about 0200, a watchstander woke me and said he saw strange lights. It took a few minutes, and then I saw a single rapid blink of blue light. There was no way to judge the distance, and I couldn't explain what it was."
"Admiral Nimitz, do we have to listen to this sorry tale? Blue lights, my ass!" complained Captain Davis.
Simpson continued, "A minute later, there was another blue flash, forty degrees right of the first one. Then the sounds started with a low noise that made no sense. Then there was a blue flash 100 degrees to the left of the others. I woke the Chief in time to see the next blue flash way to the right. The sounds were louder, but none of us could make out what they were. Then we all heard a creature roar and saw a red glow on the horizon. The next blue flash was way to our left, and the sounds were louder. Finally, we heard a woman screaming in fright, 'Help us, they're killing us.'
"That may not seem too scary in this room. Put yourselves on a tiny island in the middle of a night, subjected to unexplainable lights. You hear multiple voices screaming in pain and begging for help. Then, in the distance, we perceived what had to be a giant animal roaring and snorting. Captain Davis, the hair on the back of my neck, was standing up. The blue flashes became more frequent, every few seconds, and farther apart. The voices were screaming that they were in hell; they were trapped and tortured for eternity. I didn't know about the Chief, but I was scared shitless. Then the animal sounds became even louder, and we could briefly see some flames out on the water moving from left to right. It was coming closer and growling at us. The blue lights were blinking, and the people were screaming in pain and begging to be saved ever louder. Then we saw the animal moving at right angles to us. It was a dragon with flames out of its mouth, smoke out of its nostrils, and a glowing red eye. We only saw it for a couple of seconds. I prayed to God that it wouldn't see us – that it would keep moving away. Somehow, the blue lights were getting closer together, and the screams were from hundreds of people. My nerves were raw when the dragon reappeared closer to us and turned toward us! It was spewing flames and loudly roaring as we saw both flashing red eyes coming right at us. I pulled out my pistol, started firing, and kept firing until all my magazines were empty. The Chief fired the machine gun until it was empty. The lights went out, and the voices stopped - the dragon was gone. Both the Chief and I had waded into

the low surf. I don't know about him, but I was in the ocean up to my waist because I'd pissed myself."

Davis snorted, "That's not gonna work on the Japs. They don't understand English."

"Sir, I found out later that we also had the voices and screams recorded in Japanese."

Comdr. Joe Rochefort rushed into the room, waving several pages over his head. He was surprised to see all of the men quiet and looking stunned. Could they already know what he was going to say?

Admiral Nimitz broke the silence. "What is it, Joe? What do you have for us?"

"Admiral, a message from Selfridge, they've sunk the carrier! They sank the carrier!"

Pandemonium broke out in the makeshift Command Center. Men were jumping up and down, thumping each other on the back, yelling and whooping. Finally, Layton and Rochefort came forward with broad smiles and shook Nimitz's hand. Hardly any of the celebrating naval officers noticed Captain Davis slowly get to his feet, walk over to Lt. (jg) Simpson, and congratulate him. Nimitz waited for them to settle down.

"Comdr. Rochefort, please read us the entire message," instructed Nimitz.

"Yes sir."

"From USS Selfridge to CinCpacflt

SECRET.

1. Shokaku Class carrier attacked at 0325 Midway time, roughly bearing 345 and 215 miles from Wake Island. Carrier sunk at 0332. Selfridge has six dead and ten wounded.
2. No sign of Service Group. Selfridge retiring.
3. Regards to Lt. (jg) Simpson, your plans worked.

SECRET."

Several officers, including Nimitz, went over to Lt. (jg) Simpson and congratulated him. Nimitz returned to his desk and spoke, "Gentleman, it's been quite a day, or I should say night. Mr. Simpson, what is the reference to the Service Group?"

"The Service Group ships, sir, three destroyers, two tankers, two troopships, two freighters, and a seaplane tender. Comdr. Pelham had hoped to strike them while retiring from the main body of Orange Force."

"I see. What can we expect that Selfridge will be doing at this point?"

"They will be running to the west, then north, then east to recover the fast boat crews and sail east as far as possible before dawn. If they survive Jap aircraft attacks, they will meet up with Neosho northeast of Wake Island, refuel, and transit to Midway."

Nimitz spoke to the group, "The main action is over. I don't want to tie this talented group up any longer. Mr. Simpson and the Destroyer Desk, set up in the far corner, and carry on. Everyone listen up; although Selfridge's message was SECRET, the fact that a Jap carrier has sunk is TOP SECRET. You won't share that fact with anyone outside of this room. There are darn good reasons for this, and it will be the end of your career if you don't follow my order. We will now have a moment of silence to honor the Selfridge dead and our brave comrades on Corregidor."

Onboard Selfridge, they secured Operation Freddy and charged westward to open the distance from the main Orange Force. It was still dark, and At All Ahead Full, she was throwing up a high, bright stern wave visible for several miles. Captain Pelham was out on the Port Bridge Wing, getting some air when the JOOD came to fetch him.

"Sir, Radar reports two air contacts astern at twelve miles, closing."

David entered the Bridge, "OOD slow us down to Two Thirds, and One Third in steps. Have the JOOD watch the stern wave. 2MC Talker, report two enemy aircraft attacking from astern on all circuits."

Pelham selected ALL on his 1MC, "Now hear this, two enemy aircraft attacking from astern. All hands back to your GQ stations, in three minutes, maintain ZED throughout the ship." He mashed the GUNS button on his phone panel, and Kindall answered. "Guns, we're slowing and will turn starboard to unmask all guns. We don't know the aircraft's altitude. Radar will call the bearing and range to you. Put up some flares and knock them down with barrage fire. Captain out."

David spoke in a loud voice, "Radarman, make sure you have good comms with Gun Directors and feed them contact data every thirty seconds."

The 2MC Talker reported, "Sir, Signals standing by for instructions."

David mashed the ALL button on his 1MC and transmitted, "Secure the Signals Bridge. All personnel not manning guns or Freddy go below decks. Prepare for anti-aircraft fire, prepare for Operation Freddy, that is all."

The Radarman reported, "Good comms with Guns sir. Contacts dead astern, range fifteen miles, two aircraft closing in formation."

"Very well. Helmsman, set your rudder twenty degrees right, steady on course 350. When we're at one-third, OOD turns on the bubble machine and takes off forty turns. You have the conn."

David heard two shots from the aft guns. A few seconds later, two flares started drifting down behind the aircraft. Lookouts reported that they were float aircraft. David breathed a sigh of relief. They were probably off one of the Jap cruisers. Floaters were slower and lightly armed compared to the enemy's carrier aircraft. He was surprised to see Nagy and Akemi enter the Bridge. Nagy's arm was bandaged and in a sling.

David punched GUNS on his phone and got Lt. Kindall. "Guns, pop a couple more flares. We're slowing, turning ninety degrees starboard, and switching on the bubble machine. They will have trouble seeing us until you start firing. What range do you plan on?"

Kindall replied, "Since they're floats, we will take them at two miles, and hope they stay together."

"Roger that, Captain out. Radarman report targets."

"Two aircraft bearing right 080 relative, range four miles, speed one four-zero, lookouts estimate altitude is 2,000 feet."

"Very well. JOOD, confirm that damage control has removed Mr. Wolfe's body.

David waved Nagy and Akemi over to his chair as the JOOD confirmed that the body was gone.

"Michael, how is your arm?" Pelham inquired

"The doc says I should have a full recovery. It stings a little now. He gave me a shot of whiskey for the pain. He's patching the wounded up as fast as they arrive below."

"Outstanding. Petty Officer Akemi, you did good. Thank you. If you two go out on the starboard wing and plug your ears, you should get quite a show looking astern."

The OOD reported, "Steady three-five-zero, sir. Speed One Third minus forty turns. Bubble machine on."
"Good job, OOD. After the first barrage, turn on Freddy for five seconds."
There was a thunderous noise as all eight five-inch guns fired fragmentation shells at the floatplanes. The Talker reported that one aircraft was plummeting into the ocean in flames. David saw Freddy breathing fire and smoke on the bow. Six seconds later, the big guns were firing again. When Radar reported that the target was opening, Pelham called Check Guns. He thought, *They aren't afraid to die, but losing his wingman and the sight of a real live fire-breathing dragon was too much for that flyboy – wonder what he's going to tell his Captain?* Nagy and Akemi came back into the Bridge all smiles.
"Mr. Gehres, what's the estimate on our distance from Orange Force?"
"Captain, estimate eight miles."
"Very well. OOD, take us port to two-seven-zero, All Ahead Full."

David was getting a debrief from Nagy on how they had fooled the carrier and pitted the Jap destroyers against their cruiser when Radar called out, "Surface contact bearing right five degrees, range sixteen miles, course and speed unknown."

David slowed the ship to One-Third and clambered down from his chair. He motioned for Nagy and Akemi to follow him to the Radar plot. Once there, the Radarman pointed out a faint echo on the screen. He said he'd bet a month's pay that it was a ship's mast and predicted that the return would blossom as it got closer and the radar hits reflected off its superstructure. David had Lt. Gehres plot the single ship on the tactical board. He called the Engineer and asked him to bring the latest fuel figures to the Bridge. Then he asked the 1st Lieutenant, XO, and Guns to come to the Bridge. David went into his cabin and used the head. He sat down on his bunk and desperately wanted to lie down and sleep. His right thumb and hand were twitching, his shoulders ached, and his legs felt like lead. He could barely keep his eyes open. Finally, after a few brief moments of rest, he willed himself to get back on his feet and went into the Bridge by his chair.

Noting that it was still pitch black outside, David spoke to the ship on the 1MC. "This is the Captain. Our Gunnery Division shot down one plane, and the other one turned tail and ran after seeing Freddy. Well done. Men, this is a lull in the fighting with no aircraft on the Radar. Set modified condition YOKE. Chiefs and Petty Officers have the men move around the ship to use the heads and get a snack on the mess deck. You have about thirty minutes to rotate them, let them rest, come up on deck for some fresh air, or take a nap at their stations. All hands get a good swig of water. Only the sick and wounded can be in their bunks. The smoking lamp is lit below the main deck until GQ. That is all."

Pelham was getting quite a crowd on the Bridge. The XO, Guns, and 1st Lieutenant arrived first. David had them look at the Radar plot showing two surface contacts. Then, David reminded the OOD to rotate his men as Lt. Comdr. Haldane came on the Bridge, breathing heavily.
"Officers gather round. We've had one heck of a night. We're on one hell of a roll. I thought to take our winnings and leave the casino until five minutes ago. I believe we've found the Service Ship Group lagging about twenty miles behind the main force. They're on our nose at fifteen miles and heading right at us. Engineer, what is our fuel situation?"

"Captain, we've used forty-five percent of our fuel. We can sail about 3,000 miles at Standard Cruise Speed or 1,000 miles at Full Ahead."
"Thank you, Engineer. How are the power plants and your men?"
"Sir, everything has been holding together. The men have been magnificent. We're rotating them topside for some fresh air."
"Good. Mr. Gehres, how many miles to the tanker rendezvous?"
"Sir, if we engage these ships on a single pass, it would be forty-five miles to pick up the fast boats and a total of 310 to meet up with Neosho."
Haldane admonished, "Captain, if we have trouble tanking or she doesn't make it, we may fall short of Midway at high speeds."
"Noted, Engineer. Mr. Jones, what's the material condition of the ship?"
"Pretty good, sir. We patched up some bullet holes in the port side. The aft quad-Bofors is out. I see you have damage to your port wing door."
"Guns?"
"Captain, other than the Bofors, we're in good shape. We'll have two .50s aft by daylight. We have plenty of ammo. The crews sure needed this break. Thanks," Lt. Kindall replied.
"Men, I felt tired and decided everybody should take a break. It has been a long night. Our situation is to either engage the Service Group and run or run north now without seeing us. Mr. Nagy and Petty Officer Akemi, what are our chances of convincing her three destroyers to detach at high speed to join the main body?"

Nagy and Akemi conversed in a mixture of English and Japanese words while David drew the anticipated formation on the board.
"Hold your answer until I cover this formation. We expect to see the three destroyers out in front, and the seven service ships in two columns behind the lead destroyer, probably half a mile apart. Mr. Nagy, what say you?"
"Captain, we think it's possible. We have the cruiser's call sign, and we know one challenge and reply. Akemi orders the lead destroyer to make all speed to aid a sinking carrier and take his escorts with him. It should work. It's a cruiser CO giving orders to a smaller ship."
"Good, I like it. Mr. Gehres, what is your recommendation?"
"Skipper, let's take them on. It's like a football game. We may be a few points ahead now, but it won't hurt to rack up a few more scores before the final whistle."
"Engineer, what do you think?"
"We've pushed our luck far enough, Captain. It may not go well with the Jap destroyers; then it's three against one. Some of those service ships also have deck guns. One shell in the wrong place, and we're toast. Then, there's the question of fuel that I've already raised."
David showed his irritation. "Yes, yes. Mr. Jones, your thoughts?"
"The ship is in full fighting trim Captain. We put out a couple of fires. As Admiral King said, let's give them hell."
"XO?"
"Whatever you say, Captain, I'm ready to back your play."
"No XO, you must weigh in. I want your opinion."
"Skipper, if we think about the men, the risk to their lives, I see it as a tossup. So sorry, I remain neutral."

Steward Jackson and a Messman came on the Bridge with two large trays of sandwiches and jugs of iced tea and lemonade. David directed the OOD to slow to All Ahead Slow and told everyone to dig into the food while he and the XO went out on the Port Bridge Wing. David put his back to the ten-knot breeze, closed his eyes, and took three or four deep breaths. "XO, we're going after that Service Group even if it kills us. I hope we put down at least one of the troopships. For Christ's sake, at Pearl, their planes were strafing our boys while they were swimming for shore. I was not kidding about being tired. I'll tell you and no one else that I'm just about out of gas. Seeing Don Wolfe killed and having his blood spurt on my shirt and pants hit me hard. You're going to take Selfridge against those service ships for training purposes. At least that's what I'm going to say. Rely on Nagy and Akemi to get rid of the escorts. Fall in leading thc formation until the destroyers clear. Speed up, reverse, and give them hell. You have a big advantage, you can see them on the Radar, but they can't see you. Make one pass between the columns, swing around to the north, and hit them again. Gehres knows the sail plan for Midway. Wake me if we have enemy aircraft closing."

They went back into the Bridge, and David announced that the XO would be leading the attack on the service ships, and as CO, he'd monitor the fight from Secondary Conn. He explained that the XO would be getting a ship soon; it was time for him to get direct combat experience. David grabbed a sandwich and took a swig from the lemonade jug before leaving the Bridge. Down on the Weather Deck, he talked and joked with several men before going below to the Wardroom. Lawson was caught up in treating the wounded and gave him the Butcher's List - six dead and eleven wounded. Six men of the eleven were fit for duty; he had sent them back to their GQ stations.

Captain Pelham walked through the mess decks, shook hands with several men, and pocketed a couple of cookies on his way to Secondary Conn. Once there, he reviewed each man's job and checked the comms setup. David briefed them on the expected formation ahead, the latest plan of attack, and not to wake him unless the ship was being shelled or there were closing enemy aircraft. Finally, he lay down on the XO's cot, took out a cookie, and looked at the doctor's handwritten list. Three of the six dead men were Amjaps. *Damn,* he thought, *I hate to lose those young boys. Shit, there will be hell to pay when we get back to Pearl!*

David Pelham missed the tricky situation when the lead Jap destroyer refused to leave station until Akemi pulled rank and threatened him with a court-martial. He slept through the bombardment of the service ships, and when Selfridge sailed north at full speed, and then east to pick up the fast boat crews. The two fast boat crew climbed the cargo netting and scuttled their craft. The third speedboat was nowhere in sight.

Selfridge's Captain finally stirred three hours after hitting the Army cot. Secondary Conn personnel had kept the noise down as he slept. David woke from a dream in which a railroad engine had run over both his legs. He reached down, and they were still there, but they wouldn't move! A wooden frame member on the cot had cut off circulation to his legs, and they were asleep. Two men helped him to his feet and held him upright. He could see it was well after sunrise and decided to shower and shave in his Inport Cabin.

Nagy and Akemi were asleep as he quietly went into the head and cleaned up. He came out of the head in his skivvies and found a fresh uniform, hot coffee, and doughnuts left by Jackson. David dressed and made his way to the Bridge. After he was announced, the XO greeted him and shook his hand. David went over to the Radar plot and asked if there were

any airborne contacts. The XO briefed him that two single aircraft went on either side of Selfridge at fifteen to twenty miles after dawn. No planes had closed on the ship, so they had not woken him up.
"Mr. Gehres, estimated distance from Orange Force carrier force if you please."
"Ninety-five miles, sir."
"Very well. XO, status of fast boats?"
"We recovered two crews and scuttled the boats, no sign of the third."
"Very well. OOD, what is our course and speed?"
"Sir, course zero-eight-five, speed thirty-nine knots at All Ahead Full. Section 2 has the watch."
"Very well, XO, did you send a message reporting the service ship action?"
"Yes sir, here it is."
David read:
"SECRET
From: USS Selfridge
To: CinCpacflt
Subject: After Action Report for Orange Force Service Group Engagement
1. (S) After ordering the enemy destroyer escorts to detach and speed to the assistance of a sinking carrier, Selfridge engaged seven (7) support ships at 0355 Midway time, roughly 245 miles and a bearing of 310 from Wake Island.
2. (S) Steaming down the center of the two-column formation, Selfridge shells exploded one ship believed to be a tanker and set three (3) others on fire. At the end of the column, Selfridge reversed starboard to the northeast and brought all guns to bear on the last ship of the port column, a seaplane tender that listed hard to port and is believed to have been sunk.
3. (S) Selfridge took the remaining ships under fire as the enemy formation disintegrated into vessels moving in all directions. Solid multiple hits were observed on three ships, and a fourth believed to be a troopship was set on fire.
SECRET

David was impressed. "Wow, XO, good shooting. It's a bit wordy. We can work on that. You are relieved. I have the conn. Get some food and some rest. OOD, what is the ship's material condition?"
"Modified YOKE, sir. We stood down from GQ two hours ago, and two watches are sleeping."
"Outstanding. Mr. Gehres, show me a plot of those outbound aircraft with the times. Radarman pay extra attention to your northeast sector. Sing out for any faint echo."
"Radar aye aye sir!"
He punched his phone's GUNS button and got the Assistant Gunnery Officer, "Guns this is the Captain; if we get a snooper it will come from the northeast. We will go to GQ and set Nippon. Captain out."
He punched his ENG 1 button, and Chief Romano answered, "Chief, any reason we can't go to Flank?"
"No sir, we're turning what the Bridge asked for. Commander Haldane is asleep, do you want him to call back?"

"No Chief, let him rest while he can. Thank you."
"OOD All Ahead Flank, you have the conn. JOOD double up the forward port quarter lookouts. Go out and tell them we're expecting aircraft from that direction."
Gehres brought him a plot from 0650. It showed two aircraft tracking outbound on either side of the ship at twenty miles. Pelham thought, *Luck of the Selfridge, the first search aircraft from the carrier missed her from either side.*
"Mr. Gehres, what do you think these aircraft were doing?"
"Searching, sir."
"For what?"
"I suppose for us sir. We gave them quite a licking!"
"No, Mr. Gehres, they don't care about a dinky destroyer. They believe that there's a Yank aircraft carrier in the area. They may believe that it struck Wake and maybe even them. They're scared to death of that mythical carrier. What do you think those planes will do during their search?"
"I dunno, sir. I suppose they will come back to their carrier."
"That's it, they go out 200 miles or more, turn ninety degrees port or starboard, and return over a new area. You tracked two planes on their way out, but at least twelve planes were launched, one out each thirty-degree radial around the compass rose. So, Mr. Gehres, two planes will return to their home base that might detect us. Where will they be coming from?"
"Ohhh, I see what you mean. We've been moving east for two hours. We may run into planes coming back from the northeast or east. That's why you increased the lookouts and warned the Radarman. I see, I see."
"Think ahead, Mr. Gehres. Project your tactical picture ahead and choose your action."
"Sir, possible aircraft dead ahead twenty-four miles and closing," warned the Radarman

David took Gehres over to the display, where the Radar operator showed them a faint echo along the ship's course.

David issued orders. "OOD reduce speed to Full, then to Two Thirds after two minutes, then to One Third after two minutes. JOOD, call away GQ, and Operation Nippon, Condition ZED will be set in five minutes. Talker, tell-all circuits enemy aircraft at twenty-two miles on the bow and closing."

The Jap aircraft frustrated David. It closed to five miles at about 5,000 feet altitude but refused to come closer. It was a D3A Type 99 carrier bomber with fixed wheels and a crew of two. None of Selfridge's Nippon tricks could entice the Val pilot to fly closer to the ship. After it circled twice and headed southwest, David secured from Operation Nippon and set modified Condition Yoke. He had the deck crew remove the phony prow and rounded stern pieces and throw them over the side. Other men painted over the Jap markings and detached the heavy canvas masking the forward stack.

The honor of raising the American flag went to Petty Officer Akemi, assisted by Amjaps. Surely the Val had reported their position. They would be over a hundred miles away before the carrier could strike them. Would the Jap commanders even care about a single destroyer that might be one of theirs? David put the Selfridge back at All Ahead Flank and went below to talk to his wounded crewmen.

An hour later, he had his answer when Radar detected several aircraft at thirty miles closing from astern. Doc Lawson reported that he had treated all the initial wounded. David promised to return.

Gehres called away GQ with the Marine snipers on the Bridge Wings. David had the OOD steer for a rain squall about eight miles ahead. Lookouts identified the planes as A6M Zero fighters or Zekes carrying a bomb on each wing. At four miles, the ships' aft five-inch guns fired a wall of flak that downed one of the five planes. David dropped the ship's speed to All Ahead Two Thirds to give his gunners a more stable platform. The rainstorm·was one mile ahead, and the Zeros were two miles out when they made a mistake. Their fastest attack was against Selfridge's stern, but they flew on the ship's starboard side before turning inbound and starting their bombing run. All of Selfridge's big guns were brought to bear and exploded the trailing aircraft - now there were three. David stepped out on the Starboard Bridge Wing as the three aircraft peeled off to attack his ship from six thousand feet. He had the JOOD hold the Bridge door open so the Helmsman could hear his commands. He waited for an eternity until the lead Zero pointed directly at the ship and passed two thousand feet.

"Helmsman, prepare for starboard turn at full right rudder, OOD warn the crew. Helmsman, signal One Third on the starboard engine, All Ahead Full on port engine. Hard starboard turn now, now, now."

David watched the A6M Zero pilot try to adjust his aim before releasing two bombs. Finally, the pilot pulled off, and the ordinance exploded in the water behind the ship's stern. The second Zeke held his eggs and pushed his nose down to strafe the ship. Selfridge's forward Bofors blew off his tail just as he opened fire. Now there were two. The third aircraft in the formation raked the ship with hundreds of twenty-mike-mike shells. The leader waited for his wingman to join while Selfridge ran for the squall line at All Ahead Flank. The two Zeros were setting up for another bombing run as the ship slipped into a curtain of rain and disappeared. David went back into the Bridge as two Marines were carrying another out.

"How bad is he hurt?" David asked

The Marine had tears in his eyes. "He's dead sir. Took one through the chest."

David said, "I'm very sorry. Take him below to the doctor. OOD, hold your course, reduce speed to All Ahead Slow. Move along the edge of the squall in a clockwise direction. Radarman, what can you tell me?"

"No contacts sir. Probably blocked by the rain."

"Very well. Ops, what do you recommend?"

"We can outwait them, Skipper. They have to return to base at some point."

"Yes, I agree, but how long will this rain last?"

"Sir, I don't know."

"Mr. Gehres, neither do I. It could be half an hour, a few minutes, or even less. If it goes poof, we have two very angry Zeros on our ass - one that still has his bombs. We need something more aggressive than just hiding. Always go at the enemy and take the fight to them, Mr. Gehres. When you let up, they've time to rest, to think, and surprise you in ways you won't like. If we pop in and out of the rain, we may gain an advantage over them. They have to keep turning to stay in our area. That should help us to see them before they see us."

"Yes sir."

"OOD, ease out of the rain." On the 1MC, "All hands, stand by. We're going to pop out of this rain and see if we can take a shot at these Zekes. We will be able to fire forward, aft, and off the port side. That is all."

Selfridge slowly moved out of the heavy rain into a light mist as the OOD skirted the east side of the squall. There were no aircraft in sight and none on the Radar.

"I think they've gone sir," Gehres guessed.

"Mr. Gehres, be careful of conclusions that match your desires. Those pilots are full of revenge. If we see them on Radar retiring, I might agree with you. But, OOD, keep skirting around the squall."

A few minutes later, "Radar has a contact at ten miles opening."

Gehres shouted, "They left. We beat them!"

"Radarman, can you tell if it's one contact or two?" David asked.

"No sir."

"Very well. Mr. Gehres, that could be both of them, or it could be another aircraft passing by. It could be one of them had to leave because of fuel, or he's a decoy. OOD, keep skirting the rain."

Suddenly, the 2MC Talker sang out, "Lookouts report aircraft dead ahead, two or three miles, altitude estimated 2,000 feet."

David ordered, "Helmsman, set your rudder forty-five degrees left, steady on course 330." Then, on the 1MC, "All hands, we have a single bandit off the starboard side. Shoot him before he can bomb us."

The forward guns started firing, making a tremendous racket, and the aft guns joined in as soon as they were unmasked. David went out on the starboard wing where Gunny Giles used his spotter's shoulder like a gun rest.

The Zero pilot was banking left toward the ship as David gave him the finger and yelled, "Kill the bastard, Gunny! Kill him now!"

Shells were exploding all around the rapidly closing aircraft. He was coming right at the ship. The pilot was slumped forward; his plane was trailing flames and smoke. It just kept coming at them as the .50s riddled it with bullets. David and the two Marines watched in transfixed horror as the pilotless plane crashed into the ocean close to the ship, and aircraft pieces impacted against Selfridge. The propeller cartwheeled up through the starboard weather deck safety wires and smashed into Selfridge's forward torpedo station. There were flames and smoke from one of the tubes. David had the 2MC Talker call in the fire, and the OOD conn the ship downwind. They were moving away from the squall line, and Radar had no contacts to report.

"Ops, there is still a very slim chance that the last Zero is on the other side of the rainstorm. So we'll run with the wind until the fire is out. Then we pick up the beat to join with Neosho before dark so we can take on fuel."

"Yes sir. We can always take on fuel in the morning."

"Quite right, Mr. Gehres, but as you advance to command, you need to see fueling the same as a head call."

"Skipper, I'm not sure what you're saying."

"Whenever you get a chance to use a head, use it. Same for fuel. Take it when you can get it, always."

"Yes sir. Is your head still available?'
David smiled. "Affirmative."
The 2MC Talker reported, "Sirs, Damage Control reports torpedo fire is extinguished, and fire watch set."
"Very well. OOD, turn to new heading 085, All Ahead Full. Have Ops refine the intercept course to our tanker rendezvous when he returns. I will check the damage, talk to the XO, and drop down to sickbay. If Radar has no contacts in ten minutes, secure from GQ and set Modified Condition YOKE. You have the conn.

It took two days for Selfridge to shepherd the tanker back to Midway. First, David had Yeoman Brown collect a signed affidavit from each crew member that they knew sinking the Jap carrier was classified SECRET, and they wouldn't discuss it with anyone other than ship's company. Next, Selfridge was topping off her fuel from USS Neosho as two sections of men went ashore for a beer bust on the lagoon's beach. Other ships and the Navy base lent them liberty boats to transfer the men back and forth from the shore. From the Bridge, David could see several of the crew swimming in the lagoon and splashing each other when his Comms Officer handed him a message.

From: CinCpacflt
To: Commanding Officer, USS Selfridge (DD-357)
Subj: Sailing Orders
SECRET
1. (S) Depart Midway and report to Pearl Harbor Ammunition Pier ASAP.
2. (S) Plan to rearm, repair, and depart within 72 hours.
3. (U) Provide repair requirements to Pearl Harbor Navy Yard before departure.
4. (S) Congratulations on Slingshot. Well done!
Admiral Chester W. Nimitz
SECRET

Mitsubishi A6M "Zero" with wing tips folded for carrier operations

Source: National Museum of the United States Air Force

Chapter Thirty-five

RETURN TO HOMEPORT – Pearl Harbor, Oahu

May 1942

After reading Nimitz's message, Captain Pelham had the XO join him on the Bridge. Boats MacGregor reported that the refueling would take another three to four hours. David and Lt. Comdr. Lehman wanted to comply with CinCPac's orders but knew the crew needed to blow off some steam. They decided to spend the rest of the day preparing Selfridge for sea and send the third and last section ashore for a few hours. The word was passed to all Divisions that they would sail after supper at 1900. Lt. Gehres, in his new role as Operations Officer, ran the numbers and concurred that they would arrive at the Pearl ammo pier at dawn on Thursday, 21 May. The XO was tasked with sending a message to the Navy Yard on their need for a replacement Bofors AA gun, a set of torpedoes tubes, and other minor repairs. David went back to composing condolence letters to the families of his dead crew members. It was the hardest part of his job.

Two days later, Selfridge was tied fast to the Pearl Harbor shipyard pier by 0700. The XO monitored the Bridge team during the channel transit and the approach to the berth while their captain roamed around the ship and talked to several crewmen. Doc Lawson recommended which crewmen should be transferred to the Naval Hospital. No doubt, there would be ambulances on the pier to take their seven dead bodies to the hospital Morgue. They had chosen to put them on ice in canvas shrouds instead of burying them at sea. David thanked the Army Captain and warned him that they were going back into combat. Lawson said he was the ship's doctor and planned to stay onboard. David shook his hand and agonized out loud that he and the XO would have to see the dead Amjaps' families and try to explain their loss while thinking, *I can't send anyone else to see the families. It's all on me. I directed the recruiting. I took them into harm's way, and I got them killed.*
His thoughts were interrupted by a 1MC announcement, "Captain has a visitor on the Quarterdeck."

David made his way topside to the gangplank Quarterdeck, where the OOD was talking to Lt. Comdr. Layton. Layton said he, Rochefort, and Lt. (jg) Simpson wanted to debrief Slingshot in more detail, and perhaps Lt. Comdr. Wolfe should come along to HQ. Pelham told him that Wolfe was dead. He put out a call for Lt. Gehres, and the three of them rode in a staff car to CinCPac Fleet Headquarters. David and Lt. Gehres were heartily congratulated by Comdr. Rochefort and their shipmate Lt. (jg) Simpson in Layton's office.

They spent over two hours reviewing every detail of the operation and the lessons learned. David asked Layton why Selfridge was ordered to be clear of the Wake Island area by 2300 on the night of the attack. Layton told him that Nimitz had ordered an Army Air Force bombing strike on the island. Finally, Layton looked at his watch and informed David that he was on Admiral Nimitz's schedule in ten minutes. Layton escorted him to CinCPac's waiting room and wished him good luck. When Nimitz's double doors opened, the Admiral motioned for David to join him. Nimitz was not smiling. Once inside, the Admiral introduced Pelham to Captain Roger Kelly, who was the senior JAG officer for the Fourteenth Naval District.
Nimitz spoke first. "Commander Pelham, let me congratulate you on the operational success of Slingshot. The performance of yourself and your crew was in keeping with the highest

standards of our naval service. Captain Kelly is here because you've given my staff and Admiral Bloch a complicated legal situation. Captain."

"Commander Pelham, I must warn you that anything you say today can be used against you in a court-martial proceeding. Is it true that you had civilians on your ship and knowingly took them into a combat environment?"

"No sir, everyone but Michael Nagy, a CinCPac employee, was enlisted in the Navy."

Kelly said, "Is it true that some of these new enlistees were killed and injured?"

"Yes sir, three were killed and three injured manning a Bofors antiaircraft gun."

Kelly continued his questioning, "Some of them were underage, were less than the age of majority? Is that correct?"

"Yes sir, two of them were seventeen years old and had their parents' written permission to join the Navy."

"Join the Navy, how could they join the Navy?" Kelly asked.

"Sir, I used the long-standing prerogative of a Navy Captain to recruit men for his ship, to sign them on, and induct them into the Navy. It's a tradition that I believe goes back to the Revolutionary Navy. Each of these fifteen young men in question signed formal enlistment papers, were sworn in as Seaman Recruits by myself, and understood that their obligation included the risks of combat at sea. They were volunteers, and I must say they played a large role in our ship returning home intact."

"You knew they were of Japanese descent and that our military services were prohibited from recruiting them, yet you went ahead with some kind of trumped-up enlistment?"

"Yes sir, I had to do that. They were an integral part of a vital scheme to fool the enemy, to make them think that our ship was part of their Navy, the Imperial Japanese Navy."

"I'm told that you also have an IJN sailor onboard your ship, is that true?"

"Yes sir, we enlisted him in the Navy, swore him in, and he also helped fool the enemy - more than twice. Captain, I'm responsible for doing everything in my power to protect my ship and crew. The Navy may think I broke some rules, but I know you will find the paperwork in order. The wounded men are being treated at the Naval Hospital, and the families of the deceased young men should receive the $10,000 death insurance payment."

Captain Kelly was red-faced. "Admiral Nimitz, I've heard enough. I'm not going to sit here and argue with this sea lawyer. My staff will research his claims about the powers of Navy ship captains. This is a mess. It will take months to interview everyone involved and file charges."

"Yes Captain, I agree. The less said about this matter, the better. Please report your findings to Admiral Bloch. In the meantime, Captain Pelham and I have a war to fight. Thank you, Captain Kelly. I look forward to your report. Dismissed."

Kelly left, and when the doors were closed, Nimitz leaned back in his chair, put his feet up on his desk, and said with a half-smile, "Well, Mr. Pelham, what do you have to say for yourself?"

"Admiral, no disrespect intended; I did what I had to do to survive the Jap air superiority. Without the American Japanese, and Petty Officer Goro Akemi, we would've been sunk south of Wake and never got a shot at Wake or the carrier. When you have the time, Lt. Comdr. Layton has all the details, especially how Michael Nagy and Akemi got the Jap ships shooting at each other and tricked them so we could reach our optimum firing position."

"How is Michael Nagy?"
"He has an arm wound that will heal. He deserves a medal. He convinced me that Akemi didn't have enough English understanding to react quickly on the Signal Bridge. Michael refused to leave the ship and deserves major credit for our success."
"I want your promise that he and the young men will be off your ship when Selfridge sails."
"Yes sir, I promise. What can you tell me about our next at-sea period?'
"Layton will give you the nitty-gritty details. We believe the H8K Emily long-range flying boats are refueling at French Frigate Shoals from Jap submarines. We don't want them surveilling our carriers during the coming battle over Midway. Our special friend in San Francisco has confirmed other intelligence that Admiral Nagumo is returning with four carriers the first week in June. Layton thinks you can shut down the H8K aircraft and subs at the Shoals using your gyro glider to patrol the area. Once the battle between the carriers is joined, your mission will shift to surveilling Admiral Kondo's Invasion and Covering Force that we expect to attack from the west."
"Admiral, can we do more than surveil Kondo's force?"
"Not at first. If we're losing, we need to know where the amphibious force is going, Midway or Hawaii. If we're winning, we need to know if he keeps sailing east, or has turned back to the west and is retiring. If he does the latter, you're cleared to harass his service ships."
"Yes sir, thank you sir. The fighting 357 won't let you down. We're loaded for bear and will be ready to sail in two days."
"Good. Get the details from Layton. What are you going to do about the boys' families?"
"The XO and I'll see them this afternoon. We will be the first to tell them in person about their sons. Do you think I should mention the $10,000?"
Nimitz stood up and shook David's hand while he ushered him to the door. "Pelham, don't promise what you cannot deliver. Good luck."

David made an appointment with Lt. Comdr. Layton for the following morning. Outside, he found Yeoman Brown and Lt. Gehres waiting for him in the ship's car. They went back to Selfridge, dropped Gehres, and picked up Lt. Comdr. Lehman. The visits to the three Amjap families were as gut-wrenching as David and the XO had expected. It had to be done and done correctly with sympathy and respect for the living. They did their duty.

David knew Betty White was at Tripler Hospital and had Brown drive around the parking lot until they found her car. The driver's door was unlocked. After giving the XO her address and the phone number in case of an emergency, Pelham told Lehman to meet him at 1000 in Layton's office in the morning and bring Lt. Gehres. He got into the passenger seat of Betty's car and promptly went to sleep.

Betty woke him up a couple of hours later. "Hi there, sailor. Looking for a good time?"
"Yes Ma'am. If you could introduce me to any sex-starved nurses at this hospital, I'd be much obliged!"
She got into the driver's seat and punched him in the arm. "Where to, my Captain?"
"Betty, I've had a rough day. I want an ice cream cone. I want to drink some fresh, cold milk. I want real scrambled eggs for dinner, and bacon or ham, no Spam!"
"We can stop at a drug store and pick up some ice cream; I have the rest. My roommate is over in Kaneohe for a couple of days. How long will you be here?"
"Sorry, I can't say."

"Damn, that short of a time. What else do you really want? Did you miss me?" Betty asked coyly.
"My dear, I want a long, hot shower where I can keep the water on for the whole time."
She shifted the car into high gear on Moanalua and placed her free hand on his thigh. "Would you like one of those sex-starved Army nurses to wash your back?"
David smiled and nodded. "That sounds delightful."
Betty stroked his thigh. "What about your front? You probably need some attention there, right?"
"Even more scrumptious and delightful. Betty. On second thought, maybe we should skip the ice cream."

EPILOGUE

On Sunday, 25 May 1942, this article appeared in several major U.S. newspapers:

Japan Loses Aircraft Carrier to Denizens of the Deep?

There are rumors floating among sailors of the Pacific Fleet that the Japs had an aircraft carrier sunk in the Central Pacific amidst unusual circumstances. Some sailors think the ship set sail, or was sunk on a Friday which is considered a bad luck day. Others maintain that King Neptune, ruler of the seas, and the evil pirate Davy Jones joined air, surface and subsurface forces of the U.S. Navy to destroy the Jap carrier in only a handful of minutes. Fleet sailors in Hawaii maintain that a newly launched enemy carrier was sent directly to Davy Jones' Locker during May as payback for Pearl Harbor. A spokesman at the Navy Yard in Washington D.C. declined to comment on the rumors saying, "We cannot confirm or deny these tales. We can say we will team with any country, tribe, or entity to hurl the Japanese back to their home islands. Also, we do not have any major combatants equipped with flashing blue lights."

ACKNOWLEDGMENTS

Every author needs encouragement and feedback on their draft product. I have been fortunate to have several friends read the advance copy of this book and help me avoid some glaring errors. Good friends Bob Burton and Dard Price have continued to comment on my work and support my efforts. High School buddy Hal Goldback has weighed in from Colorado on gun and ammunition terminology and provided enthusiastic support. Finally, RV friend Chip Dashiell still listens to my trials and tribulations and cheers me on.

My wife of five decades plus two, Clare Law, has read the draft manuscript and made recommendations. We both re-read the first proof from Amazon to ferret out several more corrections and improvements.

Once again, my editor, James Oliveri (yeeditor@cox.net or www.thumbtack.com), has provided a quantum leap forward in creating a more professional paperback product. Robert Reno supplied cover material from WeareIT, Las Cruces, NM.

And then there are family and friends who keep asking for the next book, making this author feel that his time on the keyboard has not been wasted. Below is a family book signing party in Seattle for Volume One, "World War II Transport Pilot."

Sadly, our family has lost my brother-in-law Jack Nicholson (standing, second from right), a great husband to my sister (standing, second from left), and a Korean War hero. A couple of years later, my sister also passed away.

Back row (left to right) Nephew Ken, Sister Pat, Wife Clare, Nephew Tom, Sister In Law Janet, Brother In Law Jack, Niece Karla

Front row (left to right) Brother John, Author, Great Niece Jana

NOTE ABOUT THE AUTHOR

This historical fiction novel is Jim "Lawman" Law's second book in the Pacific War Series. The first volume features the adventures of Captain Harrold "Rats" Mueller, USMC, and his aviation crew during the outbreak of the Second World War. "World War II Transport Pilot" also highlights the sixteen-day siege of Wake Island before it surrendered to the Japanese.

James W. Law was born in Seattle, Washington, two months before the Japanese attack on Pearl Harbor. During his senior year in high school, he joined the Navy and advanced from Airman Recruit (E-1) to Commander (O-5) during a 24-year career.

Petty Officer Law graduated from the University of Washington in 1963 on the same day as his older brother, John K. Law, who also served in the Naval Reserves. After four months of pre-flight training and Aviation Officer Candidate School (AOCS) in Pensacola, Jim Law continued active duty as an Ensign. He then completed Naval Flight Officer school and training as an Airborne Intercept Controller (AIC) and Combat Information Center Officer (CICO). Law flew over 2,000 hours in the Grumman E-2A, E-2B, and E-2C "Hawkeye" aircraft. During his Navy career, he completed two combat cruises to Vietnam and five other carrier cruises to the Far East, Southern Pacific, and the Persian Gulf.

Lieutenant Commander Law was Operations Officer on the world-famous VAW-114 Hormel "Hawgs" during two cruises to the Far East on the USS Coral Sea (CVA-43). After his second shore tour working on E-2A/B/C software, Commander Law returned to sea duty as Executive Officer and Commanding Officer of VAW-117. This Hawkeye squadron flew off the USS Ranger (CV-61). He retired in 1983 and served as a defense contractor on several Navy command and control research and development projects for twenty-three years.

JWL has been married fifty-seven years to the former Clare Amelia Stewart and is the proud father of two children: Ryan, Amanda, and grandad to five delightful grandchildren: James, Amelia, Bradley, Connor, and Christopher.

VOLUME THREE COVER PAGE

Find more adventures of Commander David A. Pelham, USN, and Captain "Rats" Mueller, USMC. Learn about the design, competition, and construction of WWII torpedo boats. Find new facts about the Midway Island battle. ***Search Amazon for: "CDR James W. Law" and "World War II PT Boat Captain"***

Volume Four

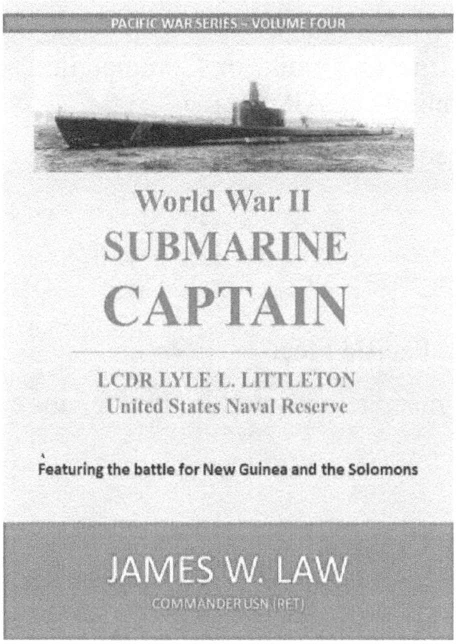

The adventures of a "90-day wonder" Naval Officer and his fight to join the "Silent Service." WWII submarine crew training, combat patrols, boat construction characteristics, and early problems with the Mark 14 torpedoes. MacArthur's challenges in Australia and Nimitz's at Guadalcanal. Plus, Rats Mueller and USS Selfridge operations "Down Under."

VOLUME TWO - GLOSSARY

AA – Anti-Aircraft

AAF – Army Air Force

ADF – Automatic Direction Finding, aircraft radio receiver provides a magnetic bearing to the transmitter on land or a ship

ARD – Assembly and Repair Department, Navy aircraft maintenance and modification organization

ASAP – As Soon As Possible

Aux – Auxiliary

Avgas – Aviation gas

Bird – Slang for aircraft or plane

BMC – Bosuns Mate Chief

BOQ – Bachelor Officer Quarters

BuAer – Navy's Bureau of Aeronautics

Bulkhead – Navy term for a wall

BuNav – Navy's Bureau of Navigation

CAP – Combat Aircraft Patrol (fighter aircraft)

Capt. or Captain – Navy O-6 rank, Marine and Army O-3 rank, or Commanding Officer of a ship regardless of rank. Modern abbreviation is CAPT

Comdr. or Cdr. or Commander – Navy O-5 rank, equivalent to Army and Marine Lieutenant Colonel. Modern abbreviation is CDR.

Chief – Navy Chief Petty Officer, E-7 to E-9 rank

CinCpac or CinCPac – Commander In Chief Pacific

CinCpacFlt or CinCPac Fleet – Commander In Chief Pacific Fleet

CNO – Chief of Naval Operations, administrative management and control of the Navy

CO – Commanding Officer

Colonel – Marine and Army O-6, equivalent to Navy Captain

CPO – Chief Petty Officer

D.C. – District of Columbia, slang for Washington D.C.

DesRonOne -Destroyer Squadron One, usually six to eight ships

DI – Drill Instructor

Dumbo – Nickname for USS Selfridge's gyro glider

Ensign – Navy O-1 rank, Second Lieutenant in Marines and Army

ETA – Estimated Time of Arrival

FDR – Franklin Delano Roosevelt, 32nd President of the United States

First Class – Navy E-6 rank

First Lieutenant – Marine and Army O-2 rank, Lt. (jg) in the Navy

FITREP – Fitness Report submitted quarterly

Freq – Radio frequency selection

Glasses – Slang for binoculars

Gouge – Slang for the straight word, the correct information

GPH – Gallons per hour

GQ – General Quarters, all ship's personnel man their battle stations

H.E. – High Explosives

Head – Bathroom, toilet, and shower

HQ – Headquarters

IAS – Indicated Air Speed, cockpit gauge provides pitot tube pressure

ICS – Inter-Communication System for aircraft crew communications

ID – Identify

IFR – Instrument Flight Rules at night and during bad weather.

IJN – Imperial Japanese Navy

JAG – Judge Advocate Group

Jarhead – Nickname for a Marine

JOOD – Junior Officer of the Deck

Ladder – Navy term for stairway

Lance Corporal – Marine Corps E-2 rank

Latitude – Degrees and minutes north or south of the Equator

Lt. Comdr.or Lieutenant Commander, Navy O-4 rank, Major in Marines and Army

Leatherneck – Slang for Marine

Lieutenant or Lt. – Navy O-3 rank, Captain in Marines and Army

Longitude – Degrees and minutes east or west of the Greenwich Meridian (England)

Lt. Col. or Lieutenant Colonel - Marine and Army O-5 rank, equivalent to Navy Commander

Lt. (jg) or Lieutenant Junior Grade – Navy O-2 rank, First Lieutenant in Marines or Army, Modern abbreviation LTjg

Maj. or Major – Marine and Army O-4 rank, equal to Navy Lieutenant Commander

MIT – Massachusetts Institute of Technology

MP – Engine Manifold Pressure, or Military Police

NAS – Naval Air Station

Nav – Navigation, a nickname for Navigator

NAVSTA – Naval Station

Navaids – Navigation equipment in an aircraft, ships, and on the ground

NROTC – Naval Reserve Officer Training Corps

OCS – Officer Candidate School

O' dark thirty – Slang for the middle of the night

OOD – Officer of the Deck

Ops – Operations Building, or Operations Officer

PBY – Twin-engine patrol aircraft

PC – Abbreviation for Plane Captain or Plane Commander

PIM – Planned Intended Movement, a ship's intended course and speed

PM – Program Manager

POC – Point of Contact

Rack – Navy term for bed

RADAR – **RA**dio **D**etection **A**nd **R**anging – transmit energy pulse and receive a reflective return from the target to determine range and bearing.

RADM – Rear Admiral, two-star flag rank

Revs – Slang for revolutions per minute

R&R – Rest and Relaxation

RON – Remain Over Night

RPM – Engine Revolutions Per Minute

QT – Closely held information, on the quiet

Seaman Apprentice – Surface Navy E-2 enlisted rank

Second Class – Navy E-5 enlisted rank

Second Lieutenant – Marine and Army O-1 rank, same as Ensign in the Navy

SITREP – Situation Report

Skipper – Commanding Officer of a ship, or an aircraft squadron

SOA – Speed of Advance

SOP – Senior Officer Present

SUB – Student Union Building

Sub or boat – Slang for submarine

TAD – Temporary Additional Duty

TAS – True Air Speed, Indicated Air Speed corrected for altitude but not corrected for wind

Third Class – Navy E-4 enlisted rank

USC – University of Southern California

USMCR – United States Marine Corps Reserve

USMC – United States Marine Corps – applies to an officer who has been selected for Regular status

USNR – United States Navy Reserve

USN – United States Navy, refers to an officer who has been selected for Regular status

Vis – Visibility

VIP – Very Important Person, usually a high-ranking officer or civilian

VFR – Visual Flight Rules, daytime in good weather

W&B – Weight and Balance calculations for an aircraft, it must be loaded correctly to fly safely

XO – Executive Officer, second in command

YOKE – XRAY – ZED/ZEBRA – Different levels of ship's material condition with hatches, doors, and openings closed and dogged down, or not.

Made in United States
Cleveland, OH
27 August 2026

43308113R00195